Production and Packaging of
Non-Carbonated Fruit Juices and Fruit Beverages

Production and Packaging of Non-Carbonated Fruit Juices and Fruit Beverages

Edited by
P. R. ASHURST
Dr P. R. Ashurst and Associates
Kingstone
Hereford

A Chapman & Hall Food Science Book

An Aspen Publication®
Aspen Publishers, Inc.
Gaithersburg, Maryland
1999

Aspen Publishers, Inc., is not affiliated with the American Society of Parenteral and Enteral Nutrition.

Originally published : New York : Chapman & Hall, 1995.
Includes bibliographical references and index.

About Aspen Publishers • For more than 35 years, Aspen has been a leading professional publisher in a variety of disciplines. Aspen's vast information resources are available in both print and electronic formats. We are committed to providing the highest quality information available in the most appropriate format for our customers. Visit Aspen's Internet site for more information resources, directories, articles, and a searchable version of Aspen's full catalog, including the most recent publications: **http://www.aspenpublishers.com**
Aspen Publishers, Inc. • The hallmark of quality in publishing
Member of the worldwide Wolters Kluwer group

Editorial Services: Ruth Bloom

ISBN 978-1-4419-5191-5

Printed in Great Britain
2 3 4 5

Preface

In the period of about five years since the first edition of this book appeared, many changes have occurred in the fruit juice and beverage markets. The growth of markets has continued, blunted to some extent, no doubt, by the recession that has featured prominently in the economies of the major consuming nations. But perhaps the most significant area that has affected juices in particular is the issue of authenticity. Commercial scandals of substantial proportions have been seen on both sides of the Atlantic because of fraudulent practice. Major strides have been made in the development of techniques to detect and measure adulterants in the major juices. A contribution to Chapter 1 describes one of the more important scientific techniques to have been developed as a routine test method to detect the addition of carbohydrates to juices.

Another, and perhaps more welcome, development in non-carbonated beverages during the past few years is the rapid growth of sports drinks. Beverages based on glucose syrup have been popular for many years, and in some parts of the world isotonic products have long featured in the sports arena. A combination of benefits is now available from a wide range of preparations formulated and marketed as sports drinks and featuring widely in beverage markets world-wide. A new chapter reviews their formulation and performance characteristics.

Another major trend in the area of fruit-containing non-carbonated beverages is the highly successful marketing of ready-to-drink products. Many products traditionally recognised, particularly in the United Kingdom, as concentrated fruit drinks and cordials for the consumer to dilute to taste, are now sold ready diluted. Unless aseptic packaging is used, the production of such non-carbonated products is a relatively high risk venture with microbial spoilage controlled largely by chemical preservatives. Hence, the development of aseptic packages, the machinery to handle them and the process plant to deliver the product in a suitably pasteurised form with minimal heat damage is a most important part of the growth of this market. Chapter 11 updates this area whilst a new contribution offers a valuable insight into some of the processing plant currently available.

A few of the chapters of the first edition remain unchanged and continue to provide valuable background material on the chemistry and technology of juice extraction and processing. Apart from these, most contributions, other than the new chapters, have been updated by the authors.

To conclude this second edition, a chapter has been included on the subjects of water and effluent. Water is the single most important ingredient and, indeed, the whole raison d'être for beverages. Growing pressures of many kinds on the environment have resulted in deterioration in water quality in many countries, and concern to obtain water of a suitable standard is of paramount importance in the beverage world. It is paradoxical that the effluent from beverage plants can itself be very demanding on the environment. This final chapter reviews standards and treatments for both water and effluent.

As ever, this book would be nothing without its contributors, whom I thank for their expertise, hard work and tolerance. Donald Hicks, the original editor, whose first edition was such a valuable work, has now retired. I am conscious of having to follow his act and I accept the full responsibility for the faults and weaknesses of this second edition.

I am very grateful to my many colleagues in the industry and in particular to Mrs Sue Bate for all her help in the preparation of the manuscripts.

I hope that readers from both commercial and scientific disciplines will find this work to be of interest.

P. R. Ashurst

Contributors

H. B. Castberg Food Science Department, Elopak A/S, PO Box 523, Lierstranda, Norway

P. J. Cooke RHM Research and Engineering Ltd., The Lord Rank Centre, Lincoln Road, High Wycombe HP12 3QR, UK

J. W. Downes H. Erben Ltd., Hadleigh, Ipswich IP7 6AS, UK

J. S. Dransfield Leatherhead Food RA, Randalls Road, Leatherhead, Surrey KT22 7RY, UK

G. R. Fenwick Institute of Food Research, Colney Lane, Norwich NR4 7UA, UK

M. A. Ford SmithKline Beecham Consumer Healthcare, Coleford GL16 8JB, UK

L.-B. Fredsted APV Pasilac AS, Pasteursvej, PO Box 320, DK 8600 Silkeborg, Denmark

J. Fry Korte Raarberg 51, 6231 KP Meerssen, The Netherlands

C. M. Hendrix Jr Intercit Inc., 1585 Tenth Street South, Safety Harbor, Florida 34695, USA

J. Hooper Lionsmead House, Shalborne, Marlborough SN8 3QD, UK

I. T. Johnson Institute of Food Research, Colney Lane, Norwich NR4 7UA, UK

A. G. H. Lea Reading Scientific Services Ltd., Lord Zuckerman Research Centre, PO Box 234, The University, Whiteknights, Reading RG6 2LA, UK

M. Lees Eurofins, S.A., Site de la Géraudière, CP 4001, 44073 Nantes Cedex 03, France

M. R. McLellan Department of Food Science and Technology, Cornell University, Geneva, New York, 14456, USA

G. G. Martin Eurofins S.A., Site de la Géraudière, CP 4001, 44073 Nantes Cedex 03, France

M. F. Moulton   SmithKline Beecham Consumer Healthcare, Smith-
Kline Beecham House, Great West Road, Brentford,
Middlesex TW8 9BD, UK

S. I. Norman   Dow Chemical Company, 1691 North Swede Road,
Midland, Michigan, Michigan 48674, USA

J. I. Osmundsen   Product Development, Elopak A/S, PO Box 523,
N-3412 Lierstranda, Norway

I. Paterson   7 Murrayfield Drive, Willaston, Nantwich, Cheshire
CW5 6QE, UK

E. J. Race   Welch Foods Inc., Westfield, New York 14787, USA

H. M. Rebeck   Gulf Machinery Corporation, 1600 Tenth Street South,
Safety Harbor, Florida 33572, USA

J. B. Redd   Intercit Inc., 1585 Tenth Street South, Safety Harbor,
Florida 34695, USA

P. Solberg   Department of Dairy and Food Industry, The Agricultural University of Norway, Norway

D. A. T. Southgate   Institute of Food Research, Colney Lane, Norwich
NR4 7UA, UK

Contents

3   Grape juice processing   88
M. R. McLELLAN and E. J. RACE

4   Tropical fruit juices   106
J. HOOPER

5 Growing and marketing soft fruit for juices and beverages 129
M. F. MOULTON

6 Apple juice 153
A. G. H. LEA

7   Equipment for extraction of soft and pome fruit juices   197
J. W. DOWNES

8   Processing of citrus juices   221
H. M. REBECK

9 Juice enhancement by ion exchange and adsorbent technologies   253
S. I. NORMAN

10 Processing systems for fruit juice and related products   274
L.-B. FREDSTED

1 Authentication of orange juice

J. FRY, G. G. MARTIN and M. LEES

1.1 Introduction

1.1.1 Incidence of adulteration

It may be safely assumed that fruit juice falsification is a permanent hazard of commercial life. In 1936, the US Food and Drug Administration (FDA) records orange juice being adulterated with sugar, citric acid, peel flavour and colour (Johnston and Kauffman, 1985), a form of dilution that is still found today. In 1981, a pilot survey by the FDA found minor irregularities in nearly half the orange juice firms inspected, and 3 of 13 companies were using illegal colours or pulp wash (Johnston and Kauffman, 1985). In 1982, the *Washington Post* was claiming that in one major US city 'fake juice is so common people don't remember what real juice tastes like' (Mitchell, 1982). The size of the problem is always unclear, but informed sources in 1983 were suggesting that up to 30% of the juice in the United States was adulterated to some degree, despite official assurances to the contrary. The feeling seems to be that this has since fallen to around 10%. There is some disagreement about the value of the US orange juice market, put at 3800×10^6 in 1983 by Nikdel and Nagy (1985) and, more conservatively, at about 2000×10^6 in 1987 (Doner *et al.*, 1987), but the economic value of juice adulteration is substantial.

At the time of writing, some Western European countries appear to be suffering a much greater incidence of falsified orange juice than the United States. This has led to suggestions in the press of large-scale fraud (Jones, 1988) and the allegation that 'Britain has become a bucket shop for fake orange juice' (Butler, 1988a).

The problem is, of course, not exclusively Britain's or the United States', but can affect any importing country, especially when times of shortage match bouyant demand—the situation in 1987/1988, for example.

The economic aspect of juice falsification is the principal reason for consumer concern over the practice. Juice adulteration seldom carries any risk to health. Although there is usually some decline in flavour quality involved, this can be less than that due to clumsy processing of genuine product. Nevertheless, passing off an assortment of cheap materials such as sugars, acid and colour as a costly and prized fruit juice is deceitful.

Consumers rightly expect that foods are correctly described, and deserve the best efforts of reputable suppliers and enforcement authorities to ensure that this is so. We should also be clear that our concern here is juice falsification and not contamination. The latter, whether deliberate or accidental, is the inclusion of foreign materials, often harmful to health. Juice adulteration is normally a deliberate act of attempted fraud. Although the materials employed may be foreign to the juice, they are increasingly so like the juice that they cannot be regarded as contaminants in the classical sense.

The discussion here relates to orange juice, particularly (but not exclusively) to frozen concentrated orange juice (FCOJ). This is the single largest juice product in international trade, and generally the focus of most concern. Locally, other juices may attract almost as much attention, for example apple juice in the United States. However, while most has been published on orange juice, many of the principles of detection are applicable to other juices. To assist the reader, the reference section is divided into those sources directly referred to in the text and a separate group of references, mostly for information on other juices.

1.1.2   The market for frozen concentrated orange juice

A brief picture is given here, as it illustrates the background to the literature.

The world's biggest market for FCOJ is the United States at around 5×10^6 tonnes of concentrate. The largest FCOJ producing state is Florida. This normally satisfies about half the domestic demand, although this is expected to be up to 70% for 1988 in a bumper harvest (Butler, 1988d). The United States imports FCOJ, typically 2.7×10^6 tonnes per year between 1984 and 1986, 90% of which comes from Brazil, with Mexico the next most important source and accounting for 6% of US imports in 1986.

After the United States, Western Europe is the second largest user of FCOJ. Orange accounts for 60% of all Western European consumption of juice and juice-based drinks. More than 65% of all single strength juice sales in the United Kingdom are orange juice. The most significant importers in Europe are Germany, followed by the United Kingdom and France, in a total market of 420 000 tonnes a year for 1984–1987 (Butler, 1987). Europe is also a producer of orange juice, responsible for 18% of world output, with Italy and Spain the major sources. However, European demand for orange juice has grown to such an extent that, to be self-sufficient, almost all the European orange crop would have to be juiced.

Table 1.1 shows the sources of imports to the three largest European markets. In world terms, the key juice suppliers are, for the United States, Florida and Brazil, and for Europe, Brazil followed by Israel. It should be remembered that intra-European sources such as Belgium and the Netherlands actually buy mostly Brazilian and Israeli juice for resale.

Table 1.1 Sources and supplies (tonnes) of orange juices and concentrates for the main European buyers 1987 (Butler, 1988b; c)

Source	Buyer		
	West Germany	United Kingdom	France
Belgium/Luxembourg	–	14 026	–
Brazil	139 918	53 555	18 060
West Germany	–	3 513	–
Israel	15 048	56 718	34 491
Italy	13 413	–	1 931
Morocco	3 833	–	5 624
Netherlands	7 178	6 520	–
Spain	2 181	–	3 745
United States	–	7 635	5 909
Other	16 134	7 318	20 009
Total	197 705	149 285	89 769

This concentration of the supply in the hands of a few producing nations helps to simplify the problems of authentication. However, the minority sources are sufficiently large in Europe to confuse the issue, and blending of juices from different countries also obscures some characteristics useful in checking authenticity.

1.1.3 Factors affecting adulteration

It should be made clear that responsible authorities in every producing country deplore adulteration, and most take active steps to discover and prevent it. In addition, importing countries are equally conscious of the need to ensure the quality of orange juice. Nevertheless, its value is such that even modest adulteration can yield handsome profits, and this inevitably encourages fraud, both on the part of producers and those receiving concentrate for reconstitution and packaging.

Against the economic incentive must be weighed the chance of detection. An important factor affecting the prevalence of falsification is the state of the permanent battle between fraudulent merchants and the enforcement authorities. Juice adulteration is cyclical. When a popular method of adulteration becomes reliably detectable it ceases to be practised, only to be replaced later by a different falsification.

An example of this process was the addition of 'orange sugar ash', an artificial mixture of sugar, minerals and ammonium salts, to simulate the minerals, total nitrogen content and the formol number of genuine juice (Scholey, 1974). These falsifications were combated by analysing for ammonia and by the introduction of the chloramine-T index. Adulterators then moved on to use cheap glycine or ethanolamine to boost the formol number (Scholey, 1974; Petrus and Vandercook, 1980). Both practices reduced in significance when even primitive amino acid analysis was able to reveal the additions. Adulterators subsequently turned to mixtures of amino acids to mask dilution, using protein hydrolysates or even especially concocted mixtures of amino acids, designed to mimic more closely the pattern of acids in genuine juice. Further advances in automated amino acid analysis have since reduced the chances of such additions escaping detection.

There is a further example of the cyclical process. Around the end of the 1970s, the most popular adulterant in the United States was simply sucrose. Isotopic methods of analysis began to detect some sucrose additions and the emphasis turned to pulp wash (material resulting from washing the rag and crushed juice sacs which remain after pressing, with water). The Petrus method for the detection of pulp wash then became more important (Petrus and Attaway, 1980). This depends on spectrophotometric measurement at both visible and ultraviolet wavelengths, and itself encouraged the addition of ascorbic acid, together with colorants, such as turmeric, annatto and beta-carotene, to try to defeat the detection method. Since adulteration involves the addition of water, isotopic methods were then produced capable of distinguishing orange juice water from ground water. These enjoyed some success, but there were reports of technologically advanced adulterators buying condensate water from orange concentrate factories to use as a solvent of appropriate isotopic composition for juice falsification (Doner et al., 1987).

Two general comments can be made. Firstly, it is usually difficult to detect reliably small degrees of falsification—a fact which possibly explains the emergence in 1982 in the United States of an attempt to fake juice by adding simultaneously 10% pulp wash, 10% Navel orange juice and 10% of a solution of beet sugar. Fortunately, the overall effect was enough to arouse suspicion. Secondly, detection methods have to be practised to be effective. It is not the case that the mere existence of a method will prevent the kind of fraud it detects. For example, buyers sometimes become complacent and cease to monitor their purchases adequately. Such lack of vigilance was apparent in the United Kingdom in the mid-1980s, and it was apparent that grossly adulterated concentrate was sometimes accepted without question or realistic test.

Where checks on quality become so relaxed it is hardly surprising that almost old-fashioned falsifications reappear. These involve adulteration, a term that is increasingly being used to describe the addition of materials not of citrus origin. In contrast, 'sophistication' implies the use of citrus products to extend or attempt to replicate genuine juice. The simplest adulteration is the addition of water, normally accompanied by sucrose and citric acid to maintain the soluble solids content. The use of non-orange citrus juices, orange pulp wash or other materials derived from peel, rag or pith constitutes sophistication. The terms are not fully precise, however, and are often interchanged.

Whatever the malpractice, the aim is always the same, namely to pass off as orange juice a mixture which is inferior and cheaper. The objective of most work on authenticity is, in essence, to render effective falsification so costly as to be no longer attractive.

1.1.4 Combating fake juice

There are various means to secure genuine juice. Most companies try to limit their purchases to reliable sources of good repute. Some may control their own primary sources of production, but many have to buy in the world market and need the means to establish authenticity. This is not to say that authentication is exclusively the concern of the industrial buyer. Producing countries, alive to the damage that adulteration can cause to legitimate trade, have inspection and control schemes. At the same time, the ultimate consumer pays the cost of the fraud, but looks to public enforcement authorities to protect him. However, few consuming nations lay down comprehensive standards. Among the exceptions are Germany, France and the Netherlands.

In Germany, the RSK values (e.g. Bielig et al., 1985) are the standards favoured. RSK stands for Richtwerte und Schwankungsbreiten bestimmter Kennzahlen, or guide values and ranges of specific reference numbers. The values are a remarkably extensive and thorough report on the composition of the two main citrus juices entering that country, namely orange and grapefruit, and also extend to a variety of other fruit juices. RSK values for orange juice cite some 35 physico-chemical determinations, plus a full amino acid analysis, a sensory analysis and include a separate specification for contaminants such as toxic metals. Juices are only accepted as authentic if their composition falls within the limits given.

The French employ their AFNOR standards in a similar way to the RSK values, but also include a statistical analysis of the figures. The Dutch rely on a more restricted range of 22 characteristic measurements, including some amino acids (Dukel, 1986). Although the Dutch specification is laid down by a trade association, it has the effective force of law in that country.

A greater measure of agreement on compositional standards may be expected within Europe, particularly with current emphasis on the single market in 1992. Thus the International Fruit Juice Union has established a European Working Party to agree both common standards and analytical methods for use in the Member States of the European Economic Community.

The same criticism can be levelled at all such approaches, namely that they treat nearly all analytical characteristics in isolation, and can thus be seen as a recipe for fraud. That is, they serve to define the target that the unscrupulous trader has to achieve. In addition, to allow for the fact that countries differ in the availability of sophisticated analytical instrumentation, internationally agreed methods are often not the most accurate or precise but rather the most easily applied. Likewise, the delays involved in selecting analytical methods usually ensure that those chosen are not up to date. Nevertheless, the two key features of orange juice authentication are, and are likely to remain, analysis and statistics. For reasons explained below these are becoming more closely intertwined.

Traditionally, the first recourse of the analyst faced with problems of authenticity is to seek marker compounds. These are, ideally, unique to orange juice and capable of rapid, reliable detection and quantification at low cost, while being difficult or prohibitively costly for would-be adulterators to add. Above all, the effective marker compound varies little in concentration either naturally or in response to processing.

Given such a list of requirements it is not surprising that the ideal marker compound for orange juice has yet to be found. A principal cause of difficulty lies in the need for invariant concentration. Orange juice has a varied composition depending on fruit cultivar, geographical source, season and fruit maturity at harvest. One has only to consider the differences between ripe and unripe fruit on the same tree to grasp the difficulties of finding components of small variation. Statistical methods thus acquire two roles. Firstly, they are used to describe the degree of variation in components of interest and, secondly, to characterise the complexities of juice composition in ways which assist the detection of improper compositional change.

Both analytical and statistical procedures are considered here, with the emphasis on the most useful techniques. In conclusion some new approaches to practical juice authentication are suggested.

1.2 Analysis

There is a long history of orange juice analysis for authentication. This will not be reported here because the dual advances of analytical technique and adulterators' cunning mean that numerous traditional analytical measures are now of little value. Even where a particular determination persists as a useful indicator, the older literature holds much data produced by comparatively unreliable means.

Published results demand careful scrutiny in any case, because analysts have persistently reported the composition of orange juice without reference to the soluble solids content. The latter is still most usually expressed on the Brix scale (strictly only appropriate for pure sucrose solutions) and determined with a refractometer. The concentration of orange juice as pressed is, of course, variable. To take a single example, Lifshitz (1983) quotes a range of 10.1–13.1° Brix for Israeli juice. At the same time there is little international agreement about what concentration properly constitutes single strength juice when it is made from diluted FCOJ. While some markets have legal minima (e.g. Dutch 11.2, US 11.8° Brix), others have been uncontrolled. Thus reconstituted juice of 10.5° Brix was widely available in the United Kingdom during 1988. (This particular situation is due to change as a result of the first agreements on common standards for fruit juices within the European Economic Community. These specify a minimum 11° Brix for orange juice made from concentrate, while a minimum of 10° Brix is allowed for fresh-pressed juice.) Nevertheless, the reporting of compositional variations per juice, rather than for a fixed amount of soluble solids, widens the apparent range of compositions and increases the difficulty of interpretation.

Old analytical approaches die hard. A recent reference work (Francis and Harmer, 1988) suggests that fruit content is still commonly estimated from determinations of potassium, nitrogen and phosphorus. Owing to their variability and the ease with which they can be adjusted by chemical additions, these measures are largely useless as primary indicators of adulteration. Similarly, the once-popular chloramine-T titration is now little used, while betaine, once a candidate marker compound, has proved both tedious to determine and rather variable in concentration. The measurement of oil content, also a labour-intensive procedure, may still find occasional application to confirm the addition of peel oil in flavours designed to hide a lack of genuine juice.

Methods of current interest and application are mentioned below, together with notes on some of limited use and indications of possible future developments.

1.2.1 Minerals

Influence of analytical method. Interpretation of results of mineral analyses require care. Different authors refer their figures to mass or volume, and some ash juice first where others use diluted and filtered or centrifuged serum directly. The latter technique is popular for the flame photometric determination of potassium and sodium, while atomic absorption analysis is more usual for other metals. Sodium, calcium and phosphorus are all more concentrated in the pulp than the serum of orange juice. Magnesium is about the same in both, and results for potassium are contradictory (Royo-Iranzo and Gimenez-Garcia, 1974). The method of preparation can be expected to influence results, and there is thus some support for the less convenient technique of ashing samples to avoid irreproducible fractionation before analysis.

Ash. Inorganic materials comprise some 0.4% of single strength orange juice (SSOJ), and the measurement of ash and alkalinity of ash were classical methods for assessing juice content. These parameters, much mentioned in the literature, have little direct significance in authentication because their adjustment is simple. However, more specific determinations of various minerals can be related to the ash content as an advantageous way of reducing their variability. For example, phosphorus/ash or potassium/ash are more narrowly distributed than the elements in orange juice (see e.g. Cohen *et al.*, 1983).

Alkalinity of ash, usually expressed as the alkalinity number or the number of millilitres of 1 M acid required to neutralise 1 g ash, ranges from 11 to 16 (Park *et al.*, 1983; Fry *et al.*, 1987). Extreme values merely indicate a need for more thorough investigation of the minerals present.

Potassium and phosphorus. In addition to ash, the three great classical measures to check on adulteration were potassium, phosphorus and total nitrogen. As a result, substantial tabulations of these exist for many fruits (Goodhall, 1969). Hulme *et al.* (1965) proposed an equation linking all three measures to estimate the fruit content of comminuted whole orange drinks. The ease with which these parameters can be manipulated reduces their usefulness in isolation. More detailed examinations, and the use of ratios of these components with others, show more promise. Thus the phosphorus content of orange juice has been fractionated into organic and inorganic types (Vandercook and Guerrero, 1969; Fry *et al.*, 1987) while Vandercook and Guerrero also separated and measured 'ethanol-insoluble phosphorus'. Lifshitz and Geiger (1985) further fractionated the phosphorylated compounds of orange juice, in a potentially valuable method discussed later (see section 1.2.8).

Measures of potassium are still employed. The absolute potassium content of orange juice depends on the origin of the fruit, but the element normally comprises 50% of the ash. If the juice comes from a single fruit variety of known or claimed source, then potassium can be used as a check. Some recent data are in the literature (Brause *et al.*, 1984). Overall, potassium commonly ranges from 1.2 to 2.8 g/l SSOJ, so the correct value for a juice of unknown origin is unpredictable. Normally, the analyst looks for at least a certain minimum concentration, and Brause *et al.* (1987) suggest 1.2 g/l, while Bielig *et al.* (1985) use 1.7 g/l for SSOJ at 11.8° Brix. The latter authors note that Israeli juice can have as little as 1 g/l, a figure supported by the data of Lifshitz (1983) and Cohen *et al.* (1983).

Calcium and magnesium. In Californian Navel juice, Vandercook *et al.* (1983) showed a strong correlation between potassium and both calcium and magnesium. Bielig *et al.* (1985) also observed the links between potassium and magnesium and suggested that genuine juice would have a potassium/magnesium ratio of no more than 21, together with magnesium, seldom outside the range 90–130 mg/l for 11.8° Brix juice. These figures are supported by independent data (Vandercook *et al.*, 1983; Fry *et al.*, 1987). However, it has been shown that Israeli juice, while conforming easily with the target ratio, only averages 90 mg/l magnesium (Cohen *et al.*, 1983).

Spanish orange juices range in magnesium concentration from about 80 to 120 mg/l (calculated to 11.8° Brix). However, of 29 compositional parameters measured in 46 different Spanish orange juice concentrates, magnesium concentration showed the least variation (Navarro et al., 1986).

In the case of Californian Navels, potassium and calcium were even more closely correlated than potassium and magnesium (Vandercook et al., 1983), but this has not encouraged widespread use of the former pair in a ratio. This is all the more surprising since pulp wash contains substantially more calcium than orange juice (Park et al., 1983; Brause et al., 1987), and the ratio potassium/calcium might thus reveal pulp wash. However, it is not clear whether the calcium of pulp wash originates in the orange or the wash water. The latter will be of variable composition, as will water used legitimately for the dilution of concentrates. Although interpretation of calcium values needs caution, high concentrations, say above the equivalent of 110 mg/l at 11.8° Brix, deserve investigation.

The ratio magnesium/calcium may also be useful, and shows a narrow distribution for both Israeli and Brazilian juices (Wallrauch, 1984). These sources differ in mean values, which are 0.88 and 1.42, respectively, according to Wallrauch although Cohen et al. (1983) record a mean of 0.94 for Israeli juice.

Chloride. This anion has featured in much work on Israeli juices (Cohen, 1982; Lifshitz, 1983; Cohen et al., 1983, 1984b; Wallrauch, 1984). RSK values merely indicate an expected maximum of 60 mg/l SSOJ, but admit that quantities of more than 100 mg/l can occur. Around 110 mg/l seems to be the maximum encountered in Israeli juice.

Sodium. Sodium is not normally correlated with chloride in orange juice. Sodium is measured much more frequently than chloride and is a valuable indicator for several reasons. Broadly, good quality SSOJ contains only about 10 mg/l sodium. A number of potential adulterants contain more than this. For example, pulp wash can provide more than 100 mg/l (Brause et al., 1984) and Israeli tap water 200 mg/l (Lifshitz, 1983). Excess sodium can also be derived from any addition of preservative or adulterant as a sodium salt. While the latter action would be a particularly incompetent adulteration, sodium is still a good indicator because it is normally so low in concentration. Amounts greater than 30 mg/l of 11.8° Brix juice are suspicious.

Other metals. These have been determined in orange juice, notably by McHard et al. (1980) and Nikdel and Nagy (1985). Although suggested as a means of characterising the geographical origins of juices, such measurements are not regularly used as a guide to adulteration.

1.2.2   Organic acids

Citric acid.   The prized flavour of orange juice includes a characteristic tartness. This is caused by citric acid, the second main dissolved component of the juice after sugars. Single strength juice contains about 1% citric acid. This can be estimated conveniently by titration with a strong base to an approximately neutral pH, typically 8.1 (Anon, 1968). The resultant figure is the 'titratable acidity' or 'total acidity' and includes contributions from all the acids present. Because citric is the most significant of these, total acidity is almost universally expressed as anhydrous citric acid. The exception is the German RSK value which not only involves a titration to pH 7.0, but expresses the result as tartaric acid, a material almost wholly absent from orange juice. Total acidity has long been used as a general indicator of quality, and much concentrate is traded on specifications which include Brix/acid ratio as a quality indicator. Citric acid may be determined accurately by high performance liquid chromatography (HPLC) (Jeuring *et al.*, 1979; Fry *et al.*, 1987).

Citric acid is of interest here because it comprises a third of the non-sugar dissolved solids of orange juice. Substantial adulteration is not possible without its addition. Unfortunately, the amount of the acid in genuine juice is rather variable, both between and within seasons (e.g. Clemens, 1964a, b; Cohen, 1982; Lifshitz, 1983). Commercially, citric acid is available at low cost and there is no practical obstacle to its use in adulteration. Attempts to identify the minor constituents of commercial citric acid have not demonstrated any marker compounds to betray the presence in orange juice of citric acid made in the usual way; that is, by fermentation of molasses using *Aspergillus niger.*

Isocitric acid.   However, the juice itself contains, amongst other acids, about 100 mg/l of isocitric acid. This may be determined enzymically (Wallrauch and Greiner, 1977; Anon, 1980) and is not found to any extent in commercial citric acid. The ratio citric/isocitric acid can thus serve to indicate the addition of acid to orange juice (Bergner-Lang, 1972, 1975; Buslig and Ting, 1977). Fortunately, there is some correlation between the two acids, and the ratio averages about 105 for orange juice, with 95% confidence limits of 80–125 (Cohen *et al.*, 1984b). The RSK guideline maximum is 130 (Bielig *et al.*, 1985). The figures for juices of different origins cluster about different mean values. For example, Bielig *et al.* (1985) give the mean ratio for Brazilian Hamelin and Pera variety juices as around 110, while Wallrauch reports 106 for Brazilian juice but only 80 for Israeli (Wallrauch, 1984). Bielig *et al.* (1985) also suggest that Israeli Shamouti and Valencia juices have a mean ratio of about 80. These authors quote a similar figure for Californian Navel orange juice, but this is not supported by Park *et al.* (1983) who found a value of 100. Some confusion is also caused by workers who have calculated the ratio from total acidity expressed as citric acid, a figure which overestimates the true citric acid content.

Isocitric acid is rather costly. At mid-1988 prices, each 'synthetic tonne' of 63° Brix concentrate would require some US$4000–5000 for isocitric acid alone to achieve the RSK recommended minimum of 70 mg/l after dilution to single strength. Thus, although the ranges are rather broad, the ratio still has a place in current methods.

Ascorbic acid. Ascorbic acid, or vitamin C, is one of the components of orange juice well known by the consumer. It can be determined by the familiar 2,6-dichlorophenolindophenol titration, or by HPLC (Dennison *et al.*, 1981; Fry *et al.*, 1987). Ascorbic acid is not stable, and its content in orange juice can be rather variable depending on the treatment of the juice. Exposure to oxygen from air and to heat both cause losses. Lifshitz *et al.* (1974) showed that 18% of the ascorbic acid in fresh Shamouti juice could be lost just on pasteurisation, with further losses on storage. Despite this, the high standards to which the industry works normally ensure that processed SSOJ contains around 350 mg/l (Lifshitz *et al.*, 1974; Bielig *et al.*, 1985).

However, because of the instability of ascorbic acid, about all that can be said about it in the context of authentication is that some should be present. RSK guidelines have a minimum of 200 mg/l SSOJ. Low concentrations indicate poor handling. Adulteration is another possible cause, but the acid is so easily added that it is of little value as an indicator of authenticity. However, ascorbic acid does contribute to the ultraviolet absorption spectrum of orange juice, and the spectrophotometric assessment of juice for authenticity is an important current technique (see Photometric methods). The variability of ascorbic acid concentration needs to be remembered therefore.

Minor acids. The numerous minor acids in orange juice are usually insignificant in authentication. Malic acid, typically about 1.7 g/l in SSOJ, features in RSK values. However, Brazilian juices, which contain more malic acid than Israeli, have tended to predominate in the RSK tables and thus weight the figures against the Israeli product (Evans and Hammond, 1989). It follows that malic acid may serve to indicate origin, but it seems too variable to be of use as an indicator of dilution. Also, it is more expensive than citric acid and is thus not a likely additive for manipulation of total acidity. If an HPLC determination of citric acid is made, malic emerges as a separate peak and can be measured with little extra effort.

Amino acids are part of the organic acid profile of orange juice, but are described later.

1.2.3 Sugars

Sugars are the main soluble solids of orange juice, but as noted by Lifshitz

(1983), total sugars should not exceed 76% of the soluble solids (expressed as Brix). Since most adulteration will involve the addition of sugars, but not a complete range of the other soluble materials of orange juice, this maximum is a convenient check on sugar addition. However, the addition of sugar is permitted in some countries such as France and Germany. The intention is to allow only limited sweetening of genuine juice (Bielig et al., 1985) but the addition does not necessarily have to be declared, and the practice can make the detection of adulteration more difficult. Accordingly, the corollary of 'total sugars', namely the 'sugar-free extract' or the sum of the non-sugar substances, is often used. This form of expression has the advantage that no confusion arises from the addition of sugar. Specifications for sugar-free extract occur in the Dutch standard (minimum 25 g/l) and the RSK values (standard value 26 g/l, minimum 24 g/l).

More useful than total sugars is the determination of the principal types present, sucrose, glucose and fructose. For this purpose HPLC (e.g. Brause et al., 1984) has largely superseded the less informative titrimetric or colorimetric determinations. Broadly, the three main sugars appear in one of two patterns. These are glucose/fructose/sucrose of 1:1:2 and 1:1:1. The former is found in orange juice from Florida, Texas and Mexico as well as in early and mid-season Californian Navels. The latter applies to Israeli and Brazilian fruit. The absolute concentrations of the sugars are typically 20–40 g/l for fructose and glucose, with 30–60 g/l for sucrose. The RSK values suggest minima of 22 and 24 g/l for glucose and fructose respectively, together with a maximum of 45 g/l for sucrose.

Although these distributions are only guidelines, they allow some judgement to be made about high concentrations of sucrose which would represent, by modern standards, a fairly primitive adulteration. The more usual additive is sucrose and invert sugar, such as a partially hydrolysed syrup. This better duplicates the profile of orange juice.

High fructose corn syrup (HFCS) is also ideal for this purpose. For separate political reasons, HFCS is not cheaply available in Europe, but is the main source of sugars for the US soft drinks industry. Fortunately, isotopic methods, discussed later, reveal HFCS in orange juice.

Occasionally juices appear which, although of correct total sugar concentration, significantly lack sucrose. Despite the high concentrations of sucrose and acid in a frozen concentrate, there is little inversion at the temperature of storage. Thus low sucrose content, together with high fructose and glucose, can indicate overheating. This is sometimes the result of clumsy processing and poor storage such that the sucrose originally present has inverted. This needs to be distinguished from a substantially adulterated product based on invert syrup. In the first case colour and flavour also suffer, and the juice deteriorates perceptibly in quality. A number of the compositional changes involved can be seen analytically and can serve to confirm that the juice is genuine although of low quality.

While the amounts of the main sugars may vary, the ratio glucose/fructose is almost constant at around 0.9. Values much above 1.0 indicate glucose addition, for example as glucose (corn) syrup. It has been alleged that too low a ratio indicates fermentation of glucose (Bielig *et al.*, 1985). Care is needed because sometimes dextrose monohydrate is used for chromatographic calibration and results may be expressed as monohydrate, whereas the ratio is calculated on the basis of anhydrous sugars.

It is reasonable to question the validity of sugar analysis when juices are heat processed in ways which must invert some of the sucrose present. There is little published, but in their work on hesperidin and pectin, Gherardi *et al.* (1980) also measured individual sugars. Their limited figures showed a 10% inversion of sucrose on pasteurisation, with a further loss of 2–7% on concentration. The losses of sucrose were roughly matched by increases in fructose and glucose. However, the quoted figures do not balance exactly, indicating some loss of monosaccharides too. This is almost certainly due to non-enzymic browning.

1.2.4 Isotopic methods

Methods which rely on the analysis of stable isotopes are important safeguards against adulteration. Such methods are routinely practised in the United States but, despite their availability at low cost through consulting laboratories, are only now being more used in Europe. Several authors provide summaries of the theory (e.g. Doner and Bills, 1981; Bricout, 1982; Parker, 1982), but it is also covered here for completeness.

Carbon stable isotope ratio analysis (CSIRA). Interest in isotopic composition originally stems from the fact that plants discriminate against the heavier isotope of carbon, ^{13}C, during photosynthesis (Bender, 1971). The largest environmental reservoir of carbon is the dissolved carbonate and bicarbonate of the oceans. All living systems contain carbon with a smaller proportion of ^{13}C than oceanic carbon. The degree of depletion differs for different groups of plants and depends on the manner in which the plant extracts carbon dioxide from the atmosphere. Plants divide into three groups according to the pathways they employ (Bender, 1971). In common with most plants, the orange uses the C3 or Calvin cycle. Sugar cane and maize utilise the second major route, the C4 or Hatch–Slack pathway. The third, and less important group, are plants such as cactus, succulents and the pineapple. These use CAM or crassulacean acid metabolism. C3 plants are more depleted in ^{13}C than C4 and the carbohydrates of the two groups can be distinguished on analysis. CAM plants use both pathways and their isotopic compositions span the complete range of the other two classes. However, the distinction between C3 and C4 is such that the carbon isotopes present can reveal the addition of a C4-derived adulterant, such as cane sugar or HFCS from maize, to a C3 plant product such as orange juice. A sample equivalent to a few milligrams of carbon is all that is required.

'Delta' notation.   Actual atomic ratios are not used to express the isotopic composition. Instead, the 'delta' notation is employed in all such work because ^{13}C represents only about 1% of carbon and its variations, although accurately measurable, are very small. A delta ^{13}C value, in effect, measures variation in parts per thousand of the carbon isotope ratio $^{13}C/^{12}C$ relative to a standard. This eliminates the cumbersome expression of small figures to many decimal places. The conventional standard for carbon is calcium carbonate from a particular fossil known as PeeDee belemnite (PDB) (Craig, 1953). Hence, results are quoted as values of $\delta\ ^{13}C$ where

$$\delta\ ^{13}C_{PDB} = \left[\frac{(^{13}C/^{12}C)_{sample}}{(^{13}C/^{12}C)_{PDB}} - 1 \right] \times 1000$$

The average $\delta\ ^{13}C$ of oceanic carbon is about 0 parts per thousand (0‰) on the PDB scale. Plants (all depleted in ^{13}C) thus have a range of negative $\delta\ ^{13}C$ values.

C3 plants fall in the range $\delta\ ^{13}C = -22$ to -33‰, while C4 types have carbohydrates with $\delta\ ^{13}C = -10$ to -20‰ (Bender, 1971; Smith and Epstein, 1971; Doner and Bills, 1981, 1982; Doner and Phillips, 1981). Orange juice has one of the narrowest distributions of values of all foods. Doner and Bills (1981) report a mean $\delta\ ^{13}C = -24.5$‰ with a standard deviation (SD) of less than 0.6‰, while Kreuger and Reesman quote -25.7‰ with SD less than 1‰. Bricout and Koziet (1987) found -25.1‰ for juice (SD = 0.9) and -25.6‰ (SD = 0.8) for pulp. These figures contrast with $\delta\ ^{13}C = -9.7$‰ (SD = 0.14) for HFCS (Doner and Bills, 1982) and between -9‰ and -12‰ for cane sugar (Bricout, 1982).

There is thus a good chance of detecting orange juice adulterated with HFCS and, as the $^{13}C/^{12}C$ ratio cannot be manipulated, the test is extremely reliable. The sensitivity can be improved further by analysis of both pulp and juice (Parker, 1982; Bricout and Koziet, 1987). This is because, although orange juices in total have some distribution of $\delta\ ^{13}C$ values, for a given sample the juice and pulp should be very similar. However, added sugars are confined to the liquid serum, thus differences of more than 2‰ between juice and pulp are cause for suspicion regardless of the overall $\delta\ ^{13}C$ value.

Sadly, sugar beet are C3 plants as are oranges. The addition of beet sugar to orange juice cannot be detected by $\delta\ ^{13}C$ measurement unless, fortuitously, the particular beet source causes the pulp and serum to differ significantly. Likewise, some sources of HFCS, such as wheat and potato, are also C3 plants and HFCS from these would also be difficult to detect by $\delta\ ^{13}C$ measurement alone.

Oxygen stable isotope ratio analysis (OSIRA).   Because of the difficulty of detecting beet invert, attention has focused on alternative isotopic analyses to reveal beet sugar. One approach relied on the determination of the $^{18}O/^{16}O$ ratio in the water of orange concentrate (Brause et al., 1984). The most abundant isotopic form of water, $H_2^{16}O$, is more volatile than, for example $H_2^{18}O$. Accordingly, the ratio of $^{18}O/^{16}O$ in ground water differs in different areas, as the isotopes fractionate between rain and atmospheric water. In addition, plants transpire $H_2^{16}O$ more readily, and so become enriched in ^{18}O relative to their water supply. The ratio $^{18}O/^{16}O$ is described by a similar notation to that for carbon, except that $\delta^{18}O$ values are given relative to a standard mean ocean water (SMOW). All orange juices have positive values and ground waters negative values. Since water is used both in the manufacture of beet invert syrup and the preparation of pulp wash, a change in $\delta^{18}O$ to less positive values reveals both these components of falsified juice. The method is, of course, only applicable to concentrates which have not been diluted with water. Brause et al. (1984) reported that 40 samples of 63–67° Brix concentrate, from a variety of sources including Californian Navels, had a mean $\delta^{18}O = +13.5\%_0$ with SD of 2.64. However, the Navels tend to be grown further inland and are thus supplied with rain already depleted in $H_2^{18}O$. Concentrated juice from Navels alone has a lower $\delta^{18}O = +9.4\%_0$ (SD = 2.84) and, excluding this variety, the mean for other types was $\delta^{18}O = +14.3\%_0$ (SD = 1.8).

The Navel is seldom used except in blends, owing to its rather bitter flavour. A typical level of Navel juice in a blend is 10%. Such a use could result in a $\delta^{18}O$ of $+10\%_0$ (for the rather extreme possibility of a base juice $2 \times SD$ lower than the mean, i.e. at $\delta^{18}O = +10.7\%_0$, blended with Navel juice $2 \times SD$ lower than its mean, i.e. at $\delta^{18}O = +3.72\%_0$). Accordingly, values of $\delta^{18}O$ below $+10\%_0$ are suspicious, but this figure could also represent a substantial addition of water to a juice of mean isotopic composition. Brause et al. (1984) used $\delta^{18}O$ of $+11\%_0$ as the minimum consistent with pure juice as 63° Brix concentrate. If Navel juice is absent, this represents $1.8 \times SD$ below the mean of the juices they tested. Brazilian juices have somewhat higher $\delta^{18}O$ than Brause et al.'s average, thus a limit of $+11\%_0$ is more than $2 \times SD$ below the mean for these. Israeli juice was not examined.

Concentrates other than 63–67° Brix (e.g. 42° Brix) are often produced by diluting the more concentrated material with water or single strength juice. The changes in $\delta^{18}O$ that such dilution causes need to be remembered. Thus Brause et al. use a lower limit of $\delta^{18}O = +1.0\%_0$ for 42° Brix concentrates.

As previously mentioned, it is a tribute to the dedication of juice adulterators that condensate water from FCOJ production, enriched in ^{18}O, has been used as a diluent to evade this method of detection (Doner et al., 1987).

Isotopes in sucrose. Doner proposed another method to avoid the problems caused by use of condensate water. He concentrated on the isotopic composition not of water, but of sucrose. He studied both $^{18}O/^{16}O$ and deuterium/hydrogen ratios, and found that North American beet sucrose had a mean $\delta^{18}O = +27.8\%_{oo}$ (SD = 2.8) and a $\delta^2H = -143\%_{oo}$ (SD = 19). Sucrose from orange juice exhibited a mean $\delta^{18}O$ of $+34.9\%_{oo}$ (SD = 2.0) with a mean δ^2H of $-27\%_{oo}$ (SD = 9). Bricout and Koziet (1987) found similar figures for orange juices mostly from Israel.

It follows from the above that δ^2H is the more sensitive measure of adulteration. Even here, however, if we consider average compositions only, up to 15% adulteration with beet sugar could still fall within $2 \times SD$ of the mean for orange sucrose. A 25% adulteration is required to reach the limit of $3 \times SD$. At this point it is still difficult to make a categorical claim that adulteration has taken place, although the result would certainly be very suspicious. The position is somewhat better if a European source of beet is postulated. Beet sucrose from colder climates is relatively more depleted in deuterium (Doner et al., 1987), and should produce a greater disturbance in the base orange value. Unfortunately, difficulties in isolating and derivatising the sucrose, which has to be converted to its octanitrate for the analysis (Doner et al., 1987), make the method unattractive as a routine screen.

Isotopes in citric acid.   Citric acid itself is produced by fermentation using the mould *Aspergillus niger* grown on various carbohydrate or even paraffin feedstocks (Roehr and Kubicek, 1981; Doner, 1985). Of the energy sources used for the fungus, cane molasses and corn syrups have different stable carbon isotope ratios from orange. Isotopic analysis has thus been suggested as a way to reveal synthetic citric acid used to adulterate lemon juice (Doner, 1985). The approach does not seem to have been extended to orange juice, possibly because citric acid produced from beet molasses or petroleum has a carbon isotope ratio too close to that of orange for the method to succeed.

1.2.5   Photometric methods

This section deals with the spectrophotometric and fluorimetric techniques developed by Petrus and Attaway (1980, 1985) from the original observations of Petrus and Dougherty (1973a, b). Intended to detect pulp wash in Florida orange juice, these methods have achieved wider use in the general characterisation of orange juices (Fry et al., 1987). The detailed procedures are available in the literature, but the method involves making an alcoholic solution of half strength orange juice, which is then left overnight to precipitate pectin. The absorption spectrum of the clarified liquid is measured between 200 and 600 nm. There are six main maxima of interest which can be used to characterise the spectrum quantitatively. Petrus also indicates that a practised qualitative assessment of the whole spectrum is valuable. Fluorescence excitation and emission spectra have been used by the same workers in conjunction with the ultraviolet and visible absorption measurements. The fluorescence spectra are mainly a check on the indications from the more easily measured absorption spectra.

The wavelengths of interest in the visible region, namely 465, 443 and 425 nm, are mainly due to absorption by carotenoids, while those in the ultraviolet, 325, 280 and 245 nm, are caused by polyphenols, flavonoids and ascorbic acid, respectively. There are peak maxima at 443, 425 and 325 nm, shoulders at 465 and 280 nm, and at 245 nm there is usually just an inflection. In contrast to orange juice, pulp wash has much weaker visible absorption which is hardly resolved into individual peaks. At the same time, its ultraviolet absorption is much stronger, with a pronounced peak at 280 nm. This enhancement of absorption, typically 1.5–2 times that of juice, is caused by the extraction of flavonoids by the pulp washing process.

The spectra can be conveniently summarised by the sum of the individual absorbances at 443, 325 and 280 nm, together with the ratio of values at 443/325 nm (Petrus and Attaway, 1980). The latter figure has the advantage of being independent of dilution, and appears to have application beyond Florida juice. For this source, however, the mean ratio for genuine juice was 0.144 (SD = 0.026) and that for pulp wash was 0.048 (SD = 0.020). It has since become widely accepted that any orange juice exhibiting a ratio of less than 0.10 is somewhat suspect, and values below 0.09 (i.e. more than $2 \times$ SD below the mean) are abnormal. Simultaneously, if the 325 nm peak is more than 1.0 absorbance units (for an analysis based on SSOJ of 12.8° Brix), sophistication with pulp wash is indicated (Petrus and Attaway, 1985). The 12.8° Brix concentration is involved because the method was developed when this was the Florida minimum requirement for reconstituted SSOJ.

The absorbance sum is also, of course, a function of the exact concentration of the juice. Under standard conditions, the absorbance sum may thus reveal dilution, such as with sugar syrups, because this malpractice should reduce the absorption throughout the relevant part of the spectrum. This is not an easy judgement to make across all juices, and Petrus' own work shows substantially different absorbances at visible wavelengths for Florida and California Valencia juices.

A further observation is that genuine juice has a peak in the ultraviolet at about 220 nm; small shifts are alleged to betray dilution or pulp wash addition. Specifically, dilution results in a slight lowering of the wavelength of maximum absorption, while pulp wash addition moves the wavelength higher. The shifts are but a few nanometres, and it seems unlikely they should be a general feature of all orange juice.

Some workers, notably the Israelis, have disputed the applicability of Petrus and Attaway's method. Saguy and Cohen (1984a) show considerable variation in the spectral characteristics of Israeli juice. The key ratio 443/325 nm had both a lower mean of 0.116 and a higher SD of 0.048 than found for Florida orange juice. Saguy and Cohen (1984b) subsequently published data for a larger number of samples. For Shamouti and Valencia juices, these showed closer agreement with the American results but still had a broader distribution. At least some of these differences may be caused by variations in extraction methods between the countries. High extraction pressures enhance ultraviolet absorption in a way akin to the addition of pulp wash (Petrus and Dougherty, 1973a, b). This is because the concentration of pulp components is increased by such extraction techniques. It is thus debatable to what extent the Israeli figures really conflict with those of Petrus and Attaway.

The fluorescence spectra serve to confirm the presence of pulp wash. Several differences in both excitation and emission spectra of pulp wash and orange juice have been summarised (Petrus and Attaway, 1985). The main effect of pulp wash presence is the appearance of a peak in the excitation spectrum at 302 nm, a diagnostic feature confirmed by Brause et al. (1987). The same technique can also reveal the use of added colours to boost the visible absorption. Artificial colours are easily detected, and natural colours are therefore preferred. Of these, turmeric has been used, but the active principle, curcumin, has a strong characteristic fluorescence. Confirmatory tests for curcumin are readily applied, for example the HPLC methods of Blake et al. (1982) and Rouseff (1988). The latter work also covers annatto.

Benzoates and sorbates, which sometimes appear in retail orange juice, interfere with the characteristic ultraviolet absorption spectrum. The possibility of their being present needs to be remembered.

Other uses of quantitative spectrophotometric data have been demonstrated by Fry et al. (1987) and are discussed later. The rather slow preparative stage of the Petrus and Attaway method is a drawback, however. Accordingly, investigations have been made of speedier approaches (Fry and White, unpublished work). These have shown that the overnight delay in the method is unnecessary. Pectin precipitates immediately on addition of alcohol to the diluted juice, and can be filtered off and the spectra measured straight away. After such treatment the visible spectra of juices are identical to those found using the conventional method. In the ultraviolet spectra, the only significant difference is that the peak at 245 nm (ascorbic acid) is lower if the sample is held overnight. This is consistent with previous observations that the 245 nm result is the least reproducible. Presumably this is due to uncontrolled oxidation by atmospheric oxygen during the originally specified preparation. Moreover, processing and storage can drastically lower the ascorbic acid concentration of freshly squeezed orange juice (Lifshitz et al., 1974). Vitamin C is thus inherently unreliable as an indicator of authenticity and should, perhaps, be omitted from consideration in this test.

Other means of shortening the preparative stages of the photometric assessment may be unsuitable. In particular, pre-treatment by filtration or centrifugation before the addition of alcohol is not recommended. Centrifugation reduces absorption at all wavelengths except 245 nm, and filtration can almost eliminate the visible absorption. These effects arise because the carotenoids, and to a lesser extent other materials, are associated with solid particles in orange juice. The highly soluble ascorbic acid is not.

1.2.6 Amino acids

Formol value. Orange juice contains significant amounts of free amino acids, typically 3–4 g/l SSOJ with a minimum around 2 g/l. Before the advent of modern amino acid analysis it was realised that these were an important component, and the total amino acids of orange juice were commonly estimated by the formol titration. Representative formol values from various sources have been tabulated (Petrus and Vandercook, 1980). The formol titration requires only simple apparatus, but provides only a single figure with no information on the component acids. Indeed, the major amino acid, proline, contributes nothing to the formol value. It is so easy to manipulate the formol value by addition of cheap amino acids, such as glycine, or even ammonium salts, that it has little value in authentication. The exception to this is where it is combined with a separate determination of proline, as suggested by Wallrauch (1974). A recent view is that the formol/proline ratio should not exceed 25, with a minimum of 575 mg/l of proline in SSOJ (Bielig et al., 1985).

Owing to the ease with which formol values can be distorted by ammonia, a corrected formol number is sometimes used. For this, ammonia is determined and the formol value reduced by the contribution calculated to be due to ammonia. As oranges mature, the ammonia content of the juice increases, but is seldom more than 25 mg/l. Addition of ammonia is indicated by a corrected formol value less than 90% of the uncorrected figure.

Methods of amino acid analysis. Full amino acid analysis has been more popular in Europe than America. The history of the method is long, and much of the older data is unreliable. There is a choice of three analytical methods, namely gas chromatography, the dedicated amino acid analyser (usually ion-exchange HPLC with post-column ninhydrin detection) and, most recent, HPLC using either *ortho*-phthaldialdehyde (OPA) derivatives or following pre-column derivatisation with phenylisothiocyanate (PITC). OPA is mostly used to derivatise amino acids before separation, but systems are available which use OPA post-column instead. A serious disadvantage is that in neither case is proline detected because it does not react with OPA. Furthermore, use of OPA post-column retains ion-exchange chromatography as the method of separation, together with its disadvantages of lengthy analyses (typically more than 90 min) and a demand for several buffers of scrupulously controlled pH value and ionic strength.

Experience has convinced the writer that pre-column derivatisation with PITC is the best method. Special apparatus is commercially available which simplifies derivative preparation, and chromatography is both automated and sufficiently rapid that multiple replicates are easily analysed. A particular advantage is that proline has the same chromophore and approximately the same molar response as other amino acids. Proline is thus detected as effectively as primary amino acids, which is not the case with alternative liquid-phase analysers. A typical PITC-based system has been evaluated in comparison with a conventional analyser by White et al. (1986).

The amino acid profile of orange juice includes both major and minor components. With detector sensitivities set to accommodate the former, the latter tend to be poorly quantified, even with modern integrators. Thus, a further advantage of the rapid HPLC method is that two analyses can be made using different detector sensitivities (or, within the limits of good chromatography, different sample sizes) to achieve better quantification of the minor acids. At the same time, the elution programme can be adjusted to obtain improved separations in crowded areas of the chromatogram.

Despite the above, most work in Europe still uses the dedicated ninhydrin-based analyser, and gas chromatography also has some supporters. Whatever technique is chosen will involve substantial capital expenditure and require skilled staff. However, the comprehensive results obtainable are valuable in detection of the most sophisticated falsifications.

The analysis is not wholly straightforward. Current attempts to harmonise methods of orange juice analysis within the EEC have shown that while laboratories are usually self-consistent, large discrepancies (coefficients of variation of 25% and more) can occur between them. Some of these difficulties seem to be associated with variable quality of calibration standards. For this reason it would seem useful to have an international reference standard. In addition, normalisation of results on a mass basis may be more useful than their quotation as absolute concentrations. Because of the importance of amino acid analysis in medical work, it has become customary to use millimoles per litre as the unit for concentration. However, much food analysis is reported in units of mass, and units of milligrams per litre are thus also common.

Amino acids in orange juice.   Full tabulations of the amino acids in orange juice are in the modern literature (Lifshitz, 1983; Ooghe, 1983; Bielig et al., 1985; Fry et al., 1987). Some amino acids are also mentioned in the Dutch standard (Dukel, 1986). Proline is the most abundant and, although the exact order thereafter may vary, orange juice is also rich in arginine, serine, asparagine, gamma-aminobutyric acid (GABA) and aspartic acid. Of these proline and serine are quite costly and pose a problem for the would-be adulterator. Alanine, glutamic acid and glycine are low cost but minor components of orange juice and, therefore, additions are easily detected. Likewise the use of protein hydrolysates to furnish a range of amino acids at low cost is betrayed by the presence of acids not common in orange juice. In particular, isoleucine, leucine, tyrosine and phenylalanine are introduced by protein hydrolysates. Each of these is naturally present at less than 1% of the total amino acids of orange juice (Ooghe and de Waele, 1982).

The last four amino acids mentioned above, together with valine and methionine, occur in orange pulp and peel at concentrations more than twice those in juice. Measurement of just amino acids, however, fails to detect peel extracts even at 10–20% addition (Evans and Hammond, 1989).

The eight main amino acids of orange juice account for more than 95% of the total. Thus some simplification can be achieved by normalising these acids to 100%. Table 1.2 shows the effect of this on otherwise quite wide ranging results quoted by various authors.

With a decline in the cost of single amino acids, some scope may return for their use in carefully made additions to disguise dilution. There are two ways to combat this, although both still require some supporting research. These approaches, described below, involve either improved statistical analysis or a different chemical analysis.

Firstly, amino acid analysis can be coupled with pattern recognition analysis (see also section 1.3). This was attempted by Fry et al. (1987), but proved unsatisfactory then because insufficient data were available concerning the sources of the juices these workers were able to use. In other applications, the combination of pattern recognition analysis with analytical data was very successful. A further study of the technique applied to amino acids would thus be useful.

While this comprehensive statistical analysis remains to be done, there have been several moves to use ratios of amino acids as characteristic properties of orange juice. Where two amino acids are correlated, this makes falsification more difficult because both absolute concentrations and the relationships between them need to be normal. This idea stems from Ooghe (1981), who worked with both fruit and commercially available juices although, unfortunately, he did not reveal their origins. He selected, initially on the basis of their occurrence in several fruits, the amino acids aspartic acid, glutamic acid, proline, alanine, GABA, arginine and the sum of serine + asparagine. In orange juice these account for more than 95% of the total. (Serine and asparagine are often summed because they are not well separated on the ion-exchange analyser when sodium buffers are used, and lithium buffers double the analysis time.) The approach was expanded (Ooghe and de Waele, 1982) to list a variety of ratios of amino acids, together with their 95% confidence limits. The latter were based on the authors' own analyses of various fruit juices purchased in different EEC countries over a three-year period. The origins of the juices were claimed to be varied, but were not stated. It was recommended that, where a ratio fell outside the limit, further tests should be done. These involved a simplified chi-squared test to express all the amino acid results for one sample in a single statistic. This was then compared with a limiting value representing a 99.5% confidence limit. If this critical value was exceeded a third test, an application of the variance ratio or

Table 1.2 Profile of the main amino acids in orange juice, normalised to 100% by mass (recalculated from the original sources shown)

Amino acid	Ooghe (1981) Mixed	Ooghe and de Waele (1982) Mixed	Lifshitz (1983) Israel	Bielig et al. (1985) Mixed	Fry et al. (1987) Israel	Fry et al. (1987) Brazil
Proline	32.2	28.3	36.2	28.1	28.7	32.3
Arginine	20.5	27.9	20.7	24.6	23.5	26.7
Serine + asparagine	18.0	17.6	16.2	21.5	20.6	17.9
GABA	12.0	9.1	9.8	8.4	8.8	8.5
Aspartic acid	9.9	10.3	9.4	9.8	9.6	7.9
Alanine	4.1	2.9	3.7	3.5	4.0	2.7
Glutamic acid	3.3	3.9	4.0	4.1	4.8	4.0

F-test, was suggested. The procedures and the theory are well described in the reference.

This whole approach was criticised by Cohen and Fuchs (1984), who showed that, in the 1981–1982 season, unprocessed Israeli juices would nearly all fail the suggested criteria for ratios. Similarly, for the seasons 1975–1976 to 1981–1982, almost all Israeli orange juice would fail the chi-squared test. Even when only Israeli data were used to form the standards, comparability between seasons was claimed to be poor. However, in the absence of common reference materials or calibrants, it is difficult to say how many of these differences are due to analytical variations.

Moreover, it is of interest here that amino acid analysis can also help to indicate the presence of pulp wash or peel products. For example, Ooghe's figures show that such materials contribute to relatively high levels of aspartic acid and alanine, together with relatively low arginine (Ooghe and de Waele, 1982). The ratios on which Israeli juices matched least well were GABA/arginine (Israeli ratios too high, i.e. low arginine) and arginine/proline (Israeli ratios too low, also indicating low arginine). For ratios involving aspartic acid or alanine, the numbers of juices failing the criteria are distributed in a way which suggests that some might contain peel or rag components.

Thus the Israeli observations may be associated at least in part with different extraction practices. It would be unwise to reject Ooghe and de Waele's work without examination of this factor, but there are no published data. Various other ratios of amino acids have been proposed by different workers and summarised (Ooghe, 1983).

The second prospective advance in amino acid analysis relies on modern chromatographic techniques to resolve mixtures of optical isomers. The natural amino acids in orange juice are all L-isomers, and there is no evidence that this is changed by racemisation during processing. However, low cost synthetic acids are usually racemic mixtures of D and L forms; even protein hydrolysates contain some D acids formed by racemisation during hydrolysis. Detection of D acids should thus prove adulteration. Gas chromatography has been applied to orange juice to demonstrate that D-aspartic acid could easily be detected in that matrix (Ooghe, 1983; Ooghe et al., 1984). A novel HPLC method has been developed which separates both D- and L-proline from orange juice and provides a detection limit for the D acid of about 10 mg/l (Duchateau et al., 1989). While this method offers the possibility of easy analyses for other D acids, proline is, unfortunately, an exception: it is produced commercially by an enzymic process which yields the L form. Costly racemisation of a DL mixture is thus not required for proline.

If adulterators are compelled to use only L amino acids, the cost of these should still be an effective barrier to falsification. Thus, based on mid-1988 prices for a few kilograms of each amino acid, it would cost about US$1500/tonne to fake the main amino acids of 63° Brix concentrate. This assumes the addition of just enough proline, arginine, serine, GABA and asparagine to meet the Dutch minimum compositional requirements. The costs of other ingredients, such as sugar, acids (particularly isocitric) and colour, would be extra. This figure for artificial material compares with the very high prices for FCOJ of around US$2200/tonne in April 1988 (Butler, 1988d), and more typical prices of only US$800/tonne throughout 1986. Proline accounts for more than half the cost of the synthetic amino acids in the above calculation.

1.2.7   Methods of limited application

The title of this section is not meant to imply that the methods described are ineffective. Rather it is a collecting point for approaches which are not well classified elsewhere. Some are, indeed, outmoded and included only for completeness. Others either have yet to become widely used or are specific indicators of a single foreign material.

Chloramine-T index.   This falls in the category of measurements often still encountered, but of questionable value. It is significant that this parameter is omitted from the otherwise comprehensive RSK values.

Chloramine-T is an oxidising agent. Its degree of reaction with orange juice, in competition with potassium iodide, is a measure of reducing substances other than sugars. It is estimated by addition of a known excess of chloramine-T in the presence of the iodide, followed by back titration with thiosulphate (e.g. Jones, 1979a, b; Cohen et al., 1984b). Israeli juices typically have a mean value around 12 ml 0.01 N chloramine-T/ml SSOJ (Cohen et al., 1983; Lifshitz, 1983; Cohen et al., 1984b). The chloramine-T index of Florida juices is reported by Maraulja and Dougherty (1975), who noted that harder squeezing of oranges raised the value. This is consistent with the finding of higher chloramine-T values for peel than the corresponding juice (Cohen et al., 1984b). The index may thus give some indication of pulp wash addition (Wucherpfennig and Franke, 1966).

Betaine.   The determination of the amino acid betaine, trimethylaminoacetic acid, also called lycine (sic), has also become unpopular. First proposed as an indicator of juice in drinks, betaine averages 720 mg/l at 11° Brix (Lewis, 1966). It also appeared to have a narrow distribution of values, with a coefficient of variation of 6.2%. Methods for its determination were collaboratively tested (Rogers, 1970) and an AOAC method adopted (Anon, 1975). However, other workers found much wider variation in betaine content. Coffin (1968) reported a coefficient of variation of 25% and Floyd and Harrell (1969) one of 12%. The mean found by Coffin was markedly lower than that given by Lewis, but this may be a reflection on his samples from the Canadian market.

Betaine can be separated during the HPLC of sugars in orange juice (Linden and Lawhead, 1975), and their separations can be further improved by using 0.005 M sodium acetate buffer (pH 5.0) in place of water in the eluant (Fry, unpublished), but the peak is small. Because it is a quaternary ammonium compound, betaine is not detected by amino acid analysers and not included in tabulations of these acids in orange juice.

Naringin. Occasionally the price of grapefruit juice falls to the point where it is a tempting adulterant for orange juice. Grapefruit juice contains a characteristic flavonone glycoside, naringin, which is responsible for the bitter taste. Naringin is easily detected, for example by HPLC (Trifiro *et al.*, 1980; Greiner and Wallrauch, 1984; Scholten, 1984; Fry *et al.*, 1987), and gas chromatography has also been used (Drawert *et al.*, 1980). The RSK values are based on the method of Davis (1947) (see also Koch and Hess, 1971; Krueger and Bielig, 1976). This does not distinguish between the various flavonoids, however.

Naringin does not occur in orange juice, where the flavonoid hesperidin dominates instead. There have been claims that 3–15 ppm naringin can appear in genuine Israeli orange juice, but a study to verify this failed to find any (Galensa *et al.*, 1986). Naringin was also absent from the peel of Israeli oranges. The peel is normally the major source of such materials. It has been suggested that more than 5 ppm naringin indicates adulteration of orange juice. Undebittered grapefruit juice typically contains about 400 mg/l, and a modern HPLC procedure can be expected to reveal about 1% grapefruit juice in orange.

Biological methods. Two procedures fall into this category, namely immunoassay and microbial assay. Immunoassay has been carried out by raising rabbit antibodies to protein isolated from orange juice (Cantagelli *et al.*, 1972). A system has also been employed for estimating the juice content of orange-based drinks (Firon *et al.*, 1978). It seems more of a curiosity than a viable procedure for authentication of orange juice.

Somewhat more useful is a bioassay reported by Vandercook *et al.* in 1976 and in subsequent papers (Vandercook and Smolensky, 1976, 1979; Vandercook *et al.*, 1980). The method depends on the growth of a fastidious microorganism, *Lactobacillus plantarum*, in orange juice. The organism requires, and therefore tests for, the presence of a complex array of nutrients, including both major and minor components of orange juice. It should be impractical to adulterate orange juice such that the organism finds its needs met as completely as in the genuine product.

The original procedure demanded a 30-h incubation with the organism before the extent of growth was assessed turbidimetrically. The results had a coefficient of variation of 24% (Vandercook, 1977). This figure was halved in a later development of an automated method (Vandercook *et al.*, 1980) which relied on monitoring pH change during growth of a huge inoculum of the *Lactobacillus*. The latter arrangement also simplified sample preparation and cut analysis time to 2 h. There are no published results of applying the method to adulterated juices, although it has been used for both pure orange juice and orange drinks. Without further research it is impossible to assess the true value of the method. The authors hint that unspecified additives may encourage the growth of the organism, and this may limit the application.

1.2.8   Future possibilities

The following three topics are not routine methods, but show promise. They have been included to illustrate the directions in which orange juice authentication might proceed.

Fractionation of phosphorus compounds. The measurement of total phosphorus has already been mentioned as one of the classical indicators of fruit content. The measurement procedure most used is a colorimetric assay of the ash, based on the method of Fogg and Wilkinson (1958). The version of Vandercook and Guerrero (1969) is a good example. As long ago as 1961 the five main components of orange juice phospholipids were identified (Swift and Veldhuis, 1961). Subsequently, the main phosphorus-containing groups of compounds in orange juice were separated and the phospholipids further separated (Vandercook *et al.*, 1970). The hope was that the technique would reveal foreign lipids added as clouding agents or emulsifiers. Modern gas liquid chromatography (GLC) lipid analysis would seem a more effective route to that end.

The whole process of fractionation was taken a step further by the use of an automated phosphorus analyser (Lifshitz and Geiger, 1985). The apparatus comprised an HPLC for separation, coupled to a 40-place turntable combining fraction collector, ashing furnace and colorimetric assay. Some 27 peaks were produced in separations of the perchloric acid-soluble phosphorus compounds in orange juice and most were identified. The main component was inorganic phosphate. This can amount to 65% of the total phosphorus (Dyce and Bessman, 1973), which itself ranges from 100 to 200 mg P/l in SSOJ. Other peaks represented sugar phosphates, mono-, di- and tri-phosphorylated nucleotides as well as phospholipid precursors. Boiling the juice failed to disturb the results, thus giving hope that the analysis would be unaffected by processing. There was no information on the variability of the concentrations of the various components nor, indeed, of the reliability of the analysis. Nevertheless, a method which displays simultaneously so many minor components of orange juice is of potential value in characterisation. The approach does not seem to have been taken further.

Pyrolysis–mass spectrometry. Apart from work on stable isotopes, the mass spectrometer has found little application in orange juice authentication. One recent exception is the pioneering work of Aries *et al.* (1986). These workers recognised the practical difficulties of acquiring large amounts of conventional analytical data and proposed a novel solution. Pyrolysis–mass spectrometry is an analytical fingerprinting technique. It requires thermal decomposition of a sample in an inert atmosphere under controlled conditions. This yields an array of volatiles, characteristic of the original sample. Mass spectrometry of these volatiles is then carried out by passing the pyrolysis products to the ion source of a mass spectrometer. Electron impact ionisation minimises fragmentation of the products. The original workers used a quadrupole spectrometer to scan the range m/z 51–140.

The advantages of the method are that it is rapid (less than 5 min per analysis) and requires no sample preparation beyond centrifuging SSOJ and drying a few microlitres on a metal foil for analysis. However, chemical interpretation of the spectra is virtually impossible, because any peak may have originated from numerous compounds simultaneously. Although some tentative assignments were made, the authors relied on pattern recognition statistics to process the copious analytical data. This has the disadvantage that, while the method may indicate an adulterated juice, it cannot be expected to show why the juice has been rejected or what adulteration has been practised. A later publication (Evans and Hammond, 1989) acknowledges that the technique requires further refinement before it can be routinely applied.

Despite these drawbacks, interesting results were produced in separating orange juices of Brazilian and Israeli origin. The mass spectrometer generates data well suited to computing the pattern recognition analysis without the need for transcription of conventional laboratory results. The penalty is that the technique demands sophisticated and costly equipment.

The published study was essentially a pilot one. Little detailed information was available about the sources of the juices examined, and no juices known to be adulterated were tested. Discrimination between juices from the two countries was incomplete, whereas the same statistical method applied to conventional chemical analyses has had more success (Fry *et al.*, 1987). Given sufficient additional study, the approach could be very useful because it does assess the whole sample, and it may thus be able to disclose a wide range of adulterations rapidly. The technique currently finds use in one consulting laboratory as a screening method to identify juices sufficiently unusual to merit further conventional multichemical analysis.

Determination of polymethoxylated flavones. Polymethoxylated flavones (PMFs) are non-polar materials which are a notable feature of citrus peel oils. However, they occur in orange juice at a maximum of around only 5 mg/l. They were first studied because they were thought to contribute to off-flavours, probably through their oxidation (Veldhuis *et al.*, 1970). The ascorbic acid in orange juice almost certainly protects the PMFs against this, and the development of HPLC methods made their analysis easier (Ting *et al.*, 1979; Rouseff and Ting, 1979; Bianchini and Gaydou, 1980). There are five main PMFs in orange juice. In declining order of concentration these are nobiletin, sinensetin, heptamethoxyflavone, tetra-*O*-methyl scutellarein and tangeritin. There are a number of minor PMFs, and their pattern appears to be characteristic of the citrus species (Ting *et al.*, 1979).

PMFs were included in methods for the liquid chromatographic characterisation of orange juice by Perfetti *et al.* (1988). Here, the results were not quantified, partly because too few standard materials exist for calibration. Rather the peak heights from the chromatograms were used directly in a pattern recognition analysis.

It seems likely that PMF analysis may have a future if peel by-products are used to sophisticate juice. However, the measurement of PMFs will contribute little to the detection of conventional adulteration because their characteristic pattern will be preserved on mere dilution and the addition of sugars and acid.

1.3 Statistics

1.3.1 Introduction and the Gaussian distribution

The purpose of this section is to show the role of statistical analysis in orange juice authentication, and to describe how that role has increased in importance. It is not an introduction to statistical method, but some basic concepts are mentioned for clarity.

Every compound measured for juice authentication exhibits a spread of values. This spread reflects compositional differences between different varieties of fruit and differing geographical origins, as well as the season, time of harvest, fruit maturity and different styles of cultivation. Seldom acknowledged, but also included, is the analytical error of the sampling and determination. These factors combine to produce a distribution of values for any measured characteristic (variate). Simple statistical parameters may be used to describe the spread of the distribution. The most common are the standard deviation (SD) and the related confidence limits. Thus a Gaussian ('normal') distribution can be completely described by the mean and SD, where the SD is given by

$$SD = \left[\sum (x - \bar{x})^2 / (n - 1) \right]^{1/2}$$

where x = the measured value, $\bar{x}$ = the arithmetic mean of measured values and n = the number of measurements. Such a distribution also has the useful property that 95% of values fall within a range of $1.96 \times SD$ either side of the mean. Because the error is small, most writers approximate this to $2 \times SD$. Provided certain criteria are met, one can be confident that a genuine value will fall outside this range by chance in only 5% of observations. Other confidence limits can also be calculated easily. For example, the 99% limit (mean $\pm 2.57 \times SD$) and the value $3 \times SD$ (99.73% of the distribution) are both often used. Many original analytical data are tabulated as mean values together with a SD or a range comprising $\pm 1 \times SD$.

Thus, while values do fall beyond a particular confidence limit by chance, it is not likely that they will occur many SDs removed from the mean. Such distant figures may arouse suspicion that a juice has been tampered with, or may be so remote from the mean that adulteration or poor processing is almost certain. The confidence limit provides a quantitative assessment of the risk of making such a judgement incorrectly, that is, the risk of wrongly condemning a genuine juice. There is no single, universally accepted confidence limit in orange juice authentication. Some national standards for orange juice composition use the 95% confidence limits as the acceptable range (Koch, 1979).

1.3.2 Some problems of statistical interpretation

The application of confidence limits is not as straightforward as might at first appear, and this has led to some confusion in the literature. For example, there is no guarantee that a particular distribution is Gaussian. This basic factor should first be checked by other statistical tests such as Filliben's test (1975). It can be shown for any distribution that the probability of a value falling beyond the mean $\pm 3 \times SD$ is less than $1/9$ (from Chebyshev's Theorem, see e.g. Freund and Walpole, 1980). However, this compares rather poorly with the special case of the Gaussian distribution, where that probability is only $3/1000$. There is thus considerable advantage in checking that a Gaussian distribution obtains for the data in question. Fry et al. (1987) did check the distributions of their measurements of most of the common analytical parameters of orange juice. They found no evidence that these were not Gaussian, and this gives reassurance that orange juice data are, in general, suitably distributed for the statistical interpretations made. Cohen et al. (1984a) indicate that the concentration of arginine is not normally distributed in Israeli orange juice. Where such circumstances are found, it is usual to seek to transform the figures (for example by taking logarithms) to a form where the distribution is Gaussian and the statistical treatments are valid.

A second source of misunderstanding is the assumption that statistical properties determined from samples also apply to the population of all orange juices. However, there is always some error involved in estimating the population characteristics this way. For example, the population mean lies (with 95% confidence) within limits defined by

$$\text{sample mean} \pm 1.96 \text{ standard error of the mean}$$

where the standard error $= (SD/\sqrt{n})$ and SD and n are as previously defined. (This is strictly true only for a random sample of the population.) The greater problem arises from the difficulty of defining which population is being examined. This is particularly acute for those checking the authenticity of orange juices from a variety of sources. Thus there are clear demonstrations that Brazilian and Israeli juices differ significantly (Wallrauch, 1984; Fry et al., 1987), while published data on orange juices from other sources also show them to be different (e.g. Californian Navel, Park et al., 1983; Spanish, Navarro et al., 1986). Equally it can be shown that significant variations in composition arise from year to year in juice from a single area. Many of the supporting references for this concern Israeli juice (Cohen, 1982; Cohen et al., 1983; Brown and Cohen, 1983), yet in some respects Israeli juice has also been claimed not to vary much between years (Lifshitz et al., 1974). It is, therefore, necessary to be aware that published figures may have been gathered from a specific sub-population of juices only, and they may have quite limited significance outside that context.

Sometimes, the manner in which statistics have been used to judge authenticity has also been extreme. For example, Richard and Coursin (1979) suggest that the appearance of a single analytical result beyond the ranges considered normal by an expert constitutes grounds for rejection. It is clearly unsatisfactory to make such a decision on the basis of a single 95% confidence limit. This point has been pursued vigorously (Brown et al., 1981; Brown and Cohen, 1983), but not always with complete clarity. Thus these latter authors point to the work of Ara and Torok (1980), who showed correctly that the probability of rejecting a genuine juice as adulterated, on the basis of at least one result outside a 95% confidence limit, becomes rather high the more measurements are made. The probability is 5% for one measurement, 22.6% for five, 40.15% for ten and 60.2% for twenty measurements. It is thus apparently highly likely that genuine juice would be condemned on the basis of the method of Richard and Coursin. However, it is for this very reason that categorical rejections are seldom made on the basis of a single deviant result. Moreover, the figures provided by Ara and Torok are true only for completely independent variates. The components of orange juice are not independent, indeed some are closely correlated, and the probabilities calculated are thus inappropriate.

An interesting approach to this problem of interpretation has been embodied in the Dutch standards (Dukel, 1986). Some 22 variates are specified, with minimum or maximum values. Only two have to be met by all juices, namely a citric/isocitric acid ratio of not more than 130, and a glucose/fructose ratio of not more than 1.00. It is permissible for two other variates to deviate by up to 25% of the standard value each. Alternatively, three may vary, but by only 10% each. Provided all other measures meet the required standards, these deviations are acceptable. This essentially quantifies the interpretation often used by an expert in assessing orange juice authenticity, that is, several indicators are sought to confirm that a juice is not genuine.

1.3.3 Ratios as variates

The use of ratios follows from the observation that the compositional characteristics of orange juice are not all independent. Various components are strongly correlated (e.g. Vandercook *et al.*, 1983) and, where this is so, ratios have several potential advantages over the single quantities. For example, ratios are independent of dilution and are thus convenient. Ratios reflect the variation in two or more components simultaneously. Thereby they impose a greater burden on the would-be adulterator, who has to match both the absolute concentrations of the components and the relationship between them.

In general, ratios have been approached as 'extra' single values. That is, they have been treated in the same way as single quantities, and ratios with narrow distributions (low coefficients of variation) have been sought. Apart from the well-known ratios included in the Dutch standards, various others have also been mentioned already (see Minerals, Photometric methods, Amino acids). Substantial groups of ratios have sometimes been used to evaluate orange juice, such as the 9 groups of 29 ratios employed by Fischer (1973).

1.3.4 Multivariate techniques

The natural extension to using ratios is to incorporate several relationships simultaneously. The regression analysis originally suggested by Rolle and Vandercook (1963) for lemon juice is an example (see also Vandercook *et al.*, 1973). Here, citric acid was linked in an equation with total phenolics, amino acids and malic acid. Coffin's work on betaine (Coffin, 1968) also contains this approach applied to orange juice, as do the various publications of Richard and Coursin (e.g. Richard and Coursin, 1984). Jones (1979a, b) tried to develop regression equations for citrus juices, but his success was limited by the range of samples available as a base set. This has been a common criticism of regression methods. The procedures can appear to define closely the characteristics of the samples examined, but are less accurate when applied to other juices. Cohen *et al.* (1984a) note that regression techniques are not used in routine quality control of Israeli citrus juices.

In a different approach, non-parametric statistics have been used to detect adulteration (Schatzki and Vandercook, 1978). The principle involves the classification of an unknown sample (i.e. one whose composition has been measured but other characteristics are unknown) by comparison with known samples in a 'training set'. The unknown is assigned to the same group as the known sample it most closely resembles. This is also called the 'nearest neighbour' scheme. The basis of comparison was a statistic referred to as the Euclidean distance. This quantity is a single figure for each sample, calculated from all its measured variates, indicating the overall distance of these from the known samples. The pair of samples, unknown plus a known one, with the minimum Euclidean distance are the nearest neighbours. The advantage of the scheme is that no assumptions are made about the underlying distributions.

For the classification to function, two training sets are needed, namely 'genuine' and 'adulterated' juices. As is usually the case, data on the former were available, but the adulterated juices had to be simulated. The classification was then tested by removing each known juice in turn and re-presenting it to the others as an unknown. The frequency of correct classification could thus be found. Initial results were unimpressive: a 20% dilution with sugar was correctly identified in 68% of cases, while genuine juice was misclassified 38% of the time. By using multiple sampling from a particular source, and a more appropriate training set, significantly improved results were achieved. For example, 20% dilution with sugar solution was always detected and the use of pulp wash and sugar was only misclassified as pure juice 4% of the time, although 12% of pure juices were wrongly classified.

Such improvement may assist regulatory authorities investigating suspect sources, but does not assist the authentication of a lone sample of orange juice. Success with the method also depends on choosing variates which tend to reveal differences, as well as on having adequate training sets. A good example of the successful application of a nearest neighbour scheme was the work of Perfetti et al. (1988). Using analyses of a wide range of substances separated from orange juice by HPLC, they made classifications based on not one but four nearest neighbours. Of 19 adulterated juices (comprising in-lab preparations as well as fake material seized by the FDA), 18 were correctly identified. At the same time, only 1 from 80 authentic Florida juices (four varieties) was misclassified.

1.3.5   Pattern recognition analysis

The nearest neighbour scheme is an example of supervised learning. Here, the number of classes is known, and boundary conditions are sought which enable an unknown to be correctly classified. The corollary is unsupervised learning, where no assumptions are made about the number of classes. Both are included in the more general name cluster analysis, itself a part of pattern recognition analysis. This area of statistics is concerned with mathematical techniques to discern significant groupings in (usually very) large volumes of data. The raw information would normally defeat simple inspection, or the traditional approach of comparing only two or three variates, as in a graph. Excellent descriptions of pattern recognition analysis are available, both at introductory (Betteridge, 1983) and more detailed (Kowalski, 1984; Jurs, 1986) levels.

A common feature of the commercial computer programs essential to pattern recognition is that they utilise n analytical measurements per sample to assign its position in n-dimensional space. Where n is no more than 3 this can be visualised, but although it is difficult to imagine coordinates in more than three dimensions they can be calculated with ease. Similarly, the distances between such points can be calculated, regardless of the number of dimensions. The distance is used to classify nearby points as part of the same cluster. In unsupervised learning, the distance indicates the possible presence of a cluster. In supervised learning the existence of clusters is known, and the distance is used to separate and discriminate between them.

The power of pattern recognition is that many variates are examined simultaneously, where even a few tax the brain. Thus the technique may be so complex and comprehensive that juice adulteration becomes difficult.

An early example of pattern recognition for orange juice was the work of Bayer et al. (1980), but there have been few others. Aries et al. (1986) analysed their data from pyrolysis–mass spectrometry of orange juice this way. Fry et al. (1987) used and compared two pattern recognition methods in their examination of citrus juice on the UK market. However, such is its promise that pattern recognition features in the largest study of orange juice authenticity to date, namely the research of the FDA (Johnston and Kaufmann, 1984; Page, 1986). This includes one of the main pattern recognition programs, SIMCA, a form of principal components analysis. In SIMCA, a mathematical model is built to represent the systematic variance in the data. Unknown or suspect juices are then compared with the model. The disadvantage of SIMCA, particularly compared to canonical variates analysis described below, is that any replication of analytical data is not taken into account. However, samples not fitting any categories in the model can be treated as outliers, which is an advantage. SIMCA is available separately and also forms part of the well-known ARTHUR package (marketed by Infometrix Inc, Seattle, Washington, USA).

The other main pattern recognition method has been canonical variates analysis (CVA) using the generalised statistical package GENSTAT (available from Numerical Algorithms Group, Oxford, OX2 8DR, UK). Other similar programs such as SPSS (Nie et al., 1975) (available from SPSS UK Ltd (Micro Division), 9–11 Queens Road, Walton-on-Thames, KT12 5LU, UK) or BMDP (Dixon, 1975) (available from Statistical Software Ltd, Cork Technology Park, Model Farm Road, Cork, Ireland) can also be used for CVA. The approach has been described for pyrolysis–mass spectrometry by Windig et al. (1983), and for orange juice analysis by Fry et al. (1987). In contrast to SIMCA, CVA can account for replicate analyses and weigh more heavily on those which are more reproducible. For this reason CVA has been preferred by some workers. Overall, CVA also gave somewhat better results than SIMCA when the two were compared for orange juice authentication (Fry *et al.*, 1987).

Although both SIMCA and CVA demand considerable computing power and statistical skill to carry out, their results can be applied with a pocket calculator if necessary. It is possible to collect and express the data from any number of variates for a single sample in a single statistic. This has a chi-squared distribution, for which defined critical values exist. These, in turn, indicate the probability that any difference between the sample and the learning set arises by chance. Where this probability is low it is conventionally accepted that the sample is significantly different. This is essentially all any statistical method does, that is, indicate those juices which are unlikely to have come from the set used to compile the basic data. This base set needs careful choice to ensure it is appropriate to the subsequent comparisons. One advantage of CVA is that, when differences are found, it is possible to inspect the statistical analysis and locate the factors which contribute to the difference. Thus, although much information is compressed into the test statistic, the method of calculation allows the contributions of individual analytical measurements to be recovered. This eases interpretation by pinpointing the compositional factors which cause a sample to be rejected by the CVA.

1.4   Recent developments in orange juice adulteration

1.4.1   Introduction

This section is designed to provide an account of the most significant developments which have occurred since the previous edition of this book. The most important of these is perhaps the progress made in the use of isotopic analysis, and in particular the development of the SNIF–NMR® technique (site-specific natural isotopic fractionation measured by nuclear magnetic resonance). This method is used to determine deuterium content at specific sites in a molecule and has been a valuable addition to the analyst's armoury as a means of combating adulteration, either on its own, or as part of a multi-isotope fingerprinting approach.

As the unscrupulous trader has become more and more sophisticated, so specific analytical approaches, based primarily on the analytical methods described earlier in the chapter, have been developed to target specific adulteration practices. These analytical approaches can be broadly divided into four different groups.

(1) Simple physical measurements such as weight, volume, density, purity, acidity and concentration of the sample. Although these parameters provide little indication of the authenticity of the product, they serve to define the value or price of the commodity in question.

(2) Comparison of fine compositional analyses to known patterns. Typical parameters include flavour profiles, sugars, organic acids, amino acids, trace element content and so on. These, and/or the interrelations between them, are generally compared to standard ranges (AFNOR or RSK standards). Chemometrics can also be used for further interpretation. Also included in this group are the various types of fingerprinting spectroscopies, such as those based on ultraviolet and visible absorption and fluorescence measurements, or near-infrared spectroscopy, which generate data that cannot always be interpreted in terms of chemical composition. These methods rely heavily on pattern recognition analysis and their reliability depends on the quality of the database used to prepare the statistical model. The main disadvantage of this approach is that the ranges and chemometric patterns for authentic products are wide and can often be matched by adding appropriate 'cocktails' to adulterated products.

(3) Checking for components that should not be present in the juice, or that should be present in only very small quantities. Those components can include: traces of synthetic intermediates or catalysts, artificial or nature-identical flavours, D-malic acid or D-aminoacids, certain oligosaccharides in freshly-squeezed orange juice, or in concentrates which have not undergone a second heating process (a second pasteurisation). One can also look for naringin in orange juice, preservatives such as benzoates, tartaric acid when there is no grape in the product etc. If one of these components is present at a level above the natural acceptable limit, then this third approach provides indisputable proof of adulteration and when used with an element of surprise, is an excellent approach. However it is ineffective against the more skilled adulterator who uses pure or modified adulterants that leave little trace of illegal components, such as invert sugar syrups stripped of their tell-tale oligosaccharides or biotechnologically-produced L-malic acid!

(4) Finally, the fourth approach, which includes isotopic methods, follows a strategy that consists in identifying the components that make up 90% of the price of the product and then testing for the genuineness of these components. In orange concentrate, for example, total sugars make up about 80% of the soluble solids. The price of the product can therefore be directly related to the origin of the sugars. In a 65° Brix orange concentrate, if the sugars originate exclusively from the orange fruit, the concentrate may be worth over $1200/tonne. However if they are not from orange but from sugar beet or corn syrup, the price is less than one fifth of this value. The economic temptation to adulterate a concentrate by adding sugar is therefore considerable. In many cases, if components other than sugar have been falsified, this is generally only in order to hide an adulteration with added sugar. The use of CSIRA to detect the addition of a C4-derived adulterant, such as cane or corn

(maize) sugars, in orange juice has been described in section 1.2.4. However since sugar beet and oranges are both C3 plants, ^{13}C isotope measurements are unable to detect the addition of sugar beet in orange concentrate. Deuterium determinations, in particular using the SNIF–NMR® method, have filled this analytical gap.

1.4.2 The SNIF–NMR method

Table 1.3 shows the average isotopic abundance of the most common naturally-occurring isotopes of the major elements making up the majority of organic products. The heavy isotope of hydrogen is deuterium, the abundance of which is low (about 150 per million hydrogen atoms), but still measurable. As in the case of ^{13}C and ^{18}O described in section 1.2.4, isotopic fractionation effects influence the deuterium content of natural products, providing information on the botanical and even geographical origin of the plant in which the molecule has been synthesised.

Deuterium isotope ratios, (D/H) are generally determined by IRMS, or isotope ratio mass spectrometry methods, expressed as a deviation relative to SMOW (standard mean ocean water) as the international standard. In the early 1980s, the SNIF–NMR technique was developed, which used high resolution deuterium nuclear magnetic resonance spectroscopy to determine the isotope content on multiple sites in a molecule (Martin *et al.*, 1982, 1985, 1986). By measuring site-specific isotope ratios, this technique is able to provide a greater wealth of information than the overall or total isotope content available from IRMS.

The original application of the SNIF–NMR technique was to detect the practice of adding sugar to grape must before fermentation to increase the final alcoholic content of a wine, and known in France as 'chaptalisation' (Martin *et al.*, 1988; Martin, 1990). Under European Union ruling, a wine grower is allowed to add a certain amount of sugar depending on the geographical region of production and the particular climatic conditions of the year in question. The SNIF–NMR method was specifically developed to detect added beet sugar and is now the official EU method for controlling this practice.

The site-specific isotope content of a given molecular species is related to the different isotopomers present. An isotopomer can be defined as a molecule containing one heavy isotope in any of the non-equivalent sites of the species. Ethanol, CH_3CH_2OH, for example, has three monodeuterated isotopomers:

$$CH_2DCH_2OH \qquad CH_3CHDOH \qquad CH_3CH_2OD$$
$$\text{(I)} \qquad\qquad \text{(II)} \qquad\qquad \text{(III)}$$

The deuterium atom can be located either on the CH_3 site (I), on the CH_2 site (II), or on the OH site (III) of the ethanol molecule. The site-specific deuterium to hydrogen ratio is therefore the proportion of each mono-

Table 1.3 Average isotopic composition of organic products

Atom	Hydrogen			Carbon			Oxygen		
	^1H	^2H	^3H	^{12}C	^{13}C	^{14}C	^{16}O	^{17}O	^{18}O
Atomic number (and mass)	1 (1)	1 (2)	1 (3)	6 (12)	6 (13)	6 (14)	8 (16)	8 (17)	8 (18)
Proportions (%)	99.985	0.015	(a)	98.904	1.096	(b)	99.763	0.037	0.2

[a] Radioactive. Practically non-existent in natural state.   [b] Radioactive. Very little present.

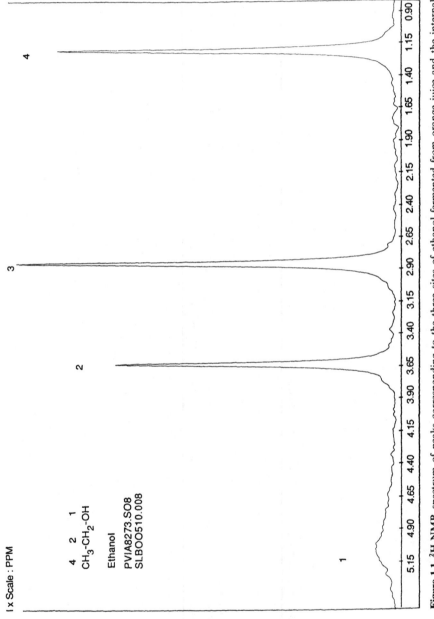

Figure 1.1 ^{2}H-NMR spectrum of peaks corresponding to the three sites of ethanol fermented from orange juice and the internal reference.

deuterated species to the fully hydrogenated molecule. In other terms, site-specific isotope ratios, R_i, can be defined as:

$$R_i = N_{hi}/P_i N_1 \qquad \text{(a)}$$

where N_{hi} is the number of molecules containing a heavy isotope in site i, and N_1 the number of light molecules. P_i denotes the number of equivalent positions in site i. $R_I = (D/H)_I$ and $R_{II} = (D/H)_{II}$ are the isotope ratios of species I and II, respectively.

An intermolecular comparison method is generally used to determine site-specific isotope ratios R_i by NMR. A measured quantity of a working standard (WS) of known isotopic content is added to the product (A) under investigation.

$$R_i^A = \frac{P^{WS} M^A m^{WS} T_i^A R^{WS}}{P_i^A M^{WS} m^A t^A} \qquad \text{(b)}$$

where M and m are the molecular weights and the masses of the products considered. t^A is the purity of A in %w/w and T the ratio of the signal intensity of isotopomer i(A) to that of the standard WS.

Figure 1.1 shows an NMR spectrum of ethanol obtained from a fermented orange juice. There are four peaks in each spectrum; the highest peak belongs to the reference or working standard, and the three other peaks each correspond to one of the monodeuterated isotopomers of ethanol. The area of each peak is proportional to the abundance of the given species, and can therefore be used to determine deuterium to hydrogen ratios at various positions in the molecule.

In wine, ethanol has been produced from the fermentation of the grape sugars. Studies have shown that, provided the conversion rate associated with the fermentation process is higher than 95%, isotopic fractionation effects can be safely neglected (Moussa et al., 1990). It has been shown that the deuterium content of the methyl site on the ethanol is directly related to the deuterium content on sites 1, 6 and 6′ of the starting sugars.

The analysis of a fruit juice using the SNIF–NMR method is not very different to that for wine once the preliminary fermentation step, to convert the fruit sugars into alcohol, has been carried out. The analysis protocol is shown in Figure 1.2. The fermentation is carefully monitored to ensure that it is complete in order to avoid isotopic fractionation.

The next step of the analysis is a very thorough computerised distillation of the hydroalcoholic solution obtained after fermentation in order to recover, with a very high yield, both the alcohol and the fermentation water. The alcohol is then measured by SNIF–NMR and by CSIRA.

The isotopic patterns of citrus, cane and beet sugars are illustrated in Figure 1.3. As discussed earlier, the $^{13}C/^{12}C$ measurement is very different for citrus and cane sugar. Addition of cane- or corn-derived sugar to citrus juices will be detected using the $\delta^{13}C$ values plotted on the vertical axis. On the

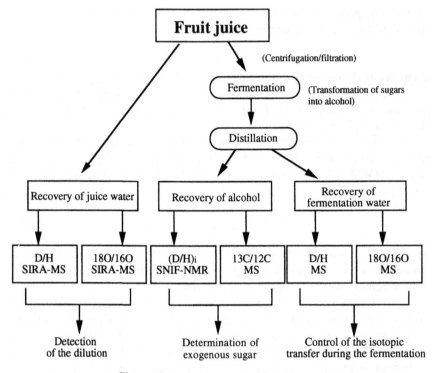

Figure 1.2 Analysis protocol of fruit juices.

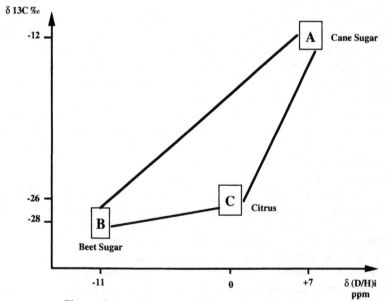

Figure 1.3 Isotopic pattern of citrus, cane and beet sugar.

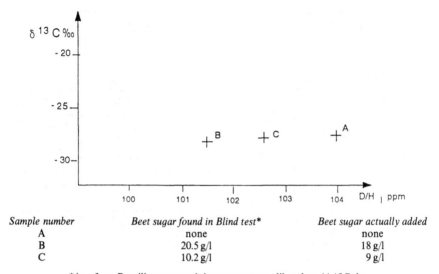

Sample number	Beet sugar found in Blind test*	Beet sugar actually added
A	none	none
B	20.5 g/l	18 g/l
C	10.2 g/l	9 g/l

* in g/l on Brazilian orange juice concentrate diluted to 11.18 Brix.

Figure 1.4 Detection of beet sugar addition by ^{2}H-NMR. Blind test performed for SGF-Germany on one authentic and two adulterated orange concentrates.

horizontal axis, the deviation of the $(D/H)_I$ ratio is plotted against the standardised value for citrus of the same origin. There is a difference of about 10 ppm between the centre of gravity of the citrus and the beet sugar populations which is highly significant, given an analytical precision of the order of 0.3. The lowering of the $(D/H)_I$ ratio will therefore indicate an addition of beet sugar to citrus juices.

Both repeatability and reproducibility of the SNIF–NMR method are good. Results of an early blind test performed for the Schutzgemeinschaft de Fruchtsaft-Industrie in Zornheim in 1988 are illustrated in Figure 1.4. The two adulterated samples were shown to have much lower $(D/H)_I$ ratios than authentic Brazilian concentrate and the amount of sugar present, calculated from the NMR spectra, was extremely close to the actual quantities added.

SNIF–NMR is now a well-known and accepted method for studying sugar added to fruit juices. It does however have other possible applications and it fits in well with an overall multi-isotopic approach.

1.4.3 Multi-isotopic fingerprint of fruit juices

Figure 1.5 gives a description of the elements that can be used to define a multi-isotopic fingerprint of fruit juices. The isotopic measurements can be carried out by IRMS or NMR, on the major components of the juice, water and sugars, or ethanol fermented from the sugars, as well as on minor components of the juice such as pulp, pectin, organic acids, vitamins, or

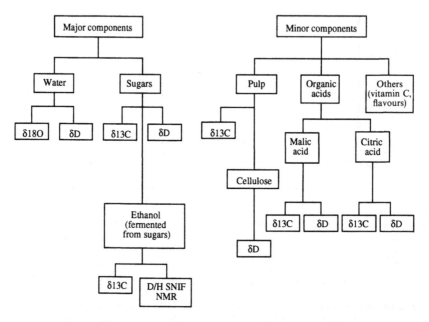

Figure 1.5 Multi-isotopic fingerprint of fruit juice.

flavours. A full isotopic fingerprint of the product can be built up, and in some cases certain interrelations can be shown to exist between the isotopic ratios of the various components. Bricout and Koziet (1987), for example, have established a relationship between the $\delta^{13}C$ measured on the sugars and the pulp of the same fruit.

The advantage of this in combating adulteration, is that it narrows the gap through which the fraudulent producer is able to squeeze. The ^{13}C content measured on the sugar of a juice containing only a small amout of cane or corn sugar may still be within the acceptable limit, but the difference with ^{13}C measured on the pulp may be abnormally high. This concept can be generalised to a great number of organic components of the juice and the vast number of measurements produced can then be used for internal normalisation, or even for verifying geographic origin.

The use of isotopic ratios to confirm a geographic origin has been explored, and the confidence intervals around the mean of the isotopic ratios measured on authentic samples from Spain, Israel, South America and South Africa have been shown to be well separated. Nikdel *et al.* of the Florida Institute of Citrus (1988) has approached the identification of geographic origin by studying trace element content in juices. Similar research has also been done on wine. A recent combination of the isotopic and trace element approaches provides further refinement in identifying the origin of wines and fruit juices (Day, 1993).

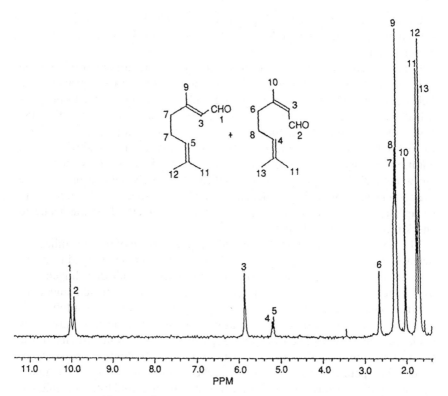

Figure 1.6 ^{2}H-NMR spectrum of citral ex-lemon grass.

1.4.4 Further applications of the SNIF–NMR method

The SNIF–NMR method also has great potential for the analysis of the authenticity of natural flavours (Martin *et al.*, 1993). Figure 1.6 shows for example the ^{2}H NMR spectra of citral. In this case it is possible to measure over a dozen isotopic ratios, making adulteration of natural fruit flavours almost impossible. In the area of fruit juice authentication, this technique may be of considerable use for proving that no flavour other than that from the named fruit has been added to a given juice.

1.4.5 The implications for orange juice authentication under the UK Food Safety Act 1990

Juice adulteration, unlike some forms of contamination, may not necessarily have adverse effects on the consumer's health. However, deliberate falsification or misrepresentation of a juice for economic gain is dealt with under Sections 14 and 15 of the UK Food Safety Act 1990, which state that it is an offence to 'sell to the purchaser's prejudice, food which is not of the nature or substance or quality demanded' or 'to falsely or misleadingly describe or present food'.

In recent years, increasingly sophisticated techniques and instrumentation have made it easier to detect a wide number of adulteration practices. In February 1991, the UK Ministry of Agriculture, Fisheries and Food published a survey carried out on 21 orange juice samples purchased in the UK. Five different methods, including those based on isotopic analyses, were carried out in a number of independent laboratories and revealed widespread addition of sugars and other ingredients in products sold as 'pure' and 'unsweetened'. The report not only confirmed a widely held suspicion as to the complacency of the UK orange juice market, it also forced the industry to sit up and take a new look at the quality control methods it was using to check incoming raw materials.

The Food Safety Act 1990 also provides a new version of the 'due diligence' defence and states that 'it shall be a defence for the person charged to prove that he took all reasonable precautions and exercised all due diligence to avoid the commission of the offence by himself or by a person under his control'. Therefore a trader whose product has been found to be adulterated and who wishes to establish the statutory 'due diligence' defence must show evidence that he had taken positive steps to ensure the authenticity of the product in question. He must show that the quality control or assurance scheme he operated was thorough and relied on the most up-to-date analytical methods. Various degrees of due diligence are required from the different levels of the processing and distribution chain, from importers to small and large retailers. In any case, though, it is not considered sufficient to rely on written warranties supplied to them by a third party. This was illustrated in the court case of January 24th and 25th, 1994, when Bradford Magistrate's Court fined a small trader £9000, plus £8000 costs, for importing and selling adulterated orange juice. An attempt to establish a due diligence defence was rejected on the grounds that the analytical tests used (a selection of compositional analyses) were known to be insufficient to detect the adulteration present.

This case has finally provided some insight into what the enforcement authoritites expect, or rather, will not accept as 'due diligence'. A 'duly diligent' scheme should sensibly include a selection of the analytical techniques described in this and the previous chapter. Whilst sugar addition remains the most economically-viable form of orange juice adulteration, isotopic techniques must be included. A sampling frequency on a random or statistical basis, calculated from raw material supplies and incorporating a certain risk factor (new or known high risk source, long term supplier, and so on) is a more cost effective alternative to a positive release system. The rate of sampling should be high enough to guarantee that an adulteration, if it exists, would be detected. In the Bradford case, the defence claimed to have tested one container out of every six supplied, by classical RSK-type analyses. This was not considered duly diligent.

At the time of writing this chapter, the European Quality Control System is being set up. Based on the subsidiarity principle advocated by the EU, the EQCS brings together established national quality control schemes under a European umbrella. This solidarity between fruit juice industries of the member states will, it is hoped, create a level playing field on a European scale which can only be to the benefit of the industry as a whole.

Editor's note

The original analytical scheme offered by Leatherhead Research Association and described in the first edition has now been superseded by a modified testing regime which incorporates a much wider range of tests including SNIF–NMR.

References

References from the text

Anon (1968) International Federation of Fruit Juice Producers Analytical Method No. 3.
Anon (1975) Section 22.055. In: *Official Methods of Analysis of the Association of Official Analytical Chemists*, 12th edn., Association of Official Analytical Chemists, Washington, DC.
Anon (1980) *Methods of Enzymatic Analysis*, Boehringer Mannheim Biochemicals.
Ara, V. and Torok, M. (1980) Statistisches Verfahren als Entscheidungshilfe bei der Beurteilung der Reinheit von Orangensaften. *Ind. Obst. Gemuseverwert.* 65, 297–300.
Aries, R. E., Gutteridge, C. S. and Evans, R. (1986) Rapid characterization of orange juice by pyrolysis mass spectrometry. *J. Food Sci.* 51(5), 1183–1186.
Bayer, S., McHard, J. A. and Winefordner, J. D. (1980) Determination of the geographic origins of frozen concentrated orange juice via pattern recognition. *J. Agric. Food Chem.* 28, 1306.
Bender, M. M. (1971) Variations in the $^{13}C/^{12}C$ ratios of plants in relation to the pathway of photosynthetic carbon dioxide fixation. *Phytochemistry* 10, 1239–1244.
Bergner-Lang, B. (1972) *Deut. Lebensm.-Rundsch.* 68, 176.
Bergner-Lang, B. (1975) On the proportions of citric acid:iso-citric acid in orange juices. *Deut. Lebensm.-Rundsch.* 71, 348–350.
Bettridge, D. (1983) Chemometrics — A buzz word explained. *Laboratory Practice* (Oct) 13, 15, 18, 21, 22.
Bianchini, J. P. and Gaydou, E. M. (1980) Separation of polymethoxylated flavones by straight-phase high-performance liquid chromatography. *J. Chromatogr.* 190(1), 233–236.
Bielig, H. J., Faethe, W., Koch, J., Wallrauch, S. and Wucherpfennig, K. (1985) Richtwerte und Schwankungsbreiten bestimmter Kennzahlen (RSK-Werte) für Orangensaft und Grapefruitsaft. *Confructa* 29, 191–207.
Blake, C. J., Porter, D. C. and Smith, P. R. (1982) The determination of added natural colours in foods. II. Turmeric. *Leatherhead Food R. A. Tech. Circ.* No. 782.
Brause, A. R., Raterman, J. M., Petrus, D. R. and Doner, L. W. (1984) Verification of authenticity of orange juice. *J. Assoc. Off. Anal. Chem.* 67(3), 535–538.
Brause, A. R., Doner, L. W. and Raterman, J. M. (1987) Detection of adulteration in apple and orange juices by chemical matrix method. *Food Prod. Manage.* 109(8), 20–21.
Bricout, J. (1982) Possibilities of stable isotope analysis in the control of food products. In: *Stable*

Isotopes, H. L. Schmidt, H. Foerstel and K. Heinzinger (eds.), Elsevier, Amsterdam, pp. 483–493.

Bricout, J. and Koziet, J. (1987) Control of the authenticity of orange juice by isotopic analysis. *J. Agric. Food Chem.* 35, 758–760.

Brown, M. B. and Cohen, E. (1983) Discussion of statistical methods for determining purity of citrus juice. *J. Assoc. Off. Anal. Chem.* 66(3), 781–788.

Brown, M. B., Cohen, E. and Volman, L. (1981) Comments on statistical methods of determining the purity of citrus juice. *Fluss. Obst.* 48, 286.

Buslig, B. S. and Ting, S. V. (1977) *Proc. Int. Citriculture* 3, 799.

Butler, D. (1987) High import requirement for citrus juices to continue. *Food News* Europe Supplement.

Butler, D. (1988a) Action on juice "falsification". *Food News* 16(14), 1.

Butler, D. (1988b) Attractive prices boosted '87 demand. *Food News* 16(8), 2.

Butler, D. (1988c) West German import surge in past year. *Food News* 16(15), 6.

Butler, D. (1988d) FCOJ tide starts turning. *Food News* 16(18), 3.

Cantagalli, P., Forconi, V., Gagnoni, G. and Pieri, J. (1972) *J. Sci. Food* 23, 905.

Clemens, R. L. (1964a) Organic acids in citrus fruits, 1. Varietal differences. *J. Food Sci.* 29, 276.

Clemens, R. L. (1964b) Organic acids in citrus fruits, 2. Seasonal changes in the orange. *J. Food Sci.* 29, 281.

Coffin, D. E. (1968) Correlation of the levels of several constituents of commercial orange juices. *J. Assoc. Off. Anal. Chem.* 51, 1199.

Cohen, E. (1982) Seasonal variability of citrus juice attributes and its effect on the quality control of citrus juice. *Z. Lebensm.-Unters. Forsch.* 175, 258–261.

Cohen, E. and Fuchs, C. (1984) Anwendung von Aminosaurenormen zum Nachweis von Fruchtsaftverdunnungen und verfalschungen: Statistische Analysen israelischer Citrussaft-daten. *Fluss. Obst.* 51(4), 169, 185–188.

Cohen, E., Hoenig, R., Sharon, R. and Volman, L. (1983) Uber die Zusammensetzung israelischer Citrussafte. *Fluss. Obst.* 50(4), 188–198.

Cohen, E., Sharon, R. and Volman, L. (1984a) Statistical methods applied in quality control and detecting adulteration of citrus juice. In: *Advances in the Fruit and Vegetable Juice Industry*, International Federation of Fruit Juice Producers (eds.), Juris Druck and Verlag, Zurich.

Cohen, E., Sharon, R., Volman, L., Hoenig, R. and Saguy, I. (1984b) Characteristics of Israeli citrus peel and citrus juice. *J. Food Sci.* 49, 987–990.

Craig, H. (1953) *Geochim. Cosmochim. Acta* 3, 53.

Davis, W. B. (1947) Determination of flavanones in citrus fruit. *Anal. Chem.* 19, 476.

Day, M. (1993) *Ph.D. Thesis*, University of Nantes.

Dennison, D. B., Brawley, T. G. and Hunter, G. L. K. (1981) *J. Agric. Food Chem.* 29, 927.

Dixon, W. J. (1975) *BMDP Biomedical Computer Programs*, University of California Press, Berkeley, CA.

Doner, L. W. (1985) Carbon isotope ratios in natural and synthetic citric acid as indicators of lemon juice adulteration. *J. Agric. Food Chem.* 33, 770–772.

Doner, L. W. and Bills, D. D. (1981) Stable carbon isotope ratios in orange juice. *J. Agric. Food Chem.* 29, 803–804.

Doner, L. W. and Bills, D. D. (1982) Mass spectrometric $^{13}C/^{12}C$ determination to detect high fructose corn syrup in orange juice: collaborative study. *J. Assoc. Off. Anal. Chem.* 65(3), 608–610.

Doner, L. W. and Phillips, J. G. (1981) Detection of high fructose corn syrup in apple juice by mass spectrometric $^{13}C/^{12}C$ analysis: collaborative study. *J. Assoc. Off. Anal. Chem.* 64(1), 85–90.

Doner, L. W., Ajie, H. O., Sternberg, L. da S. L., Milburn, J. M., DeNiro, M. J. and Hicks, K. B. (1987) Detecting sugar beet syrups in orange juice by D/H and oxygen-18/oxygen-16 analysis of sucrose. *J. Agric. Food Chem.* 35, 610–612.

Drawert, F., Leupold, G. and Pivernetz, H. (1980) Quantitative Gaschromatographische Bestimmung von Rutin, Hesperidin und Naringin. *Chem. Mikrobiol. Techn. Lebensm.* 6, 189.

Duchateau, A., Crombach, M., Munsters, B. and Droege, J. (1989) Selective analysis of proline enantiomers by high-performance liquid chromatography, paper submitted to the 13th International Symposium on Column Liquid Chromatography, Stockholm, Sweden, June 1989.

Dukel, F. (1986) Herziening Authenticiteitscriteria voor Industrieel bereid Sinaasappelsap. *Voedingsmiddelentechnologie* 19(16), 15.

Dyce, B. J. and Bessman, S. P. (1973) *Arch. Environ. Health* 27, 112.

Evans, R. and Hammond, D. (1989) Orange Juice Authenticity Testing, *European Food and Drink Review*, February 1989, pp. 41–43, 44.

Filliben, J. T. (1975) *Technometrics* 17, 111–117.

Firon, N., Lifshitz, A., Rimon, A. and Hochberg, Y. (1978) An immunoassay method for estimating the orange juice content of commercial soft drink. *Lebensm.-Wiss. Technol.* 12, 143.

Fischer, R. (1973) *Fluss. Obst.* 40, 407.

Floyd, F. M. and Harrell, J. E. (1969) *J. Assoc. Off. Anal. Chem.* 52, 1047.

Fogg, D. N. and Wilkinson, N. T. (1958) *Analyst (London)* 83, 406.

Francis, A. J. and Harmer, P. W. (1988) Fruit juices and soft drinks. In: *Food Industries Manual*, M. D. Ranken (ed.), 22nd edn, Blackie, Glasgow and London; AVI, New York, p. 252.

Freund, J. E. and Walpole, R. E. (1980) *Mathematical Statistics*, 3rd edn., Prentice Hall, Englewood Cliffs, NJ, p. 143.

Fry, J. C., White Judy, A. and Jessop Celena, A. (1987) Composition of orange and grapefruit juices. *Leatherhead Food R. A. Res. Rep.* No. 590.

Galensa, R., Ara, V. and Siewek, F. (1986) Untersuchungen zum "Naringin"-Gehalt von Orangensaften. *Fluss. Obst.* 53(9), 454–456, 471.

Goodhall, H. (1969) *The Composition of Fruits*, British Food Manufacturing Industries Research Association Sci. & Technical Survey No. 59.

Gherardi, S., Trifiro, A., Bigliardi, D. and Bazzarini, R. (1980) Effects of the technological process on the hesperidin and pectin contents of Italian oranges. *Industria Conserve* 55, 299–304.

Greiner, G. and Wallrauch, S. (1984) Naringin als Nachweis für den Zusatz von Grapefruitsaft zu Orangen- und Tangerinensaft. *Fluss. Obst.* 51(12), 626–628.

Hulme, B., Morris, P. and Stainsby, W. J. (1965) *J. Assoc. Public Anal.* 3, 113.

Jeuring, H. J., Brands, A. and Doorninck, P. (1979) Rapid determination of malic and citric acid in apple juice by HPLC. *Z. Lebensm.-Forsch.* 186, 185.

Johnston, M. R. and Kauffman, F. L. (1985) Orange juice adulteration detection and the actions of the FDA. *J. Food Qual.* 8(2/3), 81–89.

Jones, D. (1988) $700M orange juice scandal. *Today* (UK newspaper), Nov. 16, pp. 1–4.

Jones, N. R. (1979a) The authenticity of orange and grapefruit juices—interim report. *Leatherhead Food R. A. Tech. Circ.* No. 688.

Jones, N. R. (1979b) The authenticity of orange and grapefruit juices—Report II. *Leatherhead Food R. A. Tech. Circ.* No. 689.

Jurs, P. C. (1986) Pattern recognition used to investigate multivariate data in analytical chemistry. *Science* 231, 1219–1224.

Koch, J. (1979) Uber die Freien Aminosauren in Orangensaften des Handels. *Fluss. Obst.* 46, 212–216.

Koch, J. and Hess, D. (1971) *Dtsch. Lebensm.-Rundsch.* 67, 185.

Kowalski, B. R. (ed.) (1984) *Chemometrics—Mathematics and Statistics in Chemistry*, D. Reidel, Dordrecht, Netherlands.

Kreuger, E. and Bielig, H. J. (1976) *Betriebs- und Qualitaetskontrolle in Brauerei und Alkoholfreier Getraenkeinindustrie*, Verlag Paul Parey, Berlin, Hamburg, p. 368.

Kreuger, H. W. and Reesman, R. H. (1982) Carbon isotope analyses in food technology. *Mass Spectrom. Rev.* 1, 205–236.

Lewis, W. M. (1966) *J. Sci. Food Agric.* 17, 316.

Lifshitz, A. (1983) Methods to detect adulteration of citrus juice as experienced in Israel. *Food Technol. Australia* 35(7), 336–338.

Lifshitz, A. and Geiger, P. J. (1985) Phosphorylated compounds in citrus juice. *Lebensm.-Wiss. Technol.* 18(1), 43–46.

Lifshitz, A., Stepak, Y. and Brown, M. B. (1974) Method for testing the purity of Israeli citrus juices. *J. Assoc. Off. Anal. Chem.* 57(5), 1169–1175.

Linden, J. C. and Lawhead, C. L. (1975) Liquid chromatography of saccharides. *J. Chromatogr.* 105, 125.

Maraulja, M. D. and Dougherty, M. H. (1975) *Proc. Fl. State Hortic. Soc.* 88, 346.

Martin, G. J. (1990) The chemistry of chaptalisation, *Endeavour* 14, 137–143.

Martin, G. J., Martin, M. L., Mabon, F. and Michon, M. J. (1982) Identification of the origin of natural alcohols by natural abundance hydrogen nuclear magnetic resonance. *Anal. Chem.* 54, 2380–2382.

Martin, G. J., Sun, X. Y., Guillou, C. and Martin, M. L. (1985) NMR determination of absolute site-specific natural isotope ratios of hydrogen in organic molecules. *Tetrahedron* 41, 3285–3296.

Martin, G. J., Zhang, B. L., Naulet, N. and Martin, M. L. (1986) Deuterium transfer in the bioconversion of glucose to ethanol studied by specific isotope labelling at the natural abundance level. *J. Amer. Chem. Soc.* 108, 5116–5122.

Martin, G. J., Guillou, C., Martin, M. L., Cabanis, M. T., Tep, Y. and Aerny, J. (1988) Natural factors of isotope fractionation and the characterization of wines. *J. Agric. Food Chem.* 36, 316–322.

Martin, G. G., Remaud, G. and Martin, G. J. (1993) Isotopic methods for the control of natural flavours authenticity. *Flavour and Fragrance J.* 8, 97–107.

McHard, J. A., Winefordner, J. P. and Ting, S. V. (1976) Atomic absorption spectrometric determination of eight trace metals in orange juice following hydrolytic preparation. *J. Agric. Food Chem.* 24, 950–953.

McHard, J. A., Foulk, S. J., Jorgensen, J. L., Bayer, S. and Winefordner, J. P. (1980) Analysis of trace metals in orange juice. In: *Citrus Nutrition and Quality*, S. Nagy and J. A. Attaway (eds.), ACS Symposium Series 143, Washington, DC.

Mitchell, P. P. (1982) Adulterated foods. *The Washington Post*, December 12.

Moussa, I., Naulet, N., Martin, M. I. and Martin, G. J. (1990) A site-specific and multielement approach to the determination of liquid-vapor isotope fractionation parameters. The case of alcohols. *J. Phys. Chem.*, **94**, 8303–8309.

Navarro, J. J., Izquierdo, L., Aristoy, M. and Sendra, J. M. (1986) Characterisation of Spanish orange juice concentrates for detecting adulterations. In: *Fruit Juices for Europe. Advances in the Fruit and Vegetable Juice and Beverage Technology*, International Federation of Fruit Juice Producers (eds.), Juris Druck and Verlag, Zurich.

Nie, N. N., Hull, C. H., Jenkins, J. G., Steinbrenner, K. and Bent, D. (1975) *Statistical Package for the Social Sciences (SPSS)*, McGraw-Hill, New York.

Nikdel, S. and Nagy, S. (1985) Trace metals in orange juice for characterising geographical origin and detecting adulteration, Paper presented at 189th ACS National Meeting, Florida, April 29–May 3.

Nikdel, S., Nagy, S. and Attaway, J. A. (1988) Trace elements: defining geographical origin and detecting adulteration of orange juice. In *Adulteration of Fruit Juice Beverages*, S. Nagy, J. A. Attaway and H. E. Rhodes (eds), Marcel Dekker, New York.

Ooghe, W. (1981) Possibilities of the amino acid analyser in fruit juice analysis, *Proceedings of Euro Food Chem* I, Vienna.

Ooghe, W. (1983) Aminozuurnormen: een "Must" bij de Beoordeling van Industrieel Vervaardigde Vruchtesappen. *Voedingsmiddelentechnologie* 16(24), 23–26.

Ooghe, W. and de Waele, A. (1982) Anwendung von Aminosaurestandardwerten zur Aufdeckung von Fruchtsaftverdunnungen und -verfalschungen. *Fluss. Obst.* 47(11), 618–628, 632–634, 636.

Ooghe, W., Kasteleyn, H., Temmerman, I. and Sandra, P. (1984) Amino acid enantiomer separation for the detection of adulteration in fruit juices. *J. High Res. Chromatogr. Chromatogr. Commun.* Short Communication No. 10488, 7(5), 284–285.

Page, S. W. (1986) Pattern recognition methods for the determination of food composition. *Food Technol.* 104–109.

Park, G. L., Byers, J. L., Pritz, C. M., Nelson, D. B., Navarro, J. L., Smolensky, D. C. and Vandercook, C. E. (1983) Characteristics of California Navel orange juice and pulpwash. *J. Food Sci.* 48, 627–632, 651.

Parker, P. L. (1982) Uber die Anwendung des 13C/12C Verhaltnisses zum Erkennen einer Sussung von Fruchtsaftkonzentraten. *Fluss. Obst.* 49(12), 692–694, 672.

Perfetti, G. A., Joe, F. L., Fazio, T. and Page, S. W. (1988) Liquid chromatographic methodology for the characterisation of orange juice. *J. Assoc. Off. Anal. Chem.* 71(3), 469–473.

Petrus, D. R. and Attaway, J. A. (1980) Visible and ultraviolet absorption and emission characteristics of Florida orange juice and orange pulpwash: detection of adulteration. *J. Assoc. Off. Anal. Chem.* 63(6), 1317–1331.

Petrus, D. R. and Attaway, J. A. (1985) Spectral characteristics of Florida orange juice and orange pulpwash. Collaborative study. *J. Assoc. Off. Anal. Chem.* 68(6), 1202–1206.

Petrus, D. R. and Dougherty, M. H. (1973a) Spectral characteristics of three varieties of Florida orange juice. *J. Food Sci.* 38, 659–666.

Petrus, D. R. and Dougherty, M. H. (1973b) Spectrophotometric analyses of orange juices and corresponding orange pulp washes. *J. Food Sci.* 38, 913–914.

Petrus, D. R. and Vandercook, C. E. (1980) Methods for the detection of adulteration in processed citrus products. In: *Citrus Nutrition and Quality*, S. Nagy and J. A. Attaway (eds.), ACS Symposium Series 143, Washington, DC.

Richard, J. P. and Coursin, D. (1979) Utilisation de quelques methodes informatiques a l'expertise des jus de fruits. *Ind. Aliment. Agric.* 96, 89–95.

Richard, J. P. and Coursin, D. (1984) Essais de characterisation de jus d'agrumes d'Israel: application de la methode francine. In: *Advances in the Fruit and Vegetable Juice Industry*, International Federation of Fruit Juice Producers (eds.), Juris Druck and Verlag, Zurich.

Roehr, M. and Kubicek, C. P. (1981) Regulatory aspects of citric acid fermentation by *Aspergillus niger*. *Proc. Biochem.* 34–37, 44.

Rogers, G. R. (1970) *J. Assoc. Off. Anal. Chem.* 53, 568.

Rolle, B. A. and Vandercook, C. E. (1963) Lemon juice composition. III. Characterization of California–Arizona lemon juice by use of a multiple regression analysis. *J. Assoc. Off. Anal. Chem.* 58, 362–365.

Rouseff, R. L. (1988) High performance liquid chromatographic separation and spectral characterization of the pigments in turmeric and annatto. *J. Food Sci.* 53(6), 1823–1826.

Rouseff, R. L. and Ting, S. V. (1979) Quantitation of polymethoxylated flavones in orange juice by high-performance liquid chromatography. *J. Chromatogr.* 176(1), 75–87.

Royo-Iranzo, J. and Gimenez-Garcia, J. (1974) Differences in proportions of characteristic components in serum and pulp of Spanish orange juice. *Rev. Agrochim. Technol. Aliment.* 14, 136.

Saguy, I. and Cohen, E. (1984a) Evaluation of spectral characteristics of Israeli citrus products for authenticity determination. In: *Advances in the Fruit and Vegetable Juice Industry*, International Federation of Fruit Juice Producers (eds.), Juris Druck and Verlag, Zurich.

Saguy, I. and Cohen, E. (1984b) Spectral characteristics of Israeli orange products. *Z. Lebensm.-Unters. Forsch.* 178, 386–388.

Schatzki, T. F. and Vandercook, C. E. (1978) Non-parametric methods for detection of adulteration of concentrated orange juice for manufacturing. *J. Assoc. Off. Anal. Chem.* 61(4), 911–917.

Scholey, J. (1974) Detection of adulteration in citrus juices by analytical means. *Flavour Ind.* (5), 118–120.

Scholten, G. (1984) Hochdruckflussigkeitschromatographische Bestimmung von Bioflavonoiden in Citrussaften. *Fluss. Obst.* 51(10), 519–522.

Smith, B. N. and Epstein, S. (1971) Two categories of $^{13}C/^{12}C$ ratios for higher plants. *Plant Physiol.* 47, 380.

Swift, L. J. and Veldhuis, M. K. (1961) *Food Res.* 16, 142.

Ting, S. V., Rouseff, R. L., Dougherty, M. H. and Attaway, J. A. (1979) Determination of some methoxylated flavones in citrus juices by high performance liquid chromatography. *J. Food Sci.* 44(1), 69–71.

Trifiro, A., Bigliardi, D., Gherardi, S. and Bazzarini, R. (1980) *Industria Conserve* 55, 194.

Vandercook, C. E. (1977) Detection of adulteration in citrus juice beverages. *Food Chem.* 2(3), 219–233.

Vandercook, C. E. (1984) Stepwise approach to the determination of orange juice content. In: *Advances in the Fruit and Vegetable Juice Industry*, International Federation of Fruit Juice Producers (eds.), Juris Druck and Verlag, Zurich.

Vandercook, C. E. and Guerrero, H. C. (1969) Citrus juice characterization: analysis of the phosphorus fractions. *J. Agric. Food Chem.* 17, 626–628.

Vandercook, C. E. and Smolensky, D. C. (1976) Microbiological assay with *Lactobacillus plantarum* for detection of adulteration in orange juice. *J. Assoc. Off. Anal. Chem.* 59(6), 1375–1379.

Vandercook, C. E. and Smolensky, D. C. (1979) Easy method detects juice dilution. *Food Product Dev.* 13, 60–61.

Vandercook, C. E., Guerrero, H. C. and Price, R. L. (1970) Citrus juice characterization: identification and estimation of major phospholipids. *J. Agric. Food Chem.* 18, 905–907.

Vandercook, C. E., Mackey, B. E. and Price, R. L. (1973) New statistical approach to evaluation of lemon juice. *J. Agric. Food Chem.* 21, 681–683.

Vandercook, C. E., Smolensky, D. C., Nakamura, L. K. and Price, R. L. (1976) A potential microbiological assay of fruit content in orange juice products. *J. Food Sci.* 41, 709–710.

Vandercook, C. E., Lee, S. D. and Smolensky, D. C. (1980) A rapid automated microbiological determination of orange juice authenticity. *J. Food Sci.* 45(5), 1416–1418.

Vandercook, C. E., Navarro, J. L., Smolensky, D. C., Nelson, D. B. and Park, G. L. (1983) Statistical evaluation of data for detecting adulteration of California Navel orange juice. *J. Food Sci.* 48, 636–640, 655.

Veldhuis, M. K., Swift, L. J. and Scott, W. C. (1970) Fully-methoxylated flavones in Florida orange juices. *J. Agric. Food Chem.* 18(4), 590–592.

Wallrauch, S. (1984) Beurteilungsnormen, RSK-Werte—Anwendung auf israelische Orangensafte. In: *Advances in the Fruit and Vegetable Juice Industry*, International Federation of Fruit Juice Producers (eds.), Juris Druck and Verlag, Zurich.

Wallrauch, S. and Greiner, G. (1977) *Fluss. Obst.* 44, 241.

White, J. A., Hart, R. J. and Fry, J. C. (1986) An evaluation of the Waters Pico-Tag system for the amino-acid analysis of food materials. *J. Automatic Chem.* 8(4), 170–177.

Windig, W., Havercamp, J. and Kistemaker, P. G. (1983) Interpretation of a set of pyrolysis mass spectra by discriminant analysis and graphical rotation. *Anal. Chem.* 55, 81.

Wucherpfennig, K. and Franke, I. (1966) *Fruchtsaft Ind.* 11(2), 60.

Further reading

General

Bielig, H. J., Faethe, W., Koch, J., Wallrauch, S. and Wucherpfennig, K. (1984) RSK-values for apple, grape and orange juices. *Confructa* 28(1), 63–73.

Coppola, E. D. (1984) Use of HPLC to monitor juice authenticity. *Food Technol.* 38(4), 88–91.

Fang, Tzuu-Tar and Ling, Sheng-Feng (1984) Inspection of fruit juice authenticity by amino acid distribution pattern checking method. In: *Advances in the Fruit and Vegetable Juice Industry*, International Federation of Fruit Juice Producers (eds.), Juris Druck and Verlag, Zurich.

Junge, C. and Spadinger, C. (1982) Detection of the addition of L and DL malic acid in apple and pear juices by quantitative determination of fumaric acid. *Fluss. Obst.* 49(2), 46–47, 57–62.

Kreuger, D. A., Kreuger, R.-G. and Kreuger, H. W. (1986) Carbon isotope ratios of various fruits. *J. Assoc. Off. Anal. Chem.* 69(6), 1035–1036.

Kuhlmann, F. (1983) Amino acid composition of strawberry and raspberry juices. *Fluss. Obst.* 50(10), 546–548.

Maarse, H. (1987) Flavour and authenticity studies at the TNO-CIVO food analysis institute. *Perfum. Flavor.* 12(2), 45–56.

Ooghe, W. and Kastelijn, H. (1985) Amino acid analysis, a rapid, low cost, reliable screening test for evaluating commercially marketed fruit juices. *Voedingsmiddelentechnologie* 18(23), 13–15.

Roozen, J. P. and Jaussen, M. M. Th. (1981) Gas chromatographic analysis of amino acids in fruit drinks. In: *Proceedings of 1st European Conference on Food Chemistry, Vienna, 1981*, W. Baltes, P. B. Czedik-Eysenberg and W. Pfannhauser (eds.), Verlag Chemie, 1982, pp. 458–460.

Siewer, F. (1984) Isolation and identification of flavonol glycosides relevant to fruit juice adulteration. *Lebensm.-Gerichtl. Chem.* 39(1), 9–10.

Wallrauch, S. (1985) Aminosauren als Kriterium fur die Beurteilung von Fruchtsaften. *Fluss. Obst.* 52(7), 371–375, 380–381, 384.

Wrolstad, R. E., Cornwell, C. J., Culbertson, J. D. and Reyes, F. G. R. (1980) Establishing criteria for determining the authenticity of fruit juice concentrates. In: *Quality of Selected Fruits and Vegetables of North America, Las Vegas, 1980*, R. Teranishi and H. Barrera-Benitez (eds.), ACS, Washington, DC, pp. 77–93.

Apple

Bielig, H. J. and Hofsommer, H. J. (1982) On the information expressed in the amino acid spectrum of apple juices. *Fluss. Obst.* 49(2), 50–56.

Brause, A. R. and Raterman, J. M. (1982) Verification of authenticity of apple juice. *J. Assoc. Off. Anal. Chem.* 65(4), 846–849.

Burroughs, L. F. (1984) Analytical composition of English apple juice. *Fluss. Obst.* 51(8), 370–372, 393, 400–401.

Doner, L. W. and Cavender, P. J. (1988) Chiral liquid chromatography for resolving malic acid enantiomers in adulterated apple juice. *J. Food Sci.* 53(6), 1898–1899.

Doner, L. W., Kreuger, H. W. and Reesman, R. H. (1980) Isotopic composition of carbon in apple juice. *J. Agric. Food Chem.* 28(2), 362–364.

Evans, R. H., van Soestbergen, A. W. and Bistow, K. A. (1983) Evaluation of apple juice authenticity by organic acid analysis. *J. Assoc. Off. Anal. Chem.* 66(6), 1517–1520.

Mattick, L. R. and Moyer, J. C. (1983) Composition of apple juice. *J. Assoc. Off. Anal. Chem.* 66(5), 1251–1255.

Ooghe, W. (1984) Comments on the Dutch criteria for industrially prepared apple juices. *Fluss. Obst.* 51(1), 10–11, 29–31.

Smolensky, D. C. and Vandercook, C. E. (1980) Detection of grape juice in apple juice. *J. Food Sci.* 45(6), 1773–1774, 1780.

Sproer, P.-D. (1985) The importance of free amino acids in apple juice for evaluating juice identity and purity. *Fluss. Obst.* 52(1), 18–28.

Wrolstad, R. E. (1985) The chemical composition of apple juice, contaminants and adulterations. *Fluss. Obst.* 52(6), 302–307, 331–332.

Zyren, J. and Elkins, E. R. (1985) Interlaboratory variability of methods used for detection of economic adulteration in apple juice. *J. Assoc. Off. Anal. Chem.* 68(4), 672–676.

Apricot

Wallrauch, S. (1985) Standard values and ranges of specific reference numbers (RSK-values) for apricot puree (juice). *Fluss. Obst.* 52(12), 644–646.

Blackberry

Wrolstad, R. E., Culbertson, J. D., Cornwell, C. and Mattick, L. R. (1982) Detection of adulteration in blackberry juice concentrates and wines. *J. Assoc. Off. Anal. Chem.* 65(6), 1417–1423.

Blackcurrant

Beilig, H. J., Faethe, W., Koch, J., Wallrauch, S. and Wucherpfennig, K. (1984) RSK-values for blackcurrant juice. *Confructa* 28(1), 80–82.

Frank, W. (1975) The composition of blackcurrant juices and pulps. *Fluss. Obst.* 41(11), 465–473.

Nijssen, L. and Maarse, H. (1986) Authenticity studies of fruit juice concentrates using a GC-fingerprinting technique. In: *Proceedings of the 19th Symposium of the International Federation of Fruit Juice Producers, The Hague, 1986*, IFFJP (eds.), Juris Druck and Verlag, Zurich, pp. 333–355.

Nijssen, L. and Maarse, H. (1986) Volatile compounds in blackcurrant products. *Flavour Fragrance J.* 1(4, 5), 143–148.

Wald, B., Galensa, R. and Herrmann, K. (1986) Detection of adulteration of blackcurrant products with blackberry juice by determination of flavonoids with HPLC. *Fluss. Obst.* 53(7), 349–352, 353.

Cranberry

Francis, F. J. (1985) Detection of enocyanin in cranberry juice cocktail by color and pigment profile. *J. Food Sci.* 50(6), 1640–1642, 1661.

Hale, M., Francis, F. and Fagerson, I. (1986) Detection of enocyanin in cranberry juice cocktail by HPLC anthocyanin profile. *J. Food Sci.* 51(6), 1511–1513.

Hong, V. and Wrolstad, R. (1986) Cranberry juice composition. *J. Assoc. Off. Anal. Chem.* 69(2), 199–207.

Hong, V. and Wrolstad, R. (1986) Detection of adulteration in commercial cranberry juice drinks and concentrates. *J. Assoc. Off. Anal. Chem.* 69(2), 208–213.

Grape

Dunbar, J. (1982) Oxygen isotope studies on some New Zealand grape juices. *Z. Lebensm.-Unters. Forsch.* 175(4), 253–257.

Siewek, F., Galensa, R. and Herrmann, K. (1985) Detection of adulteration of grape juice and of alcoholic beverages prepared from grape juice with fig juice by determination of flavone-C-glycosides using HPLC. *Z. Lebensm.-Unters. Forsch.* 181(5), 391–394.

Werkhoff, P. and Bretschnerder, W. (1986) Gas chromatographic determination of diethylene glycol in wine, grape juice and grape juice concentrates. *Z. Lebensm.-Unters. Forsch.* 182(4), 298–302.

Grapefruit

Bielig, H. J., Faethe, W., Koch, J., Wallrauch, S. and Wucherpfennig, K. (1984) RSK-values for grapefruit juice. *Confructa* 28(1), 78–80.

Pino, J. A., Torricella, R. G., Orsi, F. and Figueras, L. (1986) Application of multivariate statistics for the quality classification of single-strength grapefruit juice. *J. Food Qual.* 9, 205–216.

Trifiro, A., Gherardi, S., Bazzarini, R. and Bigliardi, D. (1982) Bioflavonoids in grapefruit juice as authenticity indices. *Industria Conserve* 57(1), 23–25.

Lemon

Benk, E. (1980) On the composition of lemon juice. *Fluss. Obst.* 47(10), 496–499.

Petronici, C., Bazan, E. and Panno, M. (1978) On the ratio of citric acid:isocitric acid in lemon juice. *Industria Conserve* 53(2), 104–106.

Morello cherry

Bielig, H. J., Faethe, W., Koch, J., Wallrauch, S. and Wucherpfennig, K. (1984) RSK-values for morello cherry juice. *Confructa* 28(1), 84–85.

Eksl, A., Reicheneder, E. and Kieninger, H. (1980) On the chemical composition of mother juices from morello cherries of various varieties. *Fluss. Obst.* 47(10), 494–496.

Passionfruit

Bielig, H. J. and Hofsommer, H. J. (1981) Zur Analytik des Passionfruchtsaftes. *Fluss. Obst.* 48(4), 186–189.

Bielig, H. J., Faethe, W., Koch, J. and Wallrauch, S. (1984) RSK-values for passion fruit juice. *Fluss Obst.* 51(12), 622–624.

Kuhlmann, F. (1984) Recent analyses of passion fruit juices. *Fluss. Obst.* 51(2), 59–63.

Peach

Keininger, H. and Ecks, A. (1979) The distribution of free amino acids in peach pulp and commercial peach nectars. *Fluss. Obst.* 46(4), 124–131.

Pear

ALS Working Party for Fruit Products and Fruit Juices (1984) RSK-values for pear juice. *Confructa* 28(1), 74–76.

Wald, B., Galensa, R. and Herrmann, K. (1988) The detection of the addition of apple juice to pear products by the determination of flavonoids with HPLC. *Lebensm.-Gerichtl. Chem.* 42(2), 42–51.

Raspberry

ALS Working Party for Fruit Products and Fruit Juices (1984) RSK-values for raspberry juice. *Confructa* 28(1), 76–77.

Strawberry

Henning, W. (1982) Evidence of adulteration of strawberry juice concentrates. *Lebensm.-Gerichtl. Chem.* 36(3), 62–63.

2 Chemistry and technology of citrus juices and by-products

C. M. HENDRIX and J. B. REDD

2.1 Principal citrus cultivars

2.1.1 Origin of citrus

Like all citrus plants, the orange tree originated in Asia. The exact site of its appearance is hotly debated. Nevertheless, citrus fruits are said to have emerged in the east of Asia in the lands now belonging to India, China, Bhutan, Burma and Malaysia.

The first member of the citrus group to become known to European civilization was the citron, mentioned about 310 BC by Theophrastus (Hasse, 1987).

Citron seeds were said to have been identified in the excavation of ancient Babylonian ruins dating back to 4000 BC (Hendrix, 1983). An ancient Chinese manuscript of 2200 BC mentions oranges and one of AD 1178 describes 27 varieties of sweet, sour and mandarin oranges. Over many centuries, citrus was gradually spread by travelers to the near east, the Mediterranean countries, Europe and then to the new world.

Extensive classification research was done by the Japanese botanist Tyozaburo Tanaka (1931), a taxonomist at the University of Osaka. The Swedish botanist Linnaeus (1707–1778) of the University of Uppsala credited China as the definitive birthplace of citrus (Reuther *et al.*, 1967). The State of Florida, in the United States and the State of São Paulo in Brazil are the leading producing regions in the world (Hasse, 1987). Brazil became the world's biggest producer of oranges in the 1980s with more than 600 000 hectares (1.5 million acres) of citrus trees in its territory. The major reason for Florida's loss as the primary citrus producer was due to the many severe frosts (freezes) in the 1960s, 1970s and 1980s.

2.1.2 Commercial citrus regions

The citrus cultivars are grown commercially throughout the world in regions of subtropical and tropical climates where there are suitable soils and sufficient moisture to sustain the trees and hopefully insufficient periods of low temperatures to kill the trees. However, occasionally in some subtropical regions, frosts may both destroy fruit and kill the trees (Table 2.1).

Table 2.1 The major citrus producing countries (alphabetically)

Argentina	Greece	Morocco
Australia	India	Panama
Belize	Israel	Russia
Brazil	Italy	South Africa
China	Jamaica	Spain
Cuba	Japan	Trinidad
Egypt	Mexico	United States

There are a number of important citrus hybrids such as citranges (*Citrus sinensis* × *Poncirus trifoliata*), tangelos (*Citrus sinensis* × *Citrus reticulata*) and tangors (*Citrus sinensis* × *Citrus reticulata*) (Reuther *et al.*, 1967). The tangelos and tangors are important, not only to the fresh fruit industry but also to the juice processing industry. The statutes and rules of the State of Florida, USDA/AMS standards for grades and FDA standards of identity all permit the use of a small quantity of hybrid juice to enhance color and/or flavor of orange juice (USDA/AMS, 1983; FDOC, 1980) (Tables 2.2 and 2.3).

2.1.3 Citrus growing areas

Citrus is grown throughout the world in both tropical and subtropical climates. The soil types and moisture conditions as well as weather conditions are of great importance. The normal world citrus belt can range from approximately the 35° north to the 35° south latitude.

Nearly all oranges grown in the summer season from May to November in the Northern Hemisphere are non-blood oranges produced in North America, Israel and South Africa. Nearly all blood oranges used in international trade are harvested in the winter season of November to June and primarily from February to April. Most pigmented oranges are grown in

Table 2.2 The major citrus genus and species

Common name	Scientific name
Bitter (sour) orange	*Citrus aurantium*
Sweet orange	*Citrus sinensis*
Grapefruit	*Citrus paradisi*
Shaddock (pummelo)	*Citrus grandis*
Kumquat	*Fortunella* species
Lemon	*Citrus limon*
Mandarin (tangerine, satsuma, etc.)	*Citrus reticulata*
Small fruited acid limes	
(Key, West Indian, Dominican, Mexican, Sour)	*Citrus aurantifolia*
Large fruited acid limes	
(Persian, Tahiti, Bearss)	*Citrus latifolia*

Table 2.3 Important cultivars and varieties from some citrus producing areas

Country	Fruit	Varieties
Spain	Sweet oranges	Navel (Washington Navel and Thomson Navel)
		Navelina, Navelate, Salustiana, Cadeneras, Comuna, blood oranges, Verna, Valencia (late variety)
	Mandarins	Satsuma, Clementina, Common Mandarin
	Grapefruit	Marsh
	Bitter oranges	Real orange
	Lemons	Primofiore, Mesero, Fino, Verna, Verdelli
Japan	Satsuma mandarins	(*Unshiu mikan*), (*C. unshio*, Marc.) (Oct–Apr)
	Natsudaidai	(*C. natsudaidai*, Hayata) (May–Jul)
	Kawano-natsudaidai	(Feb–Apr)
	Navel oranges	
China	Orange, varieties	
	Tangerines	
Australia	Oranges	Valencia, Navel
	Lemons, Mandarins, Grapefruit	
	Bitter (Seville) oranges	
	West Indian lime	
Israel	Lemons	Washington Navel (early), Shamuti-Jaffa, Valencia (late)
	Oranges	Marsh seedless
	Grapefruit	
Italy	Oranges, Lemons,	
	Mandarins,	
	Bergamots, Sweet limes	
Morocco	Orange, varieties	
	Grapefruit	
Mexico	Oranges	Limon mexicano (*C. aurantifolia*), Villafranca, Eureka, Lisbon (*C. limon*), Persian (*C. latifolia*)
	Acid limes	Lima dulce, Atotonilco (*C. aurantifolia*)
	Sweet limes	Dancy, Honey, Murcott
	Tangerines	Minneola, Orlando
	Tangelos	Duncan, Marsh, Red Blush, Shambor
	Grapefruit	
Brazil	Oranges	Baia (Washington Navel), Baianinha, Barao, Laranja Lima (sugar orange), Brazilian Hamlin, Westin (early), Pera Rio, Pera Coroa, Pineapple (mid), Valencia, Pera Natal (late)
Russia	Lemon varieties	
	Grapefruit, Pummelo	

North Africa, Italy and Spain and one-quarter of European winter orange supplies are blood varieties (Reuther *et al.*, 1967).

The pigmented oranges are the blood oranges of the Mediterranean Basin. They differ in appearance from the common sweet oranges only because under certain climatic conditions the fruit usually exhibits pink or red coloration in the flesh and juice and on the rind. The blood oranges, in general, are characterized by a somewhat distinctive flavor which has its advocates. The coloration of the blood oranges is associated with the development of anthocyanin pigments, whereas the pink and red coloration in pigmented grapefruit is caused principally by the carotenoid pigment, lycopene. The carotene pigments especially tend to deteriorate during processing and impart a muddy color to the juice. In the Florida processing industry, frozen concentrate manufactured from such varieties as Ruby and St. Michael gives juices with a delicate punch-like flavor.

2.1.4   Effect of frost

In recent years, freezes in various citrus producing regions of the world have resulted in considerable fruit and tree loss (Hendrix, 1983). When ice crystals develop in citrus fruit, this may result in disorganization of the cell structure to an extent that depends upon the duration and temperature conditions. Normally, 4 h at $-3.3°C$ to 6 h at $-2.2°C$ is sufficient to result in damage. When the fruit is harvested shortly after the frost occurs, the composition is little affected. However, fruit that remains on the tree may suffer shrinkage and desiccation of the juice vesicles as well as a loss of weight and a reduction in juice content (Hendrix and Ghegan, 1980).

The damage to cell structures may also lead to some disturbance of the normal metabolic processes. In oranges that have been frozen, some serious consequences may result such as acid reduction, Brix/acid ratio increase, gelation/separation, and other stability problems such as increased pectinesterase activity and microbiology problems, especially resulting from pH rise as high as 4.0 or higher (Westbrook, 1957). Flavor/aroma degradation such as medicinal or stale and/or flavonoids and limonin bitterness can take place.

2.1.5   Effect of soil

Soil type, water and climate are perhaps the main factors attributed to success or failure of the production of citrus. Although citrus trees have a very wide adaptation to soils of various types, the most acceptable are those groves located on uniform sandy loams or those that have no interference with penetration of soil moisture. Those with impervious subsoils, shallow soil, and hardpans usually result in problems (Camp *et al.*, 1957). Good fertilization practices and irrigation are important with well-drained soils such as sandy and sandy loam soils which would otherwise have problems due to leaching of nitrogen and other important minerals. Some shallow soils with water aquifers near the surface also require careful growing practices. Rootstock, of course, plays a vital role and must be appropriate to soil type, water and climatic factors.

2.2 Composition and structure of citrus fruits and juices of various cultivars

2.2.1 General relationship

The demand for and acceptance of citrus fruits in the daily diet of humans is based largely on their nutritional value, flavor, aroma and other aesthetic characteristics such as color, texture and cloud. These quality factors are dependent directly on the structure and chemical composition of the fruit (Hendrix, 1984) (Figure 2.1).

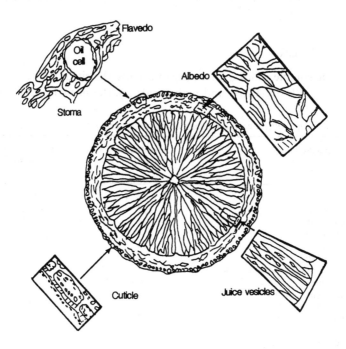

Figure 2.1 Cross-section of an orange. Adapted from P. G. Crandall, 1987.

Citrus fruits and juices make a significant contribution to and are a primary source of our daily requirement of vitamin C, and in addition, supplementary nutritional value is obtained from the amino acids, inorganic salts, carbohydrates and possibly other still unidentified factors in the edible portion of the fruit. There are also color pigments and volatile essential oils found in the peel (Beavans *et al.*, 1962).

All of these chemical constituents are a result of the combined influences of genetic regulatory mechanisms and the physical, chemical and biological environments to which the fruits are subjected during growth and after harvest.

In mature fresh citrus fruit, there are many chemical constituents and enzyme systems functioning to carry on their physiological performance without interfering with or rejecting one another. However, when the juice is extracted from the fruit by crushing, pressing, or reaming as described in chapter 8, constituent mixing occurs, which if uncontrolled, can lead to undesirable changes in flavor, color, aroma and stability. Some examples are the development and release of nomilin and naringin bitterness in immature grapefruit juice, limonin development in storage of Brazilian Baia and California Navel orange concentrates, and the gelation, separation or cloud loss of juice due to pectinesterase enzyme activity.

The complex chemical structure of orange juice is illustrated in Tables 2.4 and 2.5.

2.2.2   Organic acids

Because the soluble solids contain organic acids, the citrus fruit cultivars are classed as acid fruits. Citric and malic acids are the primary acids found in

Table 2.4 Approximate chemical composition of orange juice

Class constituents	Constituents (No.)	Proportion in total soluble solids (%)
Carbohydrates	7	76.0
Organic acids	7	9.6
Amino acids, free	17	5.4
Inorganic ions	14	3.2
Vitamins	14	2.5
Lipid constituents	18	1.2
Nitrogen bases and glutathione	5	0.9
Flavonoids	1	0.8
Volatile constituents	33	0.38
Carotenoids	22	0.013
Enzymes	12	0.0
Total	150	100.0

Adapted from USDA/ARS, Beavens *et al.* (1962).

Table 2.5 Composition of fresh orange juice

Constituent	No. of samples	Contents per 100 g	
		Range	Average
Protein (total N × 6.25) (g)	259	0.58–1.29	0.91
Amino nitrogen (g)	258	0.029–0.07	0.047
Fat (ether extract) (g)	43	0.0–0.56	0.2
Soluble solids, total (g)	4585	8.1–17.7	12.3
Sugar, total, as invert (g)	2747	6.32–14.3	9.15
Reducing (g)	1036	2.25–8.83	4.60
Sucrose (g)	1036	2.98–6.24	4.37
Acid, total (g)	4572	0.58–1.73	1.09
Malic acid (g)	9	0.10–0.17	0.15
Mineral nutrients (ash) (g)	555	0.27–0.70	0.41
Calcium (mg)	152	6.3–29.4	15
Chlorine (mg)	15	3.6–13.2	6
Fluorine (mg)	4	0.11–0.19	0.14
Iron (mg)	39	0.1–0.8	0.3
Magnesium (mg)	106	9.8–17.1	12
Phosphorus (mg)	195	8.0–30.0	20
Potassium (mg)	417	116–265	196
Sodium (mg)	258	0.2–2.4	0.5
Sulfur (mg)	57	3.5–11.3	8
Ascorbic acid (vitamin C) (mg)	2977	26–84	56.6
Betain (mg)	5	41–47	44
Biotin (mg)	12	0.0001–0.00037	0.00024
β-Carotene (mg)	132	0.23–0.28	0.12
Choline (mg)	9	7–15	11
Flavonoids (mg)	2	80–118	99
Folic acid (mg)	4	0.003–0.007	0.004
Inositol (mg)	6	170–210	194
Niacin (mg)	71	0.13–0.46	0.26
Pantothenic acid (mg)	37	0.06–0.3	0.13
Pyridoxine (B6) (mg)	12	0.023–0.094	0.038
Riboflavin (B2) (mg)	29	0.013–0.059	0.021
Thiamine (B1) (mg)	27	0.057–0.106	0.077
Vitamin B12 (μg)	5	0.0011–0.0012	0.0011

Adapted from Redd and Praschan (1975).

citrus juices. However, small amounts of other acids have also been identified. Citric acid accounts for the greatest portion of the organic acids in citrus fruit juices and is very high in the juices from lime and lemon fruit.

Citrus juice acidity is determined by titration against a known volume of juice with a standardized solution of sodium hydroxide using phenolphthalein as the indicator; the results are expressed as the percent citric acid anhydrous or monohydrate (CAMH). A pH meter may also be utilized to aid in the end point pH 8.2 determination (McAllister, 1980).

In many countries, the titratable acidity of citrus fruit juices plays a vital role in the determination of legal maturity of the fruit because external fruit color is a very poor and undependable ripeness guide.

Acidity is also an important attribute because tartness is a major factor in the acceptability of both citrus fruits and their juices. Oranges are generally acceptable at about 1.0% acid, pH 3.5; grapefruit at 1.5–2.5% acid, pH 3.0; mandarins at 0.8% acid, pH 3.5; and limes and lemons at about 5.0–6.0% acid, pH 2.2 (Lombard, 1963).

2.2.3 Carbohydrates

The primary portion of carbohydrates in citrus fruit are the three simple sugars that represent about 80% of the total soluble solids of orange juice: sucrose (49–59%), glucose (20–25%) and fructose (20–25%). This may vary somewhat with seasonal conditions, varieties and growing areas of the world. The ratios of sucrose, glucose and fructose are generally about 2:1:1 (Beavens et al., 1962).

Citrus fruits contain reducing (fructose and glucose) and non-reducing (sucrose) sugars. When citrus juices are stored, inversion of sucrose may occur with a corresponding increase in percentage of reducing sugars. The reducing and non-reducing characteristics of sugars in citrus juices are sometimes useful adulteration detection tools along with other criteria.

2.2.4 Color pigments

The color pigments are compounds generally located in the juice sacs and flavedo or outer peel and are concentrated in minute structures referred to as plastids. While the pigments of mandarin and orange may be found in the juice sacs, in some grapefruit, blood oranges, and limes, peel color may be found in the cell sap. As previously discussed, the pigments in Foster, Thompson and Marsh pink grapefruit are lycopene and β-carotene; the prime pigment in blood oranges is anthocyanin (Reuther et al., 1967).

The amounts of color pigments found in the citrus cultivars will vary with the variety, stage of maturity, seasonal variations, and even the growing region. The carotenoid pigments can be important from a nutritive standpoint due to vitamin activity.

2.2.5 Vitamins and inorganic constituents

The vitamins, generally separated into water soluble and fat soluble groups, are organic substances required in the diet for normal nutrition and health. The citrus cultivars are recognized as one of the most important sources of ascorbic acid. Scurvy was long ago associated with a lack of ascorbic acid in the diet. The use of limes and oranges in the diet of British seamen was the origin of the name 'Limey' that was given to the sailors (Nagy and Attaway, 1980).

An important factor in the acceptance of citrus fruits in the daily diet of humans is their nutritional value and especially their vitamin C (ascorbic acid) content (Sokoloff and Redd, 1949). The vitamin C of citrus juices can vary greatly depending upon such factors as soil, geographical location, climate, maturity, cultivar and other related factors.

Vitamin C stability of canned, processed juices depends mostly upon time and storage temperature. Generally, the average retention of ascorbic acid in freshly canned, processed juice is 97% and there could be a 1–2% per month loss at ambient temperature storage (Figure 2.2). Citrus juices also contain small but significant amounts of vitamins A and B and others are also known to exist (Table 2.6).

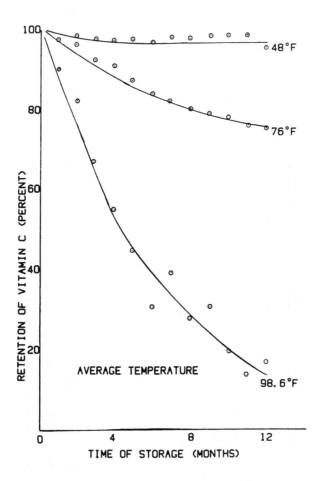

Figure 2.2 Vitamin C retention in canned orange juice. Adapted from USDA/ARS, Beavens *et al.*, 1962.

Table 2.6 Vitamin content of fresh and heat-processed canned citrus juices[a] (91–93°C)

Vitamin	Other designation	Units per 100 ml	Orange		Grapefruit		Tangerine		Lemon		Lime	
			Fresh	Processed	Fresh	Processed	Fresh	Processed	Fresh	Processed	Fresh	Processed
Provitamin A	β-Carotene	IU	190–140	2–160	[b]–21	0–15	350–420		0–2	0–2		
B$_1$	Thiamine	μg	60–145	30–100	40–100	10–50	70–120	60	30–90	40	40	
B$_2$	Riboflavin	μg	11–90	10–40	20–100	5–30	30	30	60	5–50	–[b]	–[b]
B$_6$ group[c]		μg	25–80	16–31	10–27	7–27	23	33[d]				
B$_{12}$	Animal protein factor	μg	0.0011–0.0013		0.8–1.8	0.1–2.2	1.2	1.8				
B$_c$	Folic acid, folacin, pteryl-glutamic acid	μg	1.2–3.3	1.5–3.2								
Bios I	Inositol	mg	98–210	104–178	88–150	92–112	135	147	85			
Biotin		μg	1–2.0	0.5–1.1	0.4–3.0	0.3–0.9	0.5	0.5				
C	Ascorbic acid	mg	35–56	34–52	36–45	32–45	25–50	26–31	45–60	30–58	27	
Choline		mg	10–13									
E	Tocopherols	mg	88–121									
Factor II	Pantothenic acid	μg	130–210	60–200	290	70–190						
K	Antihemorrhagic factor		0									
PABA	Para-aminobenzoic acid	μg	4									
P-P	Niacin	μg	200–300	170–300	200–220	80–200	200–220	200	100–130	80–100	100	

Adapted from USDA/ARS, Beavens *et al.* (1962).
[a] Compiled from various published data. Pasteurizing time up to 30 s.
[b] Trace.
[c] Includes pyridoxine, pyridoxal and pyridoxamine.
[d] Fresh and processed figures from different sources.

Table 2.7 Inorganic constituents of citrus fruits[a]

Constituent	Orange		Grapefruit		Lemon		Bitter orange (Seville)	
	Whole[b]	Juice[c]	Whole[b]	Juice[c]	Whole[b]	Juice[c]	Whole[b]	Juice[d]
Total ash	–	1800–4600	–	2300–4200	–	2400–2500	–	6700–8000
Potassium	920–2780	520–2840	1060–2350	670–2030	1290–3060	940–1930	1200–2290	2730–3063
Sodium	–	2–30	–	3–23	–	10–29	–	–
Calcium	–	49–206	–	–	–	31–74	–	–
Magnesium	–	50–147	–	–	–	10–60	–	–
Iron	–	0.5–6	–	–	–	1.6–10	–	–
Phosphorus	130–280	42–240	110–210	57–190	92–330	32–144	210–250	110–233
Nitrogen	1200–2460	570–1800	1000–2030	320–940	970–3050	350–840	–	–
Chlorine	–	20–55	–	24–136	–	–	–	–

Adapted from Kefford and Chandler (1970). (See this volume for full details of footnote references).

[a] In mg/kg.

[b] Extremes of ranges from Money (1964, 1966); fruit from Israel, Spain, South Africa, United States, West Indies and South America.

[c] Extremes of ranges from the following sources, recalculated where necessary: Anonymous (1967b), oranges and grapefruit from Israel; Beisel and Kitchel (1967), oranges from California and Arizona; Benk (1965), oranges from Cyprus, Greece, Israel, Italy, Morocco, Spain, Turkey, Brazil and South Africa; Birdsall et al. (1961), oranges and lemons from California; I. Calvarano (1961), oranges from Italy; I. Calvarano and M. Calvarano (1962), oranges and lemons from Italy; Dawes (1969), lemons from New Zealand; Floyd and Rogers (1969), oranges from Florida; Hart (1961), oranges from Texas and Mexico; Hopkins and Walkley (1967), oranges from Israel and Greece; Hulme et al. (1965), oranges from Mediterranean Area, Brazil and South Africa; Primo and Royo (1965a,b), oranges from Spain; Royo and Aranda (1967), oranges from Mediterranean Area, America, West Indies and South Africa; Sawyer (1963), oranges from Mediterranean Area; Sherratt and Sinar (1963), grapefruit from Cyprus, Israel, South Africa and Trinidad.

[d] From Osborn (1964).

Table 2.8 Flavonoids and flavonoid glycosides in citrus

1. Flavones

Common name of flavonoid	Chemical name or formula	Source Common name of fruit	Botanical species and variety	Part of fruit or tree	Substituents
Auranetin	3,6,7,8,4'-Pentamethoxy-flavone	Myrtle-leaf orange	Citrus aurantium var. myrifolia Ker-Gawl.		$R_3,R_6,R_7,R_8,R_{4'} = CH_3$
Nobiletin	5,6,7,8,3',4'-Hexa-methoxyflavone	Mandarin orange	C. reticulata Blanco C. tankan	Fruit	$R_5,R_6,R_7,R_8,R_3,R_{4'} = CH_3$
Rhoifolin	5,7,4'-Trihydroxyflavone-7-rhamnoglucoside	Sour orange Trifoliate orange	C. aurantium P. trifoliata	Ripe fruit Leaves	$R_5,R_{4'} = H$ R_7 = rhamnoglucose
Rutin	3,5,7,3',4'-Pentahydroxy-flavone-3-rhamno-glucoside	Satsumelo	C. paradisi × C. nobilis Lour	Fruit rind	$R_5,R_7,R_3,R_{4'} = H$ R_3 = rhamnoglucose
Tangeretin	5,6,7,8,4'-Pentamethoxy-flavone	Tangerine Dancy tangerine	C. reticulata C. tangerina Hort. ex Tanaka	Fruit rind, oil	$R_5,R_6,R_7,R_8,R_{4'} = CH_3$

Table 2.8 (cont'd)

2. Flavanones

Structure: flavanone skeleton with substituents R_2O, $OR_{3'}$, $OR_{4'}$, OR_5, R_7O.

Name	Chemical name	Common name	Species	Plant part	R groups
Citronetin	5,7-Dihydroxy-2'-methoxy-flavanone	Ponderosa, Wonder, or American Wonder lemon	C. limon (Linn.) Burm. F	Fruit peel	$R_5, R_7 = H$; $R_{2'} = CH_3$
Citronin	Citronetin-7-rhamno-glucoside	Lemon	C. limon	Fruit peel	$R_5 = H$; $R_7 = $ rhamnoglucose; $R_{2'} = CH_3$
Eriodictyol glycoside	5,7,3',4'-Tetrahydroxy-flavanone	Lemon	C. limon		$R_5, R_{3'}, R_{4'} = H$; $R_7 = $ glucose
Hesperetin	5,7,3'-Trihydroxy-4'-methoxyflavanone	Myrtle-leaf orange	C. aurantium var. myrtifolia		$R_5, R_7, R_{3'} = H$; $R_{4'} = CH_3$
Hesperidin	Hesperetin-7-rhamno-glucoside	Sweet orange	C. sinensis	Blossom petals	$R_5, R_{3'} = H$; $R_7 = $ rhamnoglucose; $R_{4'} = CH_3$
			C. kotokan Hayata	Fruit	
		Lemon	C. limon	Leaves, fruit	
		Citron	C. medica Linn.	Juice, leaves twigs, bark	
		Sweet orange	C. sinensis	Fruit	
Naringenin	5,7,4'-Trihydroxyflavanone	Shaddock or pummelo	C. tankan Hayata	Fruit	$R_5, R_7, R_{4'} = H$
Naringin (isohesperidin)	Naringenin-7-rhamno-glucoside	Sour orange	C. grandis (Linn.) Osbeck	Fruit, flowers	$R_5, R_{4'} = H$; $R_7 = $ rhamnoglucose
		Shaddock or pummelo	C. aurantium		
		Grapefruit	C. grandis		
		Trifoliate orange	P. trifoliata (Linn.) Raf.	Fruit juice, peel	
		Trifoliate orange	P. trifoliata	Fruit, leaves	
Neohesperidin	Hesperetin-7-glucorutin-oside		C. fusca Lour	Unripe fruit	$R_5, R_{3'} = H$; $R_7 = $ glucorhamnoglucose; $R_{4'} = CH_3$
			P. trifoliata	Fruit	
Poncirin	5,7-dihydroxy-4'-methoxy-flavanone-7-rhamno-glucoside	Trifoliate orange	P. trifoliata	Fruit, flowers	$R_5 = H$; $R_7 = $ rhamnoglucose; $R_{4'} = CH_3$

Adapted from USDA/ARS, Beavens *et al.* (1962).

The material remaining after all organic material has been destroyed by heating in air is ash and comprises the inorganic constituents that we recognize as minerals. Potassium accounts for by far the greatest proportion, up to 70%. However, calcium, magnesium and phosphorus are also significant and many minor elements are also found (Table 2.7).

Iron, copper, zinc, manganese and iodine are of significance in their association with basic enzyme systems concerned with human body metabolism and the low salt or sodium content is of dietetic interest associated with cardiac problems.

2.2.6   Flavonoids

The term flavonoid, as currently used, denotes a large group of chemical compounds. Chemically, flavonoids are organic compounds possessing the carbon framework of a flavone. Flavonoids are widely found in stems, roots, fruit, flowers, pollen, seeds, bark and wood. They are often found combined with carbohydrate molecules as glycosides. There are four main types of flavonoids in fruit: flavonols (such as quercetin glycosides in apple, plum and grape), anthocyanins (the red and blue pigments such as cyanidin and delphinidin and their glycosides), flavones (such as luteolin in lemon peel) and flavanones (such as hesperidin in oranges and naringin in grapefruit). In citrus fruits, the flavanones predominate whereas in colored fruit such as the black-currant and blackberry, the anthocyanins are the main form of flavonoid (Hughes, 1978) (Table 2.8).

Citrus flavonoids are of both economic and technological interest in the processing industry. Naringin provides bitterness in grapefruit juice. Hesperidin is non-bitter but can precipitate in processing equipment such as lines, heat exchangers and evaporators, and in excessive amounts it results in a low 'absence of defects' USDA grade score in orange juice. Nobiletin and tangeretin may have some effectiveness in the prevention of red blood cell adhesion (Robbins, 1977).

Almost 90% of the naringin in grapefruit may be found in the rag, pulp and albedo, and high processing pressures and other factors are responsible for its presence in grapefruit juice. Levels range from 0.02 to 0.10% depending upon maturity and variety of the fruit. Juices containing in excess of 0.07% naringin, early in the processing season, are often bitter. Florida state rules stipulate that only grapefruit juice of less than 600 ppm naringin (Davis Glycoside Test; FDOC, 1980), may be classed as US Grade A. In addition, less than 5.0 ppm limonin (a bitter limonoid), may be present in Grade A grapefruit juice.

Citrus bioflavonoids, particularly hesperidin, have been the subject of much research (Redd et al., 1962) and were referred to as vitamin P (Szent-Gyorgyi and Rusznyak, 1936). Because the substance is not now recognized as essential for health in the human diet, its vitamin terminology has been dropped by most nutritionists and researchers. There have been studies that indicate some value in reducing capillary fragility and permeability (Sokoloff and Redd, 1949).

2.2.7   Lipids

The term 'lipid' is used chemically to denote a group of substances that are water insoluble and soluble in organic solvents. Citrus lipids include simple waxes, steroids and aliphatic compounds. Lipid materials in citrus juices are closely associated with the suspended cloud matter (Nagy and Attaway, 1980). Kesterson and Braddock (1973) determined that dried citrus seeds contained 28–35% oil, 40–49% meal and 23–25% hulls. Citrus seed oil could be a potentially valuable by-product if the problems associated with bitterness and separation of peel and rag materials could be overcome. The seeds contain a bitter principle, probably limonin and/or nomilin (Beavens et al., 1962).

2.3   Operational procedures and effects on quality and shelf life of citrus juices

2.3.1   Outline of good manufacturing and processing procedures

Citrus juice in the TASTE evaporator.   During concentration the product is subjected to a high temperature short time pasteurization (99°C) for approximately 10 s. This destroys most of the microorganisms responsible for spoilage. However, practical experience shows that a small proportion of bacteria and yeasts can survive into the concentrate although their activity may not be apparent for a few days following the shock of processing (Hendrix, 1984).

Yeast spoilage can result in fermented taste and odor and the production of carbon dioxide gas bubbles. Lactic acid bacteria (lactobacillus and leuconostoc) may produce diacetyl which gives a buttermilk-like off-flavor and off-odor. Molds can produce a surface growth. Based upon practical experience of distribution of chilled orange juice, shelf life is affected by microbial load as follows (McAllister, 1980):

Organisms/ml	Shelf life (weeks)
300–700 (not more than half yeast)	3–4
5000	1–2

Concentrate handling for reprocessing and/or reconstruction
  (1) *Thawing.*   Product to be thawed should not be placed in direct sun-

light nor should steam or hot water be used on the outside of the container. Such poor handling practices may cause flavor degradation, vitamin loss, and a favorable environment for the growth of micro-organisms with a consequent reduction in shelf life.

(i) Drum or pail thawing is accomplished best in a 4°C cooler.
    (a) Pail thaw: 16–20 h
    (b) Drum thaw: 48–60 h
(ii) Product thaw at ambient temperature (24°C)
    (a) Pail thaw: 12–20 h
    (b) Drum thaw: 20–24 h
(iii) Concentrate from trailers delivered bulk will be loaded at approx. −8°C. Temperature rise will occur at the approximate rate of 2° per day in insulated tankers depending upon outside temperatures. Concentrate off-loaded to cold room tanks maintained at −8°C may be pumped to the process site at this temperature using positive displacement, sanitary pumps of proper size and horsepower.

(2) *Pumping and dumping.* Liners should be folded back over the outside of the drum to prevent outside drum contamination when dumping occurs. In no case should the product come in contact with the drum or drum dumper (premixing tank device). Sanitary hose may be used with a positive pump while being cautious to clean and sanitize all parts of the hose and pump to prevent contamination.

Sanitation or stabilization

(1) *General.* Sanitation may vary with the type of process. However, microbiological risks are present in all operations, whether aseptic, chilled, concentrated, single strength, canned, bottled, or portion pack. Good sanitation and cold temperature controls are of extreme importance, particularly with aseptic and reconstituted juices. The following operative procedures will prevent flavor degradation and/or product damage associated with enzyme activity and microbiological buildup.

(2) *Sanitizing at the end of all runs.*

(i) Rinse all equipment with clean, warm water.
(ii) Drain the system by either dismantling lines or removing valves. If lines cannot be dismantled, the removal of plug valves will permit easy inspection for cleanliness.
(iii) Use fresh cleaning solution; 2–4% NaOH or any other alkaline detergent solution is acceptable at 66°C. Recirculation of this solution through the system is standard procedure for cleaning in place (CIP). Occasional follow-up with an acid cleaner can be helpful.
(iv) Inspect the system for residue and reassemble any dismantled parts before the next run.

(3) *Sanitation of the system before beginning juice run.*
  (i) The best method for sanitation of a system is to recirculate clean, hot water (82°C) for 30 min before introducing juice into the system. If ammonia systems are in use in tanks and chillers, the refrigerant is not turned on until all hot water has been pumped from the system.
  (ii) A suitable substitute made up of 25 ppm chlorine in cold water may be utilized, but must be completely drained prior to starting the reconstitution process.
  (iii) Production areas and floors should be rinsed with water containing 20 ppm chlorine or iodine sanitizer. A thorough clean water rinse is required to avoid formation of odors and/or off-flavors.
  (iv) Joints, valve seats and form-fit gaskets are especially vulnerable to microbial buildup and should be inspected carefully.

Water for reconstitution use
  (1) It is most important that water for reconstitution be free from all contaminants such as chlorine, sulfides and iron. Even trace amounts of these elements can cause off-flavors and off-odors to develop in the juice after 2–4 days' standing. Off-flavors are even possible with the use of municipal city water or other treated waters with residual chlorine levels higher than 0.1 ppm. Charcoal filters can be used to reduce chlorine to acceptable levels provided they are checked regularly for microbiological cleanliness. Waters with high hydrogen sulfide content may be removed by spray aeration and the use of sand filters. Soft waters are preferred to hard waters as calcium bicarbonate can react with delicate flavoring substances.
  (2) Water should be chilled to between 2 and 3°C before mixing with concentrate.
  (3) Most microorganisms that originate in water do not survive at the low pH of citrus juices. Nevertheless, water should be free of contamination and a check should be routinely made for clarity, odor, taste and foreign materials. For a full account of water treatments see Hendrix (1974).

Processing of chilled high and low pulp reconstituted orange juice
  (1) Drums of concentrate, after thawing, are put into a premix dump tank and pumped by positive displacement pump to a stainless steel blending tank with slow speed agitation to minimize air entrapment. Air may result in the following problems:
    (i) Undesirable foaming;
    (ii) Improper fill of containers;
    (iii) Loss of vitamin C;

    (iv) Tendency to produce off-flavors through enzyme activity and bacterial and yeast growth.

  (2) After sufficient mixing time, the product should be checked for degree Brix and flavor, and final adjustment of Brix to 11.8° or higher should be made.

  (3) Microbial populations of 5000 to 10 000 total count per milliliter or greater on orange serum agar medium may be found in juices reconstituted from concentrates. Therefore, the juice will normally be pasteurized to destroy microorganisms and enzyme activity. Pasteurization is accomplished by heating to $85 \pm 5°C$ for 12 s in a plate pasteurizer. To avoid blockage by fruit cells, a pulpy product may require a triple tube heater or its equivalent. If product temperature drops below 70°C, the product should be considered as being insufficiently pasteurized and the operation stopped. Enzyme activity is controlled and cannot regenerate after pasteurization. However, microorganisms multiply in lines, tanks and other processing equipment downstream from the pasteurizer unless these are maintained in a thoroughly clean condition, including resanitation after any failure in pasteurization.

  (4) After pasteurization the juice is chilled; essence may be added in the holding tank before the filler to provide enhancement of flavor. Juice is then held between 1 and 3°C.

  (5) The chilled product flows by gravity to the filler. The fill is checked to determine adequacy. Cartons or other containers are sealed and dated with an expiration date. The chilled product is conveyed immediately to the cold room and not permitted to stand at ambient temperature after cartoning and palletizing.

  (6) Storage temperature of the finished containers should be between -1 and 3°C for maximum quality.

  (7) Cartons of filled product must be maintained in clean, neat appearance and a date code affixed to the carton or other package.

  (8) Juices for aseptic filling require further pasteurization. For such drinks, shelf life is limited by browning discoloration.

Finished product handling and storage

  (1) For extended shelf life of non-aseptic juice, temperatures of 2°C or below must be maintained.

  (2) Doors of delivery trucks are kept closed as much as possible to minimize heating during loading and unloading.

  (3) Stock is rotated to assure uniformity of shelf life.

  (4) All encouragement is given to the market outlet or institution to maintain low temperatures to assure good shelf life.

For further information on practical citrus methodology, refer to Murdock and Hatcher (1975).

2.4 Citrus juice flavor enhancement with natural citrus volatiles

2.4.1 Components of citrus juice flavor

Introduction. The major proportion of citrus juices now produced are extracted with a small peel oil content and are concentrated by evaporation to reduce storage and transport costs. Many of the natural flavor components are removed with the water during evaporation. A complex technology has developed to ensure that after reconstitution, the juice is as close as possible in flavor and composition to the freshly squeezed material.

The essential (i.e. essence-like) oil of the peel is mostly extracted separately from the juice by way of the emulsion leaving the Brown oil extractors or the FMC crumb process. This is referred to as 'cold-pressed oil'. As explained in

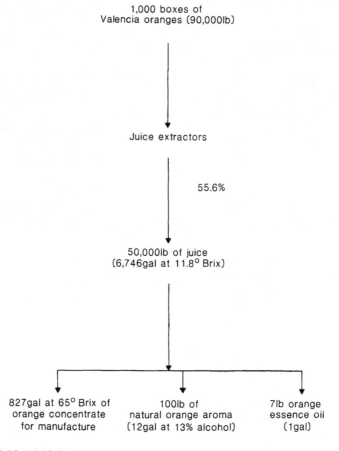

Figure 2.3 Material balance and major components of orange aroma and essence oil. Adapted from Johnson and Vora, 1983.

chapter 8, further peel oil is recovered from the de-oilers, although because of its heat treatment this has a different volatile character from cold-pressed oil. In addition, the volatiles removed with the water during evaporation are condensed separately and recovered from oil and water phases.

All these oils and aqueous distillates find application in the flavor enhancement of citrus concentrates. The materials used are recognized by legislative authorities as being a wholly natural part of the fruit juice from which they were derived and hence they require no label declaration when used to enhance its flavor and aroma (US/FDA, 1988; FDOC, 1980). The principal materials used are discussed below.

Citrus essences. The volatiles recovered during the manufacture of juice concentrates are referred to as essences. These essences divide into two groups, water soluble and oil soluble. The water soluble part is referred to as 'aroma' and the oil soluble fraction as 'essence oil'. Both parts have the characteristic 'flowery' or 'fruity' aromas that are representative of freshly squeezed juice. The quantities of juice concentrate and volatiles obtained in US processing of Valencia oranges are shown in Figure 2.3.

Water soluble aroma. This is restored to the juice to impart a 'fresh' flavor not otherwise obtainable. It contains a complex mixture of aldehydes, esters, ketones, alcohols, organic acids and hydrocarbons in a colorless aqueous phase (Table 2.9). As esters and other components in the aroma are known to hydrolyze and lose their flavor over time, aroma is stored separately from concentrate unless the concentrate is to be stored and distributed frozen.

Essence oil. This is the distilled oil fraction collected from an essence recovery unit during manufacture of concentrates and separated from the aqueous phase by decantation possibly followed by centrifugation. Although

Table 2.9 Some major components of natural orange aroma

Compound	Concentration
Acetaldehyde	600 ppm
Methanol	0.5%
Ethanol	13%
1-Propanol	100 ppm
Ethyl acetate	100 ppm
Acetal	70 ppm
Hexanal	Trace
Ethyl butyrate	50 ppm
trans-2-Hexanal	Trace
Limonene	200 ppm
Water	85%

Adapted from Johnson and Vora (1983).

Table 2.10 Some major components or orange essence oil

Compound	Concentration	Boiling points atmospheric pressure (°C)
Ethanol	0.1%	78
Ethyl acetate	50 ppm	78
Acetal	20 ppm	102.7
Hexanal	200 ppm	128
Ethyl butyrate	0.1%	121
trans-2-Hexanal	50 ppm	43 at 12 mm
α-Pinene	0.4%	156
Sabinene	0.4%	165
Myrcene	1.8%	167
Octanal	0.5%	171
d-Limonene	93.6%	178
Linalool	0.5%	198
Decanal	0.5%	208
Neral	0.2%	103 at 12 mm
Geranial	0.1%	229
Valencene	1.7%	123 at 11 mm

Adapted from Johnson and Vora (1983) and Redd (1976).

the oil fraction contains some of the volatiles from peel oils and cell oil, because of their simultaneous recovery, the oil phase contains most of the terpenes. Essence oil generally imparts the most fruity juice notes (Table 2.10). About 7 lb of essence oil is collected for each 100 lb of aqueous phase aroma (Johnson and Vora, 1983).

Cold-pressed oils. Cold-pressed oils are produced primarily for use as perfumery ingredients. Used in citrus juice, they impart the characteristic fragrance and taste of the outer part of the fruit. Data on their composition

Table 2.11 Cold-pressed citrus oil composition

Hydrocarbons
Terpenes
Sesquiterpenes

Oxygenated compounds
Aldehydes
Esters
Alcohols
Ketones
Phenols

Non-volatile residues
Waxes
Residues, coumarins, flavonoids, fatty acids, sterols, etc.

Adapted from Gunther (1949).

Table 2.12 Concentrations of classes of oxygenated compounds present in various citrus oils

Citrus oils	% (w/w) of total oil			
	Aldehydes	Esters	Alcohols	Acids
Hamlin orange	1.48	0.35	0.55	0.13
Pineapple orange	1.20	0.26	0.40	0.11
Valencia orange	1.63	0.27	0.87	0.13
Temple orange	1.87	0.42	0.63	0.30
Dancy tangerine	1.10	0.25	0.47	0.24
Orlando tangelo	0.57	1.22	0.59	0.45
Duncan grapefruit	1.80	3.26	1.06	0.39
Valencia essence oil	1.38	0.91	0.84	0.04

Adapted from Braddock and Kesterson (1976).

are shown in Tables 2.11–2.13. These oils can be folded (see below) to increase the proportion of desired flavoring materials. Fractions of these oils can provide the lighter note peel fragrances.

Oil concentrates or folded oils.   Any citrus oil may be concentrated or 'folded' by removing unwanted components. This was originally done by liquid/liquid extraction to remove excess citrus terpenes, principally *d*-limonene, but modern practice uses fractional distillation under vacuum. Reduction of the terpene content improves oxidative stability and water solubility. An oil which is reduced in volume by removal of terpenes is said to be 'folded' by the ratio of volume reduction so that a five-fold oil will have had its volume reduced by 80%. Folded oils increase fragrance with lighter peel notes without the harsh taste of citrus terpenes, referred to as 'burn' (Redd, 1988). Normally, some of both the top and bodying notes are lost during the folding process. However, newer technology has made it possible to reduce these losses (Redd and Hendrix, 1987). There are times when only highly folded oils may be used, for example, where oxidative stability or water solubility are a problem. However, folded oils are not intended to be used alone as flavorings.

Terpeneless oils.   These are an extension of the folded oils to 20-fold concentration or beyond. Being highly concentrated, they have good solubility and stability. The terpeneless oils, like all highly folded oils, cannot provide all the citrus flavor since significant flavor loss occurs during processing from the original source. Their use avoids water solubility problems and they find wide application as top notes and modifiers with single fold oils, blended oils and citrus flavor fractions.

Blended oils.   These are blends of cold-pressed oils, essence oils and certain oil fractions. They are prepared for use by bottlers of reconstituted juices as a

Table 2.13 The chemical composition of cold-pressed Valencia orange oil

Terpenes	**Alcohols**	**Ketones**
α-Thujene	Methyl alcohol	Carvone
α-Pinene	Ethyl alcohol	Methyl heptenone
Camphene	Amyl alcohol	α-Ionone
2,4-p-Menthadiene	n-Octanol	Acetone
Sabinene	n-Decanol	Piperitenone
Myrcene	Linalool	6-Methyl-5-heptene-2-one
δ-3-Carene	Citronellol	Nootkatone
α-Phellandrene	α-Terpineol	
α-Terpinene	n-Nonanol	**α,β-Dialkyl acroleins**
d-Limonene	$trans$-Carveol	α-Hexyl-β-heptyl acrolein
β-Terpinene	Geraniol	α-Hexyl-β-octyl acrolein
ρ-Cymene	Nerol	α-Heptyl-β-heptyl acrolein
α-Terpinolene	Heptanol	α-Octyl-β-heptyl acrolein
α,β-Cubebene	Undecanol	α-Hexyl-β-nonyl acrolein
α,β-Copaene	Dodecanol	α-Octyl-β-octyl acrolein
β-Elemene	Elemol	α-Heptyl-β-nonyl acrolein
Caryophyllene	cis-$trans$-2,8-p-menthadiene-1-ol	
Farnesene	cis-Carveol	**Paraffin waxes**
α,β-Humulene	1-p-Methene-9-ol	n-$C_{21}H_{44}$
Valencene	1,8-p-Menthadiene-9-ol	2-methyl-$C_{21}H_{43}$
δ-Cadinene	8-p-Methene-1,2-diol	n-$C_{22}H_{46}$
	Isopulegol	2-methyl-$C_{22}H_{45}$
Aldehydes	Borneol	n-$C_{23}H_{48}$
Formaldehyde	Methyl heptenol	3-methyl-$C_{23}H_{47}$
Acetaldehyde	Hexanol-1	n-$C_{24}H_{50}$
n-Hexanal	Terpinen-4-ol	2-methyl-$C_{24}H_{49}$
n-Heptanal		n-$C_{25}H_{52}$
n-Octanal	**Esters**	3-methyl-$C_{25}H_{51}$
n-Nonanal	Perillyl acetate	n-$C_{26}H_{54}$
n-Undecanal	n-Octyl acetate	2-methyl-$C_{26}H_{53}$
n-Dodecanal	Bornyl acetate	n-$C_{27}H_{56}$
Citral, neral	Geranyl formate	3-methyl-$C_{27}H_{55}$
geranial	Terpinyl acetate	n-$C_{28}H_{58}$
Citronellal	Linalyl acetate	2-methyl-$C_{28}H_{57}$
α-Sinensal	Linalyl propionate	n-$C_{29}H_{60}$
β-Sinensal	Geranyl acetate	
$trans$-Hexen-2-al-1	Nonyl acetate	
Dodecene-2-al-1	Decyl acetate	
Furfural	Neryl acetate	
Perillyldehyde	Citronellyl acetate	
Aldehyde A	Ethyl isovalerate	
B	Geranyl butyrate	
C	1,8-p-Menthadiene-9-yl-acetate	
D		
E	**Acids**	
	Formic	
Oxides	Acetic	
$trans$-Limonene oxide	Caprylic	
cis-Limonene oxide	Capric	

Adapted from Kesterson *et al.* (1971).

convenient single step essence addition. Essence oil is sometimes blended with peel oil to increase stability, since it lacks the natural antioxidants present in peel oil (Shaw, 1977).

2.4.2 Citrus flavor enhancement technology

This technology is best examined by considering its historical development. The procedures evolved for orange are now applied more or less to all citrus juices. For further detail on the processes described, refer to chapter 8.

During World War II, Bristow and others developed commercial production of concentrated orange juice for the army as a source of vitamin C. Flavor-wise, the product, known as 'hot pack concentrate', left a great deal to be desired because much of the essential flavor was lost during processing. The need for flavor restoration was recognized from 1942. The first attempts to correct this were by the addition of some cold-pressed orange oil to the concentrate before dilution; this gave a stable dispersion and contributed somewhat to the flavor/aroma of the reconstituted juice. It also helped in the masking of the possible oxidative 'cardboard' or 'castor oil' off-flavor that sometimes develops on storage.

In 1948, MacDowell, Moore and other Florida Citrus Commission technical associates were granted a patent for improving frozen concentrated orange juice (FCOJ) which is still in use today in some form. Flavor restoration consisted in adding to the concentrate before freezing an appropriate proportion of cold-pressed orange oil and freshly extracted, often pulpy, 'cutback' fresh juice to reduce (cut back) the 58° Brix concentrate to 42° Brix, giving a four-fold concentrated FCOJ. The main faults of this project were flavor variation and lack of adequate aroma since the loss of volatile essences during concentration had not been adequately compensated for. Also, year-round uniformity could not be wholly achieved through the frozen storage of 'pulpy cut-back juice'.

Some early work was done in Florida with essences collected from low temperature evaporators. Results were variable due to the occasional development of diacetyl and a buttermilk off-flavor resulting from lactic and other microorganism contamination.

An early essence unit designed by the Citrus Experiment Station engineers at Lake Alfred, Florida overcame this by recovering as condensate, the first 10% evaporated from the freshly extracted juice and using a fractionating still to concentrate the flavor fraction from this condensate.

In 1952, Redd and others developed flavor enhancement technology further with a vacuum concentration process for chilled juice known as the 'Sperti High Ester Restoration Process' and later with a freeze concentration process for 'pulpy cut-back juice' using centrifuges to remove ice crystals.

By 1964, the Minute Maid Company and Libby, McNeil & Libby had developed their own successful methods of essence recovery (Redd and

Hendrix, 1987). In 1967, the Pasco Packing Association started production with a system developed by C. D. Atkins and others at the Florida Citrus Commission (Jones, 1968).

In 1962, Distilkamp of Centrico, a US distributor for Westphalia Separator of Germany and McDuff of Adams Packing Company secured a patent assigned to Westphalia that recovered essence oil emulsion to give 'top flow emulsion oil'. Although this process was widely used in citrus concentrates and drinks, uniformity of oil flavor strength was always a problem (Redd and Hendrix, 1987).

Since 1960, the TASTE (thermally accelerated short time evaporator) has been developed by the Gulf Machinery Company. Due to the extreme shortness of time at maximum temperature 100°C, this allows recovery of essences with minimum change in characteristics. For a number of years, TASTE evaporators with Redd essence recovery units were operated in a number of Latin American and Caribbean countries and the essence unit design was progressively perfected to a point that allowed flavor quality grading for commercial use.

The first practical and continual day to day production of citrus essence was achieved in 1963 with a TASTE evaporator and advanced Redd recovery unit, for Suconasa, SA, at Araraquara in the State of São Paulo, Brazil. The plant was constructed and plant personnel trained in record time following the Florida freeze of the previous winter. The essence was stable and had the bouquet and aroma of freshly extracted orange juice. It was standardized, frozen and accompanied 65° Brix orange concentrate to the newly built facility of Sunny Orange, Canada Limited, Toronto, Ontario, Canada, where essence and concentrate were formulated into four-fold frozen FCOJ and citrus drinks.

The reconstituted juice received a high quality rating from the US Department of Agriculture and was purchased by almost all major private label brands from Canada and the United States. This pointed the way for the first major commercial process independent of processing season fluctuations. Because of this great success outside the United States, the Florida industry became convinced and a number of TASTE evaporators with Redd recovery units were installed and continue to be installed to the present.

2.4.3 Citrus oils and aroma and their recovery

Essential oils (cold-pressed oils).   Citrus peel essential oils are manufactured as a by-product from all citrus fruit processing: oranges, grapefruit, tangerines, tangelos, tangors, lemons and limes. Citrus oils are confined in oblate- to spherical-shaped oil glands that are located irregularly in the outer mesocarp or flavedo of the fruit. These glands or intercellular cavities have no walls of the usual type and are embedded at different depths in the flavedo of all citrus fruit. The cells surrounding the oil glands contain a colloidal aqueous emulsion of sugars, other organic materials and salts which exerts pressure on the glands (Kesterson and Braddock, 1976).

In the extraction of citrus oil, the oil is not pressed from the peel but rather the oil cells are ruptured by abrasion and the oil emerges under its own pressure to be washed away as an emulsion in water.

Historically, citrus peel oils have been expressed in Florida by seven different types of equipment: Pipkin roll; screw press; Fraser-brace excoriator; FMC rotary juice extractor; FMC in-line extractor; AMC scarifier; and Brown peel shaver. In present practice, the FMC in-line extractor is used to produce approximately 60% of the oil produced in Florida and Brown extractors the remaining 40%.

The general procedure is similar in most commercial plants: extraction or scarification to produce an oil–water emulsion is followed by a desludging centrifuge to obtain oil-rich emulsion. The oil-rich emulsion goes to a high speed 'polisher' centrifuge from which emerges an almost finished oil. In large installations, two polishers are used to minimize losses.

Essential oils are traded as liquids and so must be winterized, i.e. waxes which form insoluble deposits at low temperatures must be removed (Hendrix, 1988). For this, tall, narrow, stainless steel tanks with conical bottoms and bottom drains are in general use. A side drain is located several inches above the top of the cone to produce ample room for the wax to settle to the tank bottom. Temperature is maintained from $-23.3°C$ to $-3.9°C$, dependent upon holding time. In recent years, some companies have introduced accelerated winterization technology in which the oil is chilled rapidly using an eccentrically centered shaft scraped surface heat exchanger, sufficient to cause the formation of large numbers of wax crystal nuclei. This type of heat exchanger minimizes buildup of wax on the blades. The chilled oil is then transferred to a cold wall storage vessel for full crystallization and setting of the wax. The oil is then decanted and centrifuged and/or filtered for final removal of haze forming materials.

Care in sanitation practices is of importance to avoid bacterial contamination of citrus oils during processing (Murdock and Hunter, 1968).

Table 2.14 Optimum yield or availability of cold-pressed oils from different citrus fruit

	lb/US ton fruit
Orange	
Early and mid-season	7.0
Valencia or late-season	9.0
Grapefruit	3.5
Lemon	9.0
Lime	5.5

Adapted from FMC (1979).

The yield of cold-pressed essential oils from the principal citrus fruits is shown in Table 2.14.

De-oiler oils. As described in chapter 8, the normal operation of juice extractors results in citrus oil remaining in the juice. The level of oil in the finished juice should not exceed 0.015–0.02% oil by volume. Juice from the extractors might contain up to 0.05%. The excess oil is recovered by steam distillation. As previously stated, 1 lb of steam can potentially volatilize 0.6 lb of *d*-limonene, so to remove 0.05% it should be necessary to evaporate less than 0.1% of the juice. In practice, equilibrium conditions are not obtained. The oil is generally not present as free droplets but is more likely to be absorbed into pulp particles. In addition, not all citrus juices are the same. California orange juice, which has a higher content of suspended solids and greater viscosity than Florida juice, is much more difficult to de-oil than Florida juice.

In commercial practice, the juice is heated in a heat exchanger to approximately 71°C and then pumped tangentially into a vapor separator to form a thin film or pumped into a rapidly spinning perforated basket to form small droplets. The vapor separator is under reduced pressure, about 28 mmHg. This causes the juice to flash, reducing the temperature to about 38°C and vaporizing 6% of the juice which is generally sufficient to remove the excess oil. The amount of oil removed is controlled by the amount of flash, through temperature and vacuum. The juice can be pre-heated by the vapors to conserve energy. Normally, the condensate is cooled. The oil is separated out and the water is returned to the juice fraction.

Other oils and flavor fractions. Distillation processes are used to separate fractions of cold-pressed oil and recover and separate all other materials used in restoration of flavors to citrus juices concentrated by evaporation. Distillation involves boiling a liquid and separately condensing the vapors from it. At the boiling temperature, the sum of the vapor pressures of the components of the liquid equals the pressure at the surface of the liquid. Because of this, mixtures boil at lower temperatures than single compounds. An important application of this is steam distillation, when water already present in citrus juices, or introduced for the purpose, is boiled together with the material to be distilled. For example, in the laboratory at atmospheric pressure (approximately 760 mmHg) water and *d*-limonene, the principal component of citrus oil, boil together at 98°C contributing vapor pressures of 700 mmHg and 60 mmHg, respectively. *d*-Limonene alone boils at 177°C at atmospheric pressure. In the steam distillation, 1 lb of steam distils 0.6 lb of *d*-limonene.

Where the condensate settles into oil and water phases, these can be easily separated by decantation. However, to partition these phases into fractions for the purpose of concentrating the desired flavor notes or removing less soluble components, fractional distillation is required. This uses the principle

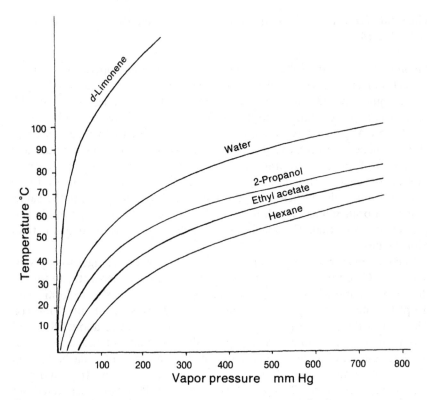

Figure 2.4 The relative vapor pressure of *d*-limonene, water and other common solvents. Adapted from Redd, 1976.

that substances with lower boiling points will evaporate earlier and faster than higher boiling components in a mixture. It is normally carried out under reduced pressure to lower the boiling temperatures and so reduce unwanted chemical changes. Also, at lower pressures the boiling points of different components are further apart so separation is easier (Figure 2.4).

A typical still and fractionating column is outlined in Figure 2.5. The column is maintained at a very constant temperature midway between the boiling point of the two components to be separated; in this way, the lower boiling, more volatile component passes over to the condenser which has a slightly lower pressure due to the condensation, while the higher boiling component refluxes back to the still.

Commercial applications of distillation to citrus production include the recovery of oils from de-oilers, recovery and separation of volatiles from the TASTE evaporators and the folding (concentration) of oils including removal of *d*-limonene.

d-*Limonene recovery.* *d*-Limonene is the principal component of the

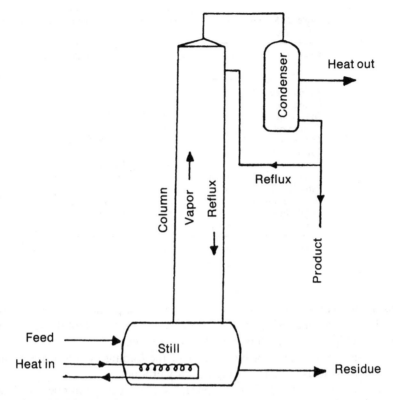

Figure 2.5 Diagram of still and fractionating column. Adapted from Redd, 1976.

essential oil recovered from citrus juices. The terms *d*-limonene, orange terpenes and stripper oil, sometimes used interchangeably, can mean different things to different people. These materials were first collected as a by-product of citrus molasses manufacture (chapter 8).

d-Limonene, the main terpene in the hydrocarbon composition (Table 2.15) of citrus essential oils finds wide and varied uses such as in the manufacture of resin adhesives, flavors, perfumes, deodorizers, soaps, shampoos, cleaning solutions, insecticide in pet shampoos and many other applications. In recent years, supply and demand has substantially increased the value of *d*-limonene and some manufacturers are diverting early season cold-pressed oils to *d*-limonene. *d*-Limonene can contribute greatly to a pollution and waste disposal problem due to a high chemical oxygen demand (COD) and it is much preferable to recover and convert to a valuable by-product.

Several machines were introduced into the industry for the specific purpose of collecting *d*-limonene. The oil can be separated out of the peel by either the FMC or Brown shaver oil systems and rather than manufacturing cold-pressed oils, *d*-limonene oil may be recovered by steam distillation. In some cases, when not carefully operated, the emulsion can be heated too high,

Table 2.15 Approximate d-limonene contents of some citrus oils.

Product	Percent limonene (v/v)
d-Limonene	> 95
Orange oil	
Grapefruit oil	
Essence oil	
Tangerine oil	
Tangelo oil	
Lemon oil	95
Lime oil, Mexican, Persian	50–55
5-fold orange, grapefruit oils	90
10-fold orange oil	80–85
25-fold orange oil	60–65
30-fold orange essence oil	1

Adapted from Braddock *et al.* (1986).

causing breakdown and hydration of the terpenes, particularly some of the bicyclic terpenes such as alpha- and beta-pinene, and some of the less stable isomers of $C_{10}H_{16}$ such as myrcene or delta-3-carene.

The d-limonene that is a by-product of the folding of cold-pressed oils, referred to as citrus terpene, is a high quality product.

More recently, the TASTE evaporators have been equipped with an internal condenser from the vent lines to draw off the d-limonene in the tube bundles and condense it in order to separate it out and prevent its being discarded with the evaporator barometric water. d-Limonene has a COD value of about 3 000 000 so even small amounts raise the chemical and biological oxygen demand (COD/BOD value) of plant effluents noticeably.

When peel is limed and pressed, a considerable portion of the oil is found in the press liquors. When these liquors are evaporated, either in the conventional molasses evaporator or in the waste heat-evaporator, d-limonene is a by-product of that operation. The product recovered from the molasses evaporators is aroma inferior to the other sources previously described.

Essence oils. Essence oils contain greater than 95% d-limonene. The aroma and flavor of these oils are quite different from other orange citrus oils, having a very fruity aroma, characteristic of fresh juice. Essence oils contain 0.5–2.0% valencene, a sesquiterpene not apparently present in other citrus oils. This terpene can be recovered and converted into nootkatone for use as a flavor enhancer (Hunter and Brodgen, 1965).

Folded oils. As previously indicated, orange and grapefruit cold-pressed oils are about 90% d-limonene. d-Limonene is a hydrocarbon with the

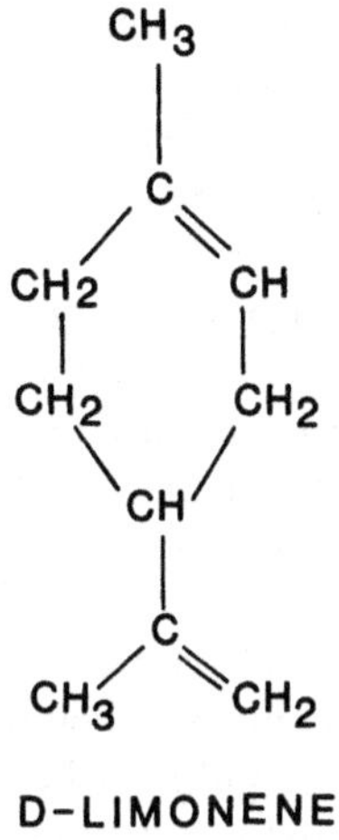

CH₃
|
C
/ \\
CH₂   CH
| |
CH₂   CH₂
\\   /
CH
|
C
CH₃ \\ CH₂

D-LIMONENE

Figure 2.6 *d*-Limonene organic formula, Adapted from Redd, 1976.

structure indicated in Figure 2.6. It boils at 178°C. It is generally conceded that *d*-limonene contributes little but harshness to the citrus flavor and the flavor of the oil can be increased and the 'bite' decreased by the removal of most of the *d*-limonene.

The folding and other fractionation of oils is surrounded by even more secrecy and mysticism than the essence and the various flavor houses have their own methods. Some processors still use solvent extraction. This is usually done by mixing ('washing') the oil with a mixture of 70% ethanol, 30% water at a controlled temperature. The mixture is then chilled and the oil which separates is removed by decantation. The ethanol/water ratio is varied to control the proportion of terpene removal.

With the advent of good mechanical vacuum pumps, the oil can now be simply and reliably separated by fractional distillation. Low boiling esters and aldehydes important to the flavor are distilled over with the *d*-limonene, and their separation and recovery is presently the subject of much research.

In designing equipment for this high vacuum work, all pipes and vessels have to be large to avoid excessive vapor speeds and ensure a maximum 2–3 mmHg pressure drop across the entire system, which must also be free of leaks.

2.5 Pectic substances and relationship of citrus enzymes to juice quality

The pectic substances found primarily in the middle lamella and in the primary cell walls of plant tissues are the cementing or cell binding materials. They are among the most abundant constituents found in the plant kingdom (Beavens *et al.*, 1962).

Pectic substances are complex organic compounds consisting of polymerized galacturonic acid, a sugar-acid related to the sugar galactose. They exhibit varied reactions and important properties such as the ability to form gels with sugars and fruit acids. Soluble citrus pectins are colloids with emulsifying properties.

On a dry weight basis, as much as 30% of orange fruit albedo may be pectin. Besides the albedo portion, the rag, core and segment membranes remaining after juice processing are a rich source of pectin. Because pectin is a cell wall constituent, relatively small amounts, 0.01–0.13%, have been reported in orange juice. Pectins contribute to viscosity and cloud stability in citrus juices.

Some viscosity in citrus juice is essential for good 'mouth feel' in flavor; however, viscosity presents technical difficulties in juice extraction, pulp washing and in concentration of pulp washed juice. Concentration above 40° Brix requires the use of pectolytic enzymes to reduce viscosity by breaking down the pectin. These enzymes are prepared especially by culturing micro-organisms, usually *Aspergillus niger*. Their safety is established by vigorously controlled tests carried out by independent bodies and they are approved for use in foods by the regulatory authorities. The juice or pulp wash is held at approximately 60°C for 2–4 h with a carefully controlled addition of enzyme. This reduces the viscosity by breaking the pectin molecules into units of lower molecular weight.

Natural pectolytic enzymes occur in citrus and other plants and are extracted with the juice. Unlike the enzymes added to reduce viscosity, which break down the chains of galacturonic acid polymer, the natural enzymes remove the methyl groups from the carboxylic acid side chains. This demethylation produces pectin substances which form gels in concentrated sugar solution and which at lower concentrations can react with calcium and magnesium naturally present in the juice and in reconstituting waters, to form insoluble flocculating precipitates. The natural pectinesterase enzyme can be deactivated by heat treatment. This is referred to as stabilization. It is common for stabilization and pasteurization to be carried out in the same operation (chapter 8).

If the pectinesterase is not deactivated in the shortest possible time after extraction of the juice, then progressive de-esterification of the pectin will occur in storage. The enzyme can facilitate demethylation even at the lowest temperature of storage, −18°C, although at these temperatures, the reaction rate is very slow. In the early days of production of frozen concentrated orange juice (FCOJ), this led to severe problems of gelling of the concentrate. Even though the heat treatment of juice during evaporation had normally destroyed the pectinesterase, the practice of 'cutting back' with freshly pressed unpasteurized orange juice to improve flavor ensured that active pectinesterase was present in the concentrate. A sugar level exceeding 50% and high acidity then provided optimum conditions for gelling with demethylated pectin.

These problems are avoided in modern practice by efficient design of the evaporator to ensure stabilization. As noted earlier, juice flavor enhancement is now carried out with components from citrus oils and volatiles thus avoiding use of unpasteurized juice.

Demethylation of pectin in citrus juice or concentrate can result in a loss of cloud stability in the finished juice. This is normally regarded as undesirable and during production operations and in purchase specifications for citrus products, pectinesterase activity is measured and/or cloud stability tested. Zero enzyme activity does not guarantee good cloud stability since nil enzyme activity at the time of testing does not guarantee that enzyme was not active for a significant period in the earlier history of the juice or concentrate.

For some products, the demethylation of citrus juices is actively promoted. Examples are in the production of completely clarified lemon and lime juice. Also, gelled citrus salads are produced by maximum pectinesterase treatment of juice before concentration.

The recent increase in the sale of freshly squeezed, non-pasteurized citrus juices relies on very short shelf life and chilled storage to avoid unacceptable cloud loss.

2.6 Effect of time, temperature and other factors on citrus products

Even when citrus juice is produced with good microbiological control and free from unwanted pectinesterase activity, chemical changes will occur on storage which can influence its shelf life. These result in flavor change due to production of furfural, hydroxymethyl furfural, α-terpineol, diacetyl, 3-methyl-2-butene-1-ol and other compounds (Nagy and Rouseff, 1986). These changes occur more rapidly in concentrate but can be minimized during storage at $-18°C$ for as long as 5 years. Single strength juices are best flash pasteurized and aseptically packed or distributed chilled. Some flavor deterioration is inevitable during hot filling.

Container type can also affect shelf life of citrus juices (Nagy and Rouseff, 1986). Vitamin and possibly flavor stability are poorer in containers that are permeable to atmospheric oxygen. Unlacquered tinned steel cans are used for orange juice since the traces of tin which will dissolve provide a reducing environment which improves color stability. However, extended storage in such cans, particularly at temperatures over 30°C, must be avoided to prevent development of off-flavors and excessive metal pick-up and corrosion which threatens can integrity.

Chilled single strength orange juice packaged in wax-coated fiberboard cartons and glass, polyethylene and polystyrene bottles have been studied by Bisset and Berry (1975). At refrigerated temperature (10°C), the glass-packed product lost about 10% of its initial vitamin C content after 4 months. However, the plastic and fiberboard-packaged juices lost 80% in only 3–4 weeks due to the effects of oxygen permeability.

References

Agricultural Marketing Services, USDA (1983) *U.S. Standards for Grade of Orange Juice*, Washington, DC.

Baker, R. A. (1980) The role of pectin in citrus quality and nutrition. In: *Citrus Nutrition and Quality*, S. Nagy and J. A. Attaway (eds.), ACS Symposium Series 143, Washington, DC.

Baker, R. A. and Bruemmer, J. H. (1972) Pectinase stabilization of orange juice cloud. *Agric. Food Chem.* 20(6), 1169.

Beavens, E. A., Bissett, O. W., Keller, J. G. *et al.* (1962) *Chemical Technology of Citrus Products*, USDA/ARS, Agricultural Handbook 98, Washington, DC.

Berry, R. E. (1973) *Commodity Stg. Manual, Code Recommended Practices for Handling Frozen Food*, Frozen Food Coordinating Committee, Washington, DC.

Berry, R. E. (1984) Recent developments in analytical technology for citrus products, *44th Annual Meeting IFT*, Anaheim, CA.

Berry, R. E., Bissett, O. W., Veldhuis, M. K. *et al.* (1971) Vitamin C retention in orange juice as related to container type. *Citrus Industry* 52(6), 12.

Bissett, O. W. and Berry, R. E. (1975) Effects of container shelf-life. *J. Food Sci.* 40, 178.

Braddock, R. J. and Kesterson, J. W. (1976) Quality analysis of aldehydes, esters, alcohols and acids from citrus oils. *J. Food Sci.* 41, 1007.

Braddock, R. J., Temelli, F., Cadwallader, K. R. *et al.* (1986) Citrus essential oils—A dossier for OSHA/MSDS. *Food Technol.* 40(11), 114–116.

Camp, A. F., Evans, R. C., MacDowell, L. G. *et al.* (1957) Citrus Industry of Florida, Department of Citrus, Citrus Commission of Florida.

FDOC (1980) *Florida Citrus Fruit Laws*, Chapter 601 (1980) Florida Statutes, FDOC, Lakeland, FL.

FMC (1979) *A Process for Determining the Efficiency of CR Citrus Oil Recovery Systems*, Citrus Machinery Division, R&D, Lakeland, FL.

Furia, T. E. (ed.) (1968) *Handbook of Food Additives*, The Chemical Publishing Co., Cleveland, OH.

Gunther, E. (1949) *The Essential Oils*, Vols I and III, Fritzsche Bros. Inc., D. Van Nostrand, Princeton, NJ.

Hasse, G. (ed.) (1987) *The Orange—A Brazilian Adv.*, Duprat & Lobe Prop. & Commun. ILDDA, Riveirao Preto, S.P., Brazil.

Hendrix, C. M., Jr. (1974) Importance of microbiology to citrus products and by-products. *Prof. Sanit. Mgt. Mag.* 4(1), 24–25.

Hendrix, C. M. (1983) *Citrus Quality Assurance and Future Industry Trends*, Fla. Sect. ASQ, Orlando, FL.

Hendrix, C. M., Jr. (1984) *Quality Assurance and Evaluation in the Citrus Industry*, Golden Gem Citrus Disp. School, Errold Est., Apopka, FL.

Hendrix, C. M., Viale, H. E., Johnson, J. D. and Vilece, R. J. (1977) Quality control, assurance and evaluation in the citrus industry. In: *Citrus Science and Technology*, Vol. 2, S. Nagy, P. E. Shaw and M. K. Veldhuis (eds.), AVT, Westport, CT.

Hendrix, D. L. (1988) The composition and yield of citrus waxes as influenced by standing temperatures, a portion of M.S. Thesis, University of Florida, Gainesville, FL.

Hendrix, D. L. and Ghegan, R. C. (1980) Quality changes in bulk standards for citrus concentrate made from frozen damaged fruit. *J. Food Sci.* 45(6), 1570–1572.

Hughes, R. E. (1978) Fruit flavonoids—some nutritional implications. *J. Hum. Nutrition* 32, 47–52.

Hunter, G. L. K. and Brogden, W. B., Jr. (1965) Conversion of Valencene to Nootkatone. *J. Food Sci.* 20(5), 876–878.

Johnson, J. D. and Vora, J. D. (1983) Natural citrus essence: production and application, *43rd Annual Meeting IFT*, New Orleans, LA.

Jones, H. A. (1968) Essences, how they develop. *Citrus World Mag.*, Lakeland, FL.

Kefford, J. F. and Chandler, B. V. (1970) *The Chemical Constituents of Citrus Fruits*, Academic Press, New York.

Kesterson, J. W. and Braddock, R. J. (1973) Citrus seed oil. *Cosmet. Perfum.* 88, 39.

Kesterson, J. W. and Braddock, R. J. (1976) *By-Products and Specialty Products of Florida Citrus*, Bull. 784, IFAS, Lake Alfred, FL.

Kesterson, J. W., Hendrickson, R., Braddock, R. J. *et al.* (1971) *Florida Citrus Oils*, Bull. 749, IFAS, Lake Alfred, FL.

Lombard, P. B. (1963) Maturity study of mandarins, tangelos, and tangors. *Calif. Citrogr.* 48, 171–176.

Marshall, M., Nagy, S. and Rouseff, R. (1985) Factors impact on the quality of stored citrus fruit beverages, *Proc. 4th Int. Flav. Conf.*, Rhodes, Greece.

Matthews, R. F. (1980) *FCOJ from Florida Oranges*, Food Science Fact Sheet, IFAS, Gainesville, FL.

McAllister, J. W. (1980) Methods for determining the quality of citrus juice. In: *Citrus Nutrition and Quality*, S. Nagy and J. A. Attaway (eds.), ACS Symposium Series 143, Washington, DC, pp. 291–317.

Moshonas, M. G. and Shaw, P. E. (1986) Analysis and evaluation of orange juice flavor components, *46th Annual Meeting IFT*, Dallas, TX.

Murdock, D. I. (1977) In: *Citrus Science and Technology*, S. Nagy, P. Shaw, and M. Veldhuis (eds.), Vol. 2, AVI Publ. Co., Westport, CT, Chap. 11.

Murdock, D. I. and Hatcher, W. S., Jr. (1975) Growth of microorganisms in chilled orange juice. *J. Milk Food Technol.* 38(7), 393–396.

Murdock, D. I. and Hunter, G. L. K. (1968) Bacterial contamination of some citrus oils during processing. *Proc. Fl. State Hortic. Soc.* 81, 242–254.

Nagy, S. and Attaway, J. A. (eds.) (1980) *Citrus Nutrition and Quality*, American Chemical Society Symposium Series 143, Washington, DC.

Nagy, S. and Nordby, H. E. (1970) The effect of storage condition on the lipid composition of commercially prepared orange juice. *J. Agric. Food Chem.* 18, 593–597.

Nagy, S. and Rouseff, R. L. (1986) Citrus fruit juices. In: *Handbook of Food and Beverage Stabilization*, Academic Press, New York, pp. 719–743.

Redd, J. B. (1976) Distillation systems. In: *The Annual Short Course for the Food Industry*, IFT, University of Florida, Gainesville, FL.

Redd, J. B. (1977) Brazilian Citrus Products Around the World, Part 7. In: *Citrus Science and Technology*, Vol. 2, S. Nagy, P. E. Shaw and M. K. Veldhuis (eds.), AVI Publ. Co., Westport, CT, Chap. 15.

Redd, J. B. (1988) The volatile flavors of orange juice, *Annual ASME Citrus Engineering Conf.*, FDOC, Lakeland, FL.

Redd, J. B. and Hendrix, C. M., Jr. (1987) *A History of Juice Enhancement*, Intercit, Inc., Safety Harbor, FL.

Redd, J. B., Hendrix, C. M., Jr. and Jefferson, J. E. (1966) *Quality Control Manual for Citrus Processing Plants*, Redd Labs Inc., Safety Harbor, FL.

Redd, J. B. and Praschan, V. C. (1975) *Quality Control Manual for Citrus Processing Plants*, Intercit, Inc., Safety Harbor, FL.

Redd, J. B., Hendrix, C. M., Jr. and Hendrix, D. L. (1987) *Quality Control Manual for Citrus Processing Plants*, Vol. I, Intercit, Inc., Safety Harbor, FL.

Reuther, W., Batchelor, L. D. and Webber, H. J. (1967) *The Citrus Industry*, University of California.

Ross, E. (1944) Effect of time and temperature of storage on vitamin C retention in canned citrus juice. *Food Res.* 9, 27–33.

Robbins, R. C. (1977) *Int. J. Vitam. Nutrition Res.* 47, 373.

Shaw, P. E. (1977) In: *Citrus Science and Technology*, Vol. 1, S. Nagy, P. E. Shaw and M. K. Veldhuis (eds.), AVI Publ. Co., Westport, CT, pp. 246–257.

Sokoloff, B. and Redd, J. B. (1949) The health angle in the consumption of canned orange juice. *The Citrus Industry Mag.*, Lakeland, FL.

Szent-Gyorgyi, A. and Rusznyak, I. (1936) *Nature*, London.

Ting, S. V. (1980) Nutrients and nutrition of citrus fruits. In: *Citrus Nutrition and Quality*, S. Nagy and J. A. Attaway (eds.), ACS Symposium Series 143, Washington, DC, Chap. 1.

US/FDA (1988) Title 21, FDA Code, Fed. Reg. 21, Part 146, Canned Fruit Juice, US Government Printing Office, Washington, DC.

Varsel, C. (1986) Citrus juice production as related to quality and nutrition. In: *Citrus Nutrition and Quality*, S. Nagy and J. A. Attaway (eds.), ACS Symposium Series 143, Washington, DC, Chap. 11.

Weast, R. C. (ed.) (1972) *Handbook of Chemistry and Physics*, 53rd edn., CRC, Cleveland, OH.

Westbrook, G. F., Jr. (1957) Effects of the use of frozen damaged fruit on the characteristics of FCOJ, Ph.D. Thesis, University of Florida, Gainesville, FL.

3 Grape juice processing
M. R. McLELLAN and E. J. RACE

3.1 History of grape juice processing in North America

The fruit juice processing industry of the United States is said to have been started by Dr Thomas B. Welch and his son Charles in Vineland, New Jersey in 1868. By applying the theory of Louis Pasteur to the processing of Concord grapes, they were able to produce an 'unfermented sacramental wine' for use in their church. By 1870 this grape product was being produced on a small scale for local church use.

By 1893, grape juice had become a national favorite beverage in the United States as thousands sampled it at the Chicago World's Fair. It was during this year that Dr Charles Welch turned his full attention to the marketing of grape juice. In 1897 a new plant location was chosen for processing operations at Westfield, New York. Some 300 metric tons of grapes were processed that year; in 1989 Welch's, now one of the largest producers of processed grape juice in the world, handled some 186 000 metric tons of grapes.

3.2 Grape cultivars

In the United States, four broad classes of grapes are grown: *Vitis lubruscana*, hybrids of the northeastern United States native grape; *Vitis vinifera*, European grapes common to California area; *Vitis rotundifolia*, the southern and southeastern Muscadine grapes; and French hybrids. Prior to the discovery of the Americas the species, *Vitis vinifera*, supplied the known world's grapes. *Vinifera* grapes are still among the most important in the world but in harsh climates these grapes cannot tolerate severe winters, diseases, and pest problems.

When selections of *Vitis lubrusca* were crossed with other grapes, new varietals were produced, such as *Vitis lubruscana*. Ephraim Bull was a horticulturalist who pioneered in this cross-breeding and selection process with native American grapes. The Concord grape, a varietal almost synonymous with grape juice in the United States, was a seedling that Bull found in his vineyard. Its parentage is still unknown. Virtually the entire unfermented grape juice industry has developed from this one cultivar.

The average production of Concords is approximately 4–5 tons per acre in eastern United States. Yields can reach 7–8 tons in western states. Cultural practices such as irrigation, pruning severity and fertilization can have a very significant influence upon the quality and quantity of grapes harvested (Morris et al., 1983).

There are now many hybrids of the native species available for use in the industry. Some of the older cultivars are: Catawba, Delaware and Niagara. The Concord grape, grown throughout the cooler regions of the United States and Canada, is still the principal grape for the industry.

Though not possessing as large a market share in unfermented grape juice, there are some Vinifera grapes processed into grape juice. They tend to be much higher in sugar and lower in acidity than the Lubruscana grapes, and consequently are not as flavorful.

The genus, Vitis, is generally considered to consist of two sub-species, Euvitis and Rotundifolia. All grape species other than muscadines fall into the Euvitis sub-genus; muscadines alone make up the sub-genus, Rotundifolia. For this reason some botanists do not classify muscadines as grapes. Some basic differences are identifiable such as clustering of the berries and pit configuration, however, muscadines are utilized in an identical fashion and treatment as the Euvitis sub-genus. Consequently, in matters of commerce and functionality, muscadines are considered grapes.

3.3 The chemistry of grape juice

The quality of grape juice can be described almost entirely by its chemistry. Its color is caused by anthocyanins, their glucosides and condensation products (Hrazdina and Moskowitz, 1981), its taste by acids, sugars and phenolics (Ribéreau-Gayon, 1968) and its aroma by a diverse mixture of volatile secondary metabolites in very low concentrations (Schreier et al., 1976). Since 1967 over 1000 research papers have been abstracted by Chemical Abstract Service describing the chemistry of grapes and grape juice. In these papers thousands of chemicals and their reactions in grapes are described. However, only a small percent of these chemicals are responsible for the quality attributes that people perceive when they drink grape juice. Table 3.1 lists the major components of grape juice, their concentrations and the quality attributes they determine.

This table shows clearly why it is possible to predict sweetness (Shallenberger, 1980), sourness and acidity (Plane et al., 1980) in grape juice by measuring certain carbohydrates and organic acids. These compounds are major constitutents of grape juice and are measurable by some very simple techniques. For many years the sugar and acid content of grapes has been used to set standards of quality resulting in new horticultural and processing techniques that modify the sugars and acids in grapes in order to optimize juice quality. These developments are the direct result of our knowledge of the

Table 3.1 The chemical composition of grape juice solutes, their amounts and corresponding quality attributes

Chemical class	Concentration (g/100 ml)	Quality attribute
Carbohydrates	20	Sweetness
Acids	1	Sourness and acid taste
Phenolics	0.1	Color and astringency
Volatiles	0.0001	Aroma

chemical causes of quality and the availability of tools to measure them. However, the volatiles that cause aroma are present in such minute quantities that most of them are still unknown and those that are known generally require the most advanced chemical and spectroscopic techniques to quantify them.

Although the methods of chemical analysis that define a quality grape will continue to develop and improve, there is another approach, sensory analysis. This involves the determination of quality attributes using methods from the sciences of behavioral psychology and sensory perception (Amerine *et al.*, 1967; O'Mahony, 1986). These methods relate the human perception of quality to food components. Instead of using chemical reactions and physical measurements, sensory analysis uses human subjects as tools to determine quality by measuring their behavior when they are subjected to the food. A review of sensory methodology as it relates to the optimization of quality attributes is given by Moskowitz (1983).

3.3.1 Carbohydrates

Carbohydrates are the most abundant component in grapes. On average, grapes have per 100 g, 6.2 g glucose, 6.7 g fructose, 1.8 g sucrose, 1.9 g maltose and 1.6 g of other various mono- and oligosaccharides (Lee *et al.*, 1970). Pectic substances which act as the intercellular cement are a mixture of numerous long chain carbohydrates and related compounds which occur in solution and/or colloidal dispersion in grape juice.

3.3.2 Acids

Tartaric acid is the predominant acid in grapes and accounts for the tartness of the juice. Tartaric acid is present as the D-isomer and malic acid as the L-isomer. Other acids present in minor amounts include citric, lactic, succinic, fumaric, pyruvic, α-oxoglutaric, glyceric, glycolic, dimethyl-succinic, shikimic, quinic, mandelic, *cis*- and *trans*-aconitic, maleic and isocitric acids (Colagrande, 1956; Mattick *et al.*, 1973). The resulting pH of grape juice can range typically from pH 3 to pH 4 depending upon variety, climate and soil conditions.

3.3.3 Mineral content

Although differences among varietals in mineral content are not typically noted, differences during maturation have been studied. Mineral anions and cations are taken up at a relatively constant rate and distributed throughout the system resulting in a relatively poor concentration in the berry compared to other parts of the vine. During maturation heavy metals increase by as much as 50%. Phosphate content increases in both the peel and the pulp (Bonastre, 1971).

3.3.4 Phenolics

The color of Concord grapes is due in large part to the anthocyanin pigments located in and adjacent to the skin. These pigments are extracted by heat and/or fermentation. Seven individual color components have been identified in Concord juice, the major contributor being delphinidin monoglucoside (Mattick et al., 1967). The phenols typically found in grapes include benzoic acids, cinnamic acids, flavonols, anthocyanidins as well as various flavans which constitute the tannin precursors (Ribéreau-Gayon, 1964; Peri, 1967).

3.3.5 Volatiles

Maarse (1984) listed 500 volatile compounds in grapes but most of these compounds occur at levels well below their detection threshold. Only a very few seem to evoke sensation and affect perception (Acree, 1981). Furthermore, the odor-active molecules have no identifiable chemical features that distinguish them from the more prevalent odorless volatile compounds. The odor-active volatiles in Lubruscana grapes come from different metabolic pathways (Figure 3.1). 2,5-Dimethyl-3(2H)-4-furanone, a sweet strawberry smelling compound, is produced from carbohydrate metabolism. Ethyl 3-mercapto-propanoate, with a foxy smell, methyl anthranilate with its Concord smell and the floral smelling 2-phenylethanol are formed from amino acids, while beta-damascenone with a rose-like aroma and linalool with the aroma of orange peel oil are derived from terpenoids. The green-leaf-like smelling (E)-2-hexenol is a lipid oxidation product.

This diversity results in a variety of processing effects on the aroma of grape products. For example, the heat treatments used in color extraction, depectinization, juice concentration and pasteurization cause beta-damascenone, vanilin, and/or methyl anthranilate to dominate the aroma of certain cultivars.

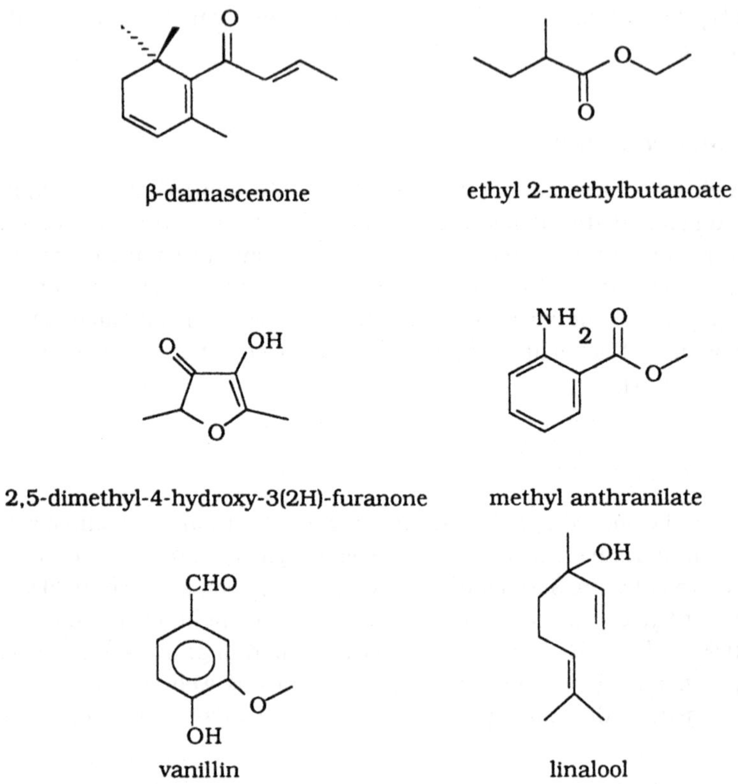

β-damascenone

ethyl 2-methylbutanoate

2,5-dimethyl-4-hydroxy-3(2H)-furanone

methyl anthranilate

CHO

vanillin

linalool

Figure 3.1 The major odor-active compounds found in *Lubruscana* grapes.

3.4 Modern grape juice processing

3.4.1 Harvesting/ripening

Grape juice consists of a natural aqueous mixture of various carbohydrates, organic acids, anthocyanins and flavor compounds. In the soluble solids of grape juice, the primary sugars are glucose and fructose. In unripe grapes, glucose accounts for as much as 85% of the sugar content. As the grape approaches full ripeness there is generally a slight excess of fructose. Sugars are manufactured in the leaves via photosynthesis and translocated to other organs, particularly the berries, as needed. It is the sugar content of the grapes which often is the basis for purchase and a primary indicator of optimal harvest time for grape juice production.

A typical level of acidity at time of harvest in Concord grapes is 1.3 g/100 ml as tartaric acid (Mattick *et al.*, 1973). This acidity generally starts quite high in the grapes and decreases as ripening takes place. Optimum ripeness is often associated with balanced levels of sugars and acids (Winkler, 1932). On occasion, tannins, color and aroma are also considered. Although acidity levels of Concord grapes at harvest are very high, cold temperature storage (32°F) of the grape juice extracted from these high acid grapes will induce a detartration and reduction of total acidity to more acceptable levels. Degrees Brix, a refractive index-based measurement of soluble solids, is the primary method used in grading grapes at harvest which are scheduled for juice production.

Mechanical harvesting has greatly improved the efficiency and speed of bringing grapes from the vineyard to the processor. Mechanically harvested grapes are handled in bulk boxes (47 × 42 × 38 inch) equipped with polyethylene liners. These boxes will hold approximately 1 ton of grapes. Studies have shown that mechanically harvested grapes actually contain fewer stems and trash than hand harvested grapes (Marshall et al., 1971). This is important due to the detrimental effect of materials other than grapes on the quality of expressed grape juice (Huang et al., 1988). These boxes can be expected to have approximately 21 inch of free run juice in them because of mechanical damage, vibration and weight of grapes from depth of load.

The Chisholm–Ryder grape harvester is a self-propelled system that straddles the rows of vines and is one of the most commonly used harvesting units. This harvester bats off the grapes and grape clusters with metal or rubber-like strips as it moves down the length of the row. The grapes fall onto a series of collector leaves located on either side of the unit underneath the vine being harvested. These rotating leaves then transfer the grapes to conveyers which carry them to the top of the harvester unit and dump them into a transfer chute for filling of a bulk bin separately transported by truck.

Bulk bins of grapes are transported to a grading station where core samples are taken from each bin and combined to form a load sample. This is used to measure the soluble solids (Brix) upon which payment for the load will be based. Grapes are typically processed within 4–6 h of picking.

Most grape juice processed in the United States is made from Concord grapes. A significant volume of literature can be found concerning grape juice processing and some reviews of the technology are also available (Pederson, 1980; Luh and Kean, 1975). A typical process outline is described below and shown in Figure 3.2

3.4.2 Stemmer/crusher operation

Grapes at the plant are first conveyed to a stemmer/crusher which removes residual stems, leaves and petioles from fruit. This unit is designed around a rotating perforated drum. The perforations are approximately 1 inch in diameter and generally cover the entire drum. In the process of traversing the rotating drum, grapes are caught by the perforated drum and knocked from the stems. The individual grapes are broken open or crushed in the process

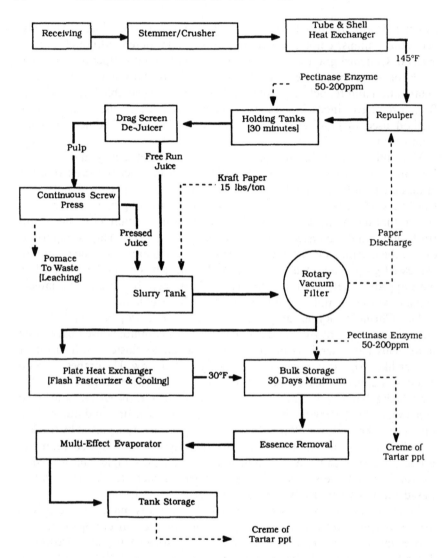

Figure 3.2 A typical process for grape juice production.

and drop through the drum. Stems and leaves, etc. continue on in the center of the drum and are discharged at the end for waste.

3.4.3 Hot-break process

Once the stemmed/crushed grapes are separated from the vines; they pass through a large bore tubular heat exchanger where they are heated to 140°F. This is known as the hot-break process and it is primarily designed to extract a large amount of color and assist in maximizing the yield of juice. To the hot grapes is added a pectolytic enzyme and in the case of the typical process shown here, Kraft (wood pulp) paper (actually spent filter media from the rotary vacuum juice filter) is added to serve as a press aid.

The addition of press aid to the mash provides firmness and exit channels for the juice (Hurler and Wey, 1984). Sterilized rice hulls, bleached Kraft fiber sheets or rolled stock and ground wood pulp are common press aids used commercially. Ideally, a rotary filter press aid should have relatively long fibers and it should be able to be separated with a minimum breakage of those fibers for maximum effectiveness.

Soluble pectin is found in the juice and is a result of pectolytic enzymes which are primarily located in the cell wall of the grape. This soluble pectin causes difficulties in pressing due to the lubrication it affords the press cake and the consequential reduction in screw press effectiveness. Typically, 50–100 ppm of a pectinase enzyme is sufficient for the de-pectinization process at this stage. De-pectinization is designed to reduce the slipperiness of the pulp and thus permits the effective use of a screw press when combined with wood pulp bulking agent. Enzyme treatment is allowed to continue for about 30 min prior to pressing. Several de-pectinizing tanks are employed so that a continuous flow may be maintained to the presses.

3.4.4 De-juicing/pressing operation

The mash is generally hot-pressed in order to maximize yields and color extraction. Hot-pressed juice is higher in total solids, tannins, anthocyanins and other matter. Control of the combined hot-break/hot-press procedures is important because the extraction of astringent materials (i.e. tannins) will increase with time and temperature of these two unit operations. Phenolic compounds contribute to both the color and the astringency of grape juice. The extraction of both the phenols and the anthocyanins is greatly affected by the degree of heating used in the hot-break process. In general, the longer and hotter this process, the more significant the extraction process.

The de-juicing or pressing operation can utilize a number of different equipment variations. These include screw, hydraulic, belt and pneumatic presses. Selection of the equipment can depend on production capacity, juice yield, desired level of automation, availability of man power, operating cost and required capital investment.

The hydraulic rack and frame press was the mainstay of the grape juice pressing operations for many years. Heavy cotton or nylon cloths were filled with a set amount of mash and then folded to produce what is called a cheese. The individual cheeses were stacked and separated by a wooden or plastic lattice work. The combined stack was then compressed using a hydraulic ram during which the juice was expressed.

The Bucher Guyer Press is a highly automated pressing system used in a batch pressing operation. Generally this system requires no press aid. The system consists of a rotatable cylinder with a hydraulic ram used for juice expression. Within the cylinder are fabric-covered flexible rubber rods. The rods have longitudinal grooves which allow the juice to transport easily to the discharge port.

The Wilmes press is a commonly used system for grape juice pressing. It is a pneumatic-based system consisting of a perforated, rotatable horizontal cylinder with an inflatable rubber tube in the center. The cylinder is filled with grape mash through a door in the cylinder wall which is rotated to the top position. After filling, the press is rotated to ensure even filling. During this rotation the air bag is filled, creating the mash compression action. The bag is then collapsed and the cylinder is rotated. The rotation and pneumatic compression of the mash is repeated many times with increasing air pressure.

The de-pectinized mash produced in the hot-break process can be passed over dejuicing screens where 50–60% free run juice is removed. With this low technology step, the mash to be pressed is reduced by approximately half. The dejuiced pulp is then pressed in (horizontal or vertical) screw presses (Figure 3.3). The typical screw press consists of a reinforced stainless steel cylindrical screen enclosing a large bore screw with narrow clearance between the screw and the screen. Breaker bars are typically located between the screw intervals in order to disrupt the compressing mash. Back pressure is provided at the end of the chamber and is usually adjustable. Typical capacities for screw presses with a 12 inch and 16 inch diameter are 5 and 15 tons/h, respectively (Bump, 1988).

Figure 3.3 De-juiced grape pulp in a horizontal screw press.

The Serpentine belt press, the Ensink and other belt presses are effective for grape juice processing (DeVos, 1970). In belt presses, a layer of mash is pumped onto the belt entering the machine. Added press aid is usually required for improved yield and reduced suspended solids. The belt is either folded over or another belt is layered on top of the one carrying the mash. A series of pressurized rolls compress the enveloped mash. Expressed juice is caught in drip pans. The cake is discharged after the last pressure roller.

The waste pomace from any of these systems may be repressed for improved yields of an additional 3–5 gallons of juice per ton leading to a total juice yield of 200–205 gallons per ton. Extraction of the pomace using hot water, in a plant equipped with concentration equipment, can recover additional soluble solids in the juice. This secondary extraction when handled with care will produce a low soluble solids liquor which will have no odor or taste defects. This low flavor impact, low soluble solid extract can be recombined (in natural proportions) with single strength juice intended for concentration.

All pressing operations must be able to deal with free run juice and in some cases handle it with special care to give the highest possible product quality. In comparing the different press types, the screw press is probably most popular; however, this is highly dependent on required yields, available capital at purchase and smoothness of integration into current operations.

Liquefaction is also a common de-juicing process though not as common with grape juice. The process generally requires a fine grinding of the fruit in preparation for enzyme treatment. Typically the mash is treated with a combination of pectinase and cellulase which induces dissolution of the cell wall leading to a potential increase in total juice yield as well as increased soluble and insoluble solids. Due to the large increase in solids, particular care must be taken to ensure that components likely to induce post-bottling haze are sufficiently reduced or removed. A method of addressing this concern is the use of further enzyme treatments typically including hemicellulases, oligomerases and glycosidases. New commercial enzyme preparations are specifically tailored for the liquefaction process and will usually produce a satisfactory liquefaction in approximately 3 h at elevated temperatures.

3.4.5 Coarse filtration

The pressed juice is merged with free run juice and accumulated in slurry tanks where the press aid, usually paper or wood fiber is added at the rate of about 15 lb/ton of grapes. The paper, which typically comes in 1 lb sheets, requires very heavy agitation to break it up and disperse it throughout the juice. Alternatively this paper may be shredded first. Having been dispersed, this paper serves as filter media for the rotary vacuum belt filtration that follows.

Prior to filtration the insoluble solids (ISS) content of Concord juice pressed with paper (measured volumetrically by centrifugation) can reach 10–12%. Without paper, pressing is extremely difficult and produces juice containing up to 30% ISS. After filtration through the belt and paper, insoluble solids will be 1% or lower. The spent paper discharging from the belt is then re-slurried into the fresh hot grapes entering the system just prior to the treatment tanks; thus the paper serves the dual function of filter media and press aid.

The importance of achieving low ISS cannot be overemphasized for two reasons: (i) high efficiency heat exchangers used for pasteurization have very small clearances between plates and tend to plug over a period of time with high levels of insolubles; (ii) any insolubles remaining will ultimately settle to the bottom of the storage tanks. The actual volume of bottoms is dependent primarily on the level of insolubles in the initial juice and the filtration of bottoms juice containing high levels of insolubles is difficult and costly.

3.4.6 Bulk storage and tartrate precipitation

A common practice is the storage of single strength grape juice in bulk storage tanks. The tanks themselves are pre-cleaned and then filled with clean municipal ice water for periods of 1–4 weeks prior to use. This water is then used for cooling processed juice during the periods of highest process demand thereby dramatically reducing the immediate refrigeration needs. Air inlets at the tops of these tanks are generally fitted with absolute sterile air filters in order to ensure no airborne contamination.

In the actual process, filtered juice is pasteurized at 185–190°F for a minimum of 1 min. The heat exchanger system utilized in this step, cycles through a cleaning every 24 h. Hot caustic, followed by 200 ppm chlorine solution, followed by 185°F hot water rinse is the usual sanitizing method. By proper flow design this cleaning approach can be used for not only the heat exchangers but also for all the feed lines to the storage tanks. Once the juice is properly pasteurized it is immediately cooled using a regenerative heat exchanger, to 30–32°F prior to storing in refrigerated tanks.

The entire system is treated much like an aseptic process. Special care is taken to avoid any dead ends or other such areas for contamination buildup. Use of heat exchangers for the cooling process is required in order to deliver juice at proper temperature to the storage tanks. Failure to reach storage temperature prior to delivery of juice to the tank will cause the tank to remain above proper storage temperature for an excessively long time.

These tanks, which typically are either stainless steel or Food Grade epoxy-lined cold-rolled steel, are not truly aseptic but nevertheless are carefully sterilized prior to introduction of any new juice. Design of the tanks allows for complete filling up to a top manhole. The manhole top is usually equipped with an ultraviolet lamp to preclude the growth of mold. Tank capacity varies typically in a range of 150 000–320 000 US gallons but can range over 700 000 gallons. Juice processed in this manner and held at 28–30°F is very stable and has been held for over 1 year without any evidence of fermentation. Obviously, stringent sanitation requirements are necessary.

Yeasts and molds occasionally are found in juice stored in this manner. Storage tanks are typically monitored on a weekly basis for any change in alcohol content. By current practices in the industry, a specific change of alcohol determined on a company by company basis is significant enough to warrant a complete re-pasteurization of the stored juice.

Bulk storage creates a large pool of grape juice, allowing for bottling of a juice with very standard characteristics which are stable over a long term bottling period. Juice stored in this bulk storage process maintains its quality over the period of a year or more. This access to large quantities of juice reserves allows processors to operate over a longer period of time, thus spreading their capital costs. Packing throughout the year helps to ensure a high quality product is reaching the consumer no matter how long after harvest the juice is purchased.

Freshly pressed Concord grape juice is naturally saturated with potassium bitartrate (cream of tartar) upon receipt. Typically, it will contain 1.1–1.2 g potassium acid tartrate/100 ml. Under refrigeration, however, the excess tartrate will precipitate until equilibrium conditions are achieved. This equilibrium is approximately 0.65–0.70 g potassium acid tartrate/100 ml. The difference precipitates out of solution and settles to the bottom of the tank. These precipitates are called argols. This is the equivalent of about 9 lb/ton of original grape juice. Unlike tartrates precipitating from wine, which tend to appear in the form of large, pure crystals (wine stones), tartrates from grape juice are much smaller and contain substantial occluded pigments and other insolubles that render them unsuitable for tartrate recovery and are therefore considered a waste product.

3.4.7 Enzyme clarification

Additional enzyme treatment is required before the juice is further processed. For concentration it is necessary to complete the de-pectinization process to preclude gelation of the concentrate once produced. For bottled juice, de-pectinization speeds and simplifies polish filtration and clarification prior to bottling. De-pectinization followed by filtration with diatomaceous earth or its equivalent prior to bottling is necessary in order to eliminate or minimize sedimentation upon long term storage after hot packing.

To complete the de-pectinization, 50–200 ppm of additional pectolytic enzyme is added with agitation. By adding the enzyme while the tank is being filled, the incoming juice velocity will provide adequate mixing and agitation. At the high level of enzyme treatment, de-pectinization and settling can be accomplished within 7–10 days at 30°F. If a longer time is available, 50 ppm will prove adequate at that temperature.

Enzymatic clarification of grape juice, as well as other juices of similar character, is based on the partial hydrolysis of soluble pectin which otherwise acts as a protective colloid. Hydrolysis of this pectin allows insoluble and fine particles to flocculate.

De-pectinization of the supernatant juice is deemed to be completed sufficiently for concentration when the apparent viscosity has been reduced to within 0.05 cP of the viscosity of a sucrose solution of the same percent water soluble solids.

3.4.8 Polish (fine) filtration

The juice is prepared for final filtration by decanting the supernatant from the settled tank bottoms which might account for as much as 8–10% of the tank volume. This is generally carried out through draw off valves at convenient locations elevated from the tank bottom.

A final filtration process is accomplished typically through the use of a diatomaceous earth filtration system. This system requires the suspension of filter media (diatomaceous earth) in the juice and subsequent passing of the juice through a pre-coated pad or plate. When done properly, this filtration process is highly efficient. Use of the correct grade and amount of diatomaceous earth for filtration can increase the rate and effectiveness of filtration. In the filtration process, suspended filter-aid deposits on a pre-coated layer as the filtration progresses. This deposition is permeable and prevents the extended buildup and subsequent clogging of the filter by the suspended solids.

There are a number of different types of pressure filters available for this application. A plate and frame filter is designed around a series of alternate stainless steel plates and frames. Juice is fed into the frames of the system and filters through the pre-coated layer of diatomaceous earth on the plate. The unfiltered juice has had filter aid added as a body feed and over the time of filtration, this body feed gradually builds up in the frames providing fresh filtration surfaces continuously. Filtered juice collects from all of the plates and exits to a hold tank.

A leaf filter is another common system where individual leaves of filtration elements are suspended in a sealable feed tank. These leaves are pre-coated and then used as filtration elements. Spacing between leaves is adjustable according to the suspended solids level. High suspended solids call for wide leaf placement. A cleaner juice will allow for a narrower spacing and thus higher filtration rates due to additional leaf accommodation.

Rotary vacuum pre-coat filters are also common in the juice industry. In these systems, a layer of pre-coat is built up on the rotary drum and then removed by a knife as the juice filters through the pre-coat. Removal of a layer of the pre-coat exposes a new clean pre-coat surface thus preserving the rate of filtration.

Although diatomaceous earth filtration is common, especially in the handling of bottom solids, ultrafiltration may offer specific advantages to the grape juice processor. Ultrafiltration membranes are constructed of various materials, however, for juice processing there are two general classes, polymeric membranes and ceramic membranes.

The polymer membranes are typically a polysulfone type membrane usually of a hollow fiber design. Characteristics of these membranes include good stability under typical conditions of juice processing, moderate costs, and as compared to the ceramic membranes, the flow rates are generally higher because of the larger filter area per cartridge unit.

Ceramic membranes are relative newcomers to the fruit juice industry. These systems are generally designed to handle higher solids concentration than the hollow fiber systems. Consequently, the major use for them is in the area of 'bottoms' filtration. After the relatively clear supernatant has been drawn off the tank, the bottoms are mixed and filtered by the ceramic system. Where the hollow fiber systems are able to obtain a filtration up to 35% spin-solids (centrifuged suspended solids), the ceramic systems are able to push that limit to 200% spin-solids. At high feed velocities (14.6 m/s) and high temperatures (50°C), very high flux rates (400–500 kg/h m²) have been reported using ceramic membranes (Padilla and McLellan, 1993).

Membranes with molecular weight cut-offs of 10 000 have been shown to dramatically reduce the level of soluble proteins in the juice to almost zero. The result is a higher degree of heat stability (Hsu et al., 1987).

3.4.9 Hot fill

Retail bottling is accomplished by passing the polished juice through a heat exchanger and raising the temperature such that the temperature of the juice filling the bottle or can is typically at a minimum of 180°F. This hot fill process is adequate for acidic beverages such as grape juice. In a glass bottle, a shelf life of high quality retention can be from 6 to 12 months, though typically juice movement through US markets to consumers will take only 3–6 months.

3.5 Process alternatives

3.5.1 Cold-pressing

White grapes and even Concords are sometimes pressed without a hot-break, preheating step. Advantages of this procedure may include a more fresh grape flavor with less astringency and usually a somewhat lighter color.

Yields can be expected to drop with this technique possibly as much as 20% (Pederson, 1980). Generally cold-pressed juices do not settle out as well and therefore may appear muddy unless full clarification and filtration steps are taken. Particularly in white grape juices, browning rapidly sets in unless countermeasures are taken, such as the addition of ascorbic acid or SO_2.

3.5.2 Aseptic process

With the approved use of hydrogen peroxide as a surface sterilant for packaging, the way was cleared for the introduction of aseptic processing. This process requires that the product be commercially sterile at the time of packaging. Additionally, the package itself must be free of any micro-organisms at time of filling, and finally the filling and sealing process must be done such that no re-contamination is possible. For a high acid product such as grape juice, an acceptable process would be achievable with the equivalent of a 190°F hot fill. Thus 2–3 min at 190°F should be adequate.

Use of high temperature continuous processes can produce a sterile product with a minimal effect on product quality because of the greater sensitivity of spoilage organisms than the quality factors to increasing temperatures. Another benefit of aseptic processing is that packaging materials are generally cheaper than those used in traditional hot fill operations.

Aseptic processing is not a new concept. It was commercially practised in Europe for milk during the 1950s. However, the widespread use of this technology has awaited the development of more suitable packaging designed for the process.

There exist a number of methods appropriate for grape juice sterilization. For fully clarified grape juice with no remaining particulate matter, sterile filtration would be effective, Membranes with a pore size small enough to exclude microorganisms ($< 0.45\,\mu m$) would be required. Alternatively heat sterilization would be required. For grape juice, this would typically require 200–212°F for 15–45 s. Equipment selection for this process would typically be a tube and shell or plate type heat exchanger.

3.5.3 Concentration

Grape juice concentration can offer significant advantages to the processor. The concentration process reduces the bulk of the juice, thereby reducing storage volume requirement and transportation costs. Additionally, concentration allows a more complete deposition of insoluble solids and tartrates. Storage of the cold concentrate is less likely to exhibit yeast growth because of the high sugar concentration; and if frozen concentrate is a primary end product, storage as a concentrate prior to canning and freezing is a logical step.

Reviews which cover the technological aspects of concentration of fruit juices have been published by Heid and Casten (1961). In depth discussion concerning fluid food evaporators (Moore and Hesler, 1963), energy use in the concentration of fluid foods (Schwartzberg, 1977), principles of freeze concentration and its economics (Thijssen, 1974) and engineering principles of aroma recovery (Bomben *et al.*, 1973) have been presented in the literature.

Grape juice is most often concentrated by evaporation of water, the major constituent of the juice. Because of possible loss of aroma constituents, the first major step is stripping of volatiles, from which aromas can be recovered. Stripping is primarily handled by partial evaporation. Alternatively, a steam stripping process could be employed.

A number of different evaporator designs can be employed for grape juice concentration. These include single pass re-circulating concentrators, single or multiple stage, single or multiple effect and whether the condenser is a surface mixing type (Heid and Casten, 1961). When describing evaporators, the word 'stage' is used to identify the flow of juice through the system. Thus single strength juice enters the first stage and final concentrate exits the last stage. Use of the term 'effect' describes the path of steam through the system. The first effect of an evaporator system receives steam from a boiler and the second effect receives steam and vapors boiled off from the first effect (Rebeck, 1976).

A Schmidt Sigmaster multi-effect evaporator in use for grape juice concentration is shown in Figure 3.4. The volatiles removed from this system are concentrated 100–150-fold and added back to the concentrate prior to storage.

Concentration of 'clean' juice streams is also possible using a combination

Figure 3.4 Schmidt Sigmaster multi-effect evaporator used to concentrate grape juice.

of reverse osmosis and evaporation. Cross-flow membrane systems are ideally suited for this application due to a self-cleaning turbulence effect. The reverse osmosis technology is effective in concentrating a low solids juice (7–8° Brix) two- or threefold. From there, evaporation technology will take over. Recently, a new reverse osmosis design has claimed effectiveness in achieving a 50–60° Brix concentrate (Cross, 1988).

3.5.4 *Sulfur dioxide preservative*

Though use of sulfites is currently discouraged in the industry, it nevertheless is a viable method of grape juice preservation. Current US law allows for the addition of up to 1000 ppm to single strength juice as a preservative, however, research has shown that when yeast contamination is not excessive, SO_2 levels at 200 ppm should be adequate to provide long term storage at low temperature (Splittstoesser and Mattick, 1981). Other detailed systems have been proposed for adding and subsequently removing sulfite from grape juice (Potter, 1979; Anon, 1978).

References

Acree, T. E. (1981) *Quality of Selected Fruits and Vegetables*, ACS Symposium Series 170, Washington, DC, p. 11.
Amerine, M. A., Pangborn, R. M. and Roessler, E. B. (1967) *Principles of Sensory Evaluation of Food*, Academic Press, New York.
Anon (1978) *Food Eng. Int.* 3, 74.
Bomben, J. L., Bruin, S., Thijssen, H. A. C. and Merson, R. L. (1973) *Adv. Food Res.* 20, 1.
Bonastre, J. (1971) In: *The Biochemistry of Fruits and Their Products*, Vol. 2, A. C. Hulme (ed.), Academic Press, New York.
Bump, V. L. (1988) In: *Processed Apple Products*, D. L. Downing (ed.) AVI, New York, Chap. 3.
Cofelt, R. J. (1965) *Calif. Agric.* 19, 8.
Colagrande, O. (1956) *Ann. Microbiol.* 9, 62.
Cross, S. (1988) In: *1988 IFT Program and Abstracts*, IFT Chicago, IL, p. 217.
DeVos, L. (1970) *Proc. Int. Fed. Fruit Juice Prod.* 10, 173.
Heid, J. L. and Casten, J. W. (1961) In: *Fruit and Vegetable Juice Processing Technology*, D. K. Tressler and M. A. Joslyn (eds.) AVI, New York, p. 278.
Hrazdina, G. and Moskowitz, A. H. (1981) In: *The Quality of Foods and Beverages*, Academic Press, New York, p. 341.
Hsu, J., Heatherbell, D. A., Flores, J. H. and Watson, B. T. (1987) *Am. J. Enol. Vitic.* 38, 17.
Huang, P. D., Cash, J. N. and Santerre, C. R. (1988) *J. Food Sci.* 53, 173.
Hurler, A. and Wey, R. (1984) *Confructa Stud.* 28, 125.
Lee, C. Y., Shallenberger, R. S. and Vittum, M. T. (1970) *New York's Food Life Sci. Bull.* 1, 1.
Luh, B. S. and Kean, C. E. (1975) In: *Commercial Fruit Processing*, J. G. Woodroof and B. S. Luh (eds.), AVI, New York, Chap. 6.
Marshall, D. E., Levin, J. H. and Cargill, B. F. (1971) *Trans. Am. Soc. Agric. Engrs.* 14, 273.
Maarse, H. and Visscher, C. A. (1984) *Volatile Compounds in Food*, TNO, Zeist.
Mattick, L. R., Shaulis, N. J. and Moyer, J. C. (1973) *New York's Food Life Sci. Bull.* 6, 8.
Mattick, L. R., Weirs, L. D. and Robinson, W. B. (1967) *J. Assoc. Off. Agric. Chem.* 50, 299.
Moore, J. G. and Hesler, W. E. (1963) *Chem Eng. Prog.* 59, 87.
Morris, J. R., Spayd, S. E. and Cawthon, D. L. (1983) *Am. J. Enol. Vitic.* 34, 229.
Moskowitz, H. R. (1983) *Product Testing and Sensory Evaluation of Foods*, Food and Nutrition Press Inc., Westport, CT.

O'Mahony, M. O. (1986) *Sensory Evaluation of Food*. Marcel Dekker, New York.

Padilla, O. I. and McLellan, M. R. (1993) *J. Food Sci.* 58, 369.

Pederson, C. S. (1980) In: *Fruit and Vegetable Juice Processing Technology*, P. Nelson and D. Tressler (eds.), AVI, New York, Chap. 7.

Peri, C. (1967) In: *IIème Symposium intern. d'Oenologie*, p. 47.

Plane, R. A., Mattick, L. R. and Weirs, L. D. (1980) *Am. J. Enol. Vitic.* 31, 265.

Potter, R. (1979) *Food Technol. Australia* 31, 113.

Rebeck, H. (1976) In: *Proceedings of 16th Annual Short Course for the Food Industry*, University of Florida, Gainesville, FL, p. 1.

Ribéreau-Gayon, P. (1964) In: *Les Composés phénoliques du Raisan et du Vin*, Dunod, Paris, p. 1.

Schreier, P. D., Drawert, F. and Junker, A. (1976) *J. Agric. Food Chem.* 24, 331.

Schwartzberg, H. G. (1977) *Food Technol.* 31, 67.

Shallenberger, R. S. (1980) *Food Technol.* 65.

Splittstoesser, D. F., Cadwell, M. C. and Martin, M. (1969) *J. Food Sci.* 34.

Splittstoesser, D. F., Kuss, F. R., Harrison, W. and Prest, D. B. (1971) *Am. Soc. Microbiol.* 21, 335.

Splittstoesser, D. F. and Mattick, L. R. (1981) *Am. J. Enol. Viticult*, 32, 171.

Thijssen, H. A. C. (1974) In: *Advances in Preconcentration and Dehydration of Foods*, A. Spicer (ed.). John Wiley, New York p. 115.

Winkler, A. J. (1932) *Calif. Agric. Exp. Stn., Bull.* 529, 1.

4 Tropical fruit juices

J. HOOPER

4.1 Introduction

Tropical fruits are the newest arrivals on the juice and fruit beverage market. With the exception of pineapple, they have only recently become established in this sector, and there is still considerable room for further development. Pineapple juice has been available commercially since 1932. The development of the production and sale of other tropical juices has been much more recent, and the market for such juices is still immature both in Europe and especially in North America.

The expansion of demand was no doubt stimulated by the increasing diversity of tropical fruits that became available in the 1980s. They were first introduced by small shops in ethnic communities and are now widely available in the major supermarkets throughout Europe and North America. This new interest encouraged more sophisticated methods of production, selection and packaging and hence the availability of higher quality and more attractive looking fresh fruit which, in turn, led to increased demand. The result was an enhanced public awareness of tropical fruits leading to a growing interest in the juices and beverages made from them. This interest was encouraged by the marketing activities of many of the leading distributors.

Apart from pineapple, the principal juices which have gained some level of popularity in Europe are guava, mango and passionfruit, although some other tropical juices are also seen. With the exception of pineapple, tropical juices tend not to be suitable for drinking on their own and are usually found either in nectars or juice-based drinks particularly, or as an ingredient in the mixed fruit drinks which have become more common in recent years.

The commercial production of tropical fruit juices has, thus, increased considerably. The method of production of juice or purée from certain tropical fruits is described below. In many cases the juices are concentrated and then packed either in frozen or aseptic form. Not all tropical juices are, however, suitable for concentration because of their physical nature. This is particularly true where the liquid expressed from the fruit is a purée with a high pulp content rather than a free flowing juice. In these cases, it is usually possible to separate the serum from the pulp but the result is frequently a poor, characterless liquid bearing little resemblance to the original fruit. Another constraint is the fact that many tropical juices are produced in relatively small quantities and unless a sufficient throughput can be generated by, for instance producing other juices in greater quantities, the installation of modern evaporation capacity would be uneconomic.

When concentration is possible, modern evaporators are employed including those produced by APV, Alfa Laval and Gulf. Much development work is also being done to improve the production of those natural volatiles which are normally lost during evaporation. In many cases these are particularly elusive but breakthroughs have been made and more are expected.

Throughout this chapter, reference to degree of concentration is given in degrees Brix rather than by volume (e.g. 3:1 or 2+1). The reason for this is twofold. In the first instance, there are no official criteria for the Brix or soluble solid content of single strength tropical juices or purées, and secondly the natural solids in the various fruits vary considerably depending on origin, variety and the time at which the fruit is harvested.

In the following, an attempt has been made to give estimates of production figures. It should be noted, however, that reliable statistical information is not generally available. Many countries lump all tropical fruit statistics together under one heading whilst others release export figures rather than those for total production. It should be understood therefore that many of the figures given are based on information privately obtained, and from the author's personal experience. They should be regarded as a guide only and are intended to give the reader some general perspective of the world production and trade in tropical juices rather than an accurate record.

4.2 Guava

The guava, *Psidium guajava* L. (Figure 4.1) is a member of the large Myrtaceae or Myrtle family which includes the common Mediterranean myrtle of mythological importance to the Greeks and the Romans, the eucalyptus group and also cloves which are the dried buds of *Syzygium aromaticum*. Guavas are the only edible fruits of this family which are processed commercially although there are other fruits such as Brazilian cherry, water rose and rye apple which are sold as fresh fruits mostly locally.

Guavas are believed to have originated in the tropical regions of South America but were well established in Asia by the early 17th century and are now common in most tropical and sub-tropical regions throughout the world.

No figures exist showing the production of fresh guava. The plant grows in the wild and is well adapted to a wide range of different conditions in both tropical and sub-tropical regions. The most important commercial regions include Florida, California and Hawaii, Central and South America, India, South East Asia, Australia and the South Sea Islands, and Northern and Southern Africa.

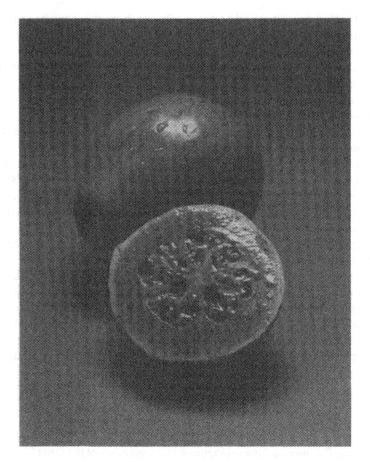

Figure 4.1 Guava, Brazil.

The fruit, which is roughly the size of a lemon, can be round or pear-shaped and grows on small trees or shrubs which bear after 3 years and are mature by 6 years. Commercial plantations are generally planted to a density of around 250 trees per hectare and when mature the yield is close to 35 tonnes of fruit per hectare. Propagation is readily achieved from seed and it reproduces rapidly enough to grow in the wild state to the extent that in some areas, such as Florida, it is almost regarded as a weed. It has a yellow outer skin and the inside is not unlike a tomato in appearance containing a large amount of small sharp seeds.

There are many different varieties and cultivars, too numerous to list here, which have been developed in different producing areas, as well as a large range of wild and low quality types. For the purposes of the fruit juice industry it is convenient to divide the different types broadly into the white-fleshed and the pink-fleshed varieties of which the Beaumont is probably the best known.

There is so far only a small, but nevertheless growing, international trade in fresh guava because of its rather limited keeping quality although the fruit is consumed in very large quantities in those areas where it is grown. Most commercial plantations are thus geared to industrial purposes where the fruit is either canned or converted into juice or purée, or used for producing jam and guava paste. There are two main problems facing the processing industry. One is the high proportion of sharp seeds in the fruit, and the other is the fact that the flowering, and hence the ripening of the fruit, occurs indiscriminately creating problems for production planning. The reason that so much work has been done to produce specialized cultivars is the need to reduce these problems and although much progress has been made, there is still some way to go in this area.

The juice or purée produced from guava has a high nutritional value and has a distinctive flavour. The total soluble solids tends to be on the low side averaging around 9° Brix. It has a Brix/acid ratio ranging from about 6 up to 20 and the pH varies between 3.2 and 4.2.

The usual method of producing guava purée is to crush the fruit into a pulp and pass it through screens to filter out the seeds and fibrous material from the skins. The juice yield is quite low at about 25–30% of the weight of fruit employed and various methods have been used including both cooking and freezing to break down the cell structure to increase yield. Cooking, usually for 30 min at 100°C, tends to spoil the flavour and freezing is expensive. More recently some processors have used enzyme treatment to give higher yields. The major difficulty in processing is caused by the seeds or seed particles. These are very sharp and if present in the purée can give it an unpleasant, gritty texture. In addition they cause excessive wear and tear on machinery, particularly the rubber parts on some filling equipment. In order to eliminate this problem it is necessary to pass the purée through a mill so that any seed particles that have not been trapped in the screens are all eliminated. This means the introduction of equipment that would not be necessary in a plant processing other types of tropical fruits.

Table 4.1 Estimated world production of guava purée 1987–1988

	Tonnes
United States (Hawaii)	5500–6000
Brazil	4000
South Africa	800–1000
Australia	600–1000
India	1500
Pakistan	100–200
Taiwan	100–200
Others	400–600

Guava purée does not concentrate easily because of its high pulp content. Although attempts have been made using assisted circulation in the evaporator it is normally available commercially only as a frozen single strength purée.

Estimated world production of guava purée in selected areas is shown in Table 4.1. Both the Brazilian and the Australian production, together with smaller amounts produced in Fiji and New Zealand are mostly consumed locally, whereas that of Hawaii, which has grown from about 150 tonnes in 1970 is consumed either locally or in America. The guava drink market has been very actively promoted in Hawaii and North America since about 1980. Dole and several dairies in Hawaii produce a guava nectar and in mainland United States, some of the first to introduce it were Libbys, Kern Foods and Hansen Foods. There is also a large market in the Middle East, supplied mainly by India and another growing market is Japan where supplies are obtained from Taiwan, India and Pakistan as well as from some production in the Philippines. The European market which is supplied mainly by South Africa, and to a lesser extent, the Philippines and Brazil, is still small, measuring a few hundred tonnes, but shows signs of growth in the field of nectars and in mixed fruit drinks.

4.3 Mango

The mango, *Mangifera indica* L. (Figure 4.2) is a member of the Anacardiaceae family which includes among its 600 or so members, the cashews and pistachios as well as some lesser known fruits such as the bauno, gandaria and ambarella from South East Asia and the umbu and hogplum from central America.

It is believed to be the oldest cultivated fruit in the world and its origin is to be found in the Indian sub-continent, most probably in the Indo-Burmese area where it is known to have existed for at least 4000 years. Today mango trees cover roughly 2 million hectares in India alone. Ancient stocks also exist throughout Malaysia and much of the Far East.

By the 16th century, the Portuguese had introduced mangoes to South Africa and Brazil and today the fruit grows freely throughout all tropical regions and many temperate zones including Florida and the Southeastern rim of the Mediterranean. Because it is so widespread, it has earned a reputation of being the 'apple of the tropics'. No production figures exist but it is known that India produces at least 5–6 million tonnes and that another million tonnes are grown in Brazil.

The fruit, which is oval or round in section and usually slightly kidney shaped, has a large flat clingstone seed, and can vary in size between that of a plum and a melon. A large fruit can sometimes weigh up to 2 kg. It grows on an evergreen tree which bears a slight resemblance to a chestnut and which in

Figure 4.2 Mangoes, Brazil.

the wild state can reach a height of 30 m. For convenience the trees are often grafted onto dwarf rootstocks. Propagation is readily achieved by both seed and grafting.

Thousands of different varieties of mango exist. One of the more important ones is the Mulgoba from which the Alfonso, Haden, Tommy Atkins and Keitt subvarieties have been derived. These grow chiefly in India, Florida, Hawaii and in various parts of Africa. Others worth mentioning include the Carabao, grown in the Far East, Bourbon, Extreme and Imperial which are popular in South America, and Kensington which is the main Australian variety, grown extensively in Queensland. All these varieties differ greatly in colour, size, composition and, above all, flavour.

Mangoes are a very popular fruit and the vast majority are consumed in the fresh state. Regretfully, as with so many tropical fruits, transport of the fruit is difficult and as a result fruit destined for export has to be picked in the green state. This means that the fresh mangoes we can obtain in Northern Europe are often very disappointing as they lack the delicate aroma and sweetness that is one of the characteristics of the tree-ripened fruit.

Much fruit is processed into pickles and chutneys, particularly in India, and there is also a well established canning industry producing slices in syrup. More recently, frozen mango cheeks have begun to gain popularity.

Mango juice or purée is very aromatic and tends to be sweet. Soluble solids average 15–17% and the Brix acid ratio can vary between 20 and 45; pH averages 4.35. There is a high content of carotene and vitamins A and C, and the level of terpenes is also very high, particularly in the part of the fruit close to the skin.

The production of mango juice or purée involves steaming the fruit for a few minutes to loosen the skin. The next stage is generally to pass the fruit through a stirring and cutting vessel where knives break the fruit up into a mixture of pulp and skin without breaking the seeds or stones which remain intact. Thereafter it passes through a sieve to remove stone and skins and on to a final pulper or centrifuge to yield a smooth purée. As with guava, concentration is difficult, and though evaporators with assisted circulation have been used to some effect, the degree of concentration which can be achieved remains low and much of the aromatic flavour is lost. Another method of concentration which has been employed with some success is to centrifuge the purée to separate the pulp from the serum. The latter can then be concentrated to about 45–50° Brix and when the pulp is returned the result is a concentrate of about 40° Brix equivalent to a threefold concentration which has much of the aroma and quality of the original purée. Despite this, most mango is still only available as a single strength purée, packed aseptically or frozen.

Mango purée is produced in most of the main growing areas and is virtually a cottage industry. Much of the production would not meet the quality standards required in Europe. It is widely used for the production of drinks in the countries where it grows, and has only gained any popularity in Europe since the 1960s. The quality of the purée varies very widely, depending

Table 4.2 Estimated world production of mango purée 1987–1988

	Tonnes
Egypt	24 000
India	22 000
Central America and Caribbean	20 000
Taiwan	8 000
Pakistan	1 000
Philippines	500
Brazil	500
Others	1 000

on the variety of fruit used to produce it, and in Europe the most sought after quality is that made from the Alfonso and related Tutapuri cultivars which are grown in India. The result is that the Indian purée commands a substantial premium over that from any other origin and although in general, supplies are very plentiful and relatively inexpensive compared to other fruit juices, there are, from time to time shortages of the best qualities. Table 4.2 gives estimated production figures for mango purée in the main areas.

The largest market for mango purée is the CIS which probably consumes around 30 000 tonnes per annum with supplies coming mainly from Egypt, Cuba and India. Almost as large is the Middle East, mainly North Yemen, Saudi Arabia and the Gulf States which draw mainly from India. As with guava the main European use for mango purée is for the production of nectars and mixed fruit drinks and consumption is particularly high amongst the Asian communities living in Europe where there is a market for imported canned mango drinks as well as a developing local production of such drinks.

Figure 4.3 Passionfruit, Kenya.

4.4 Passionfruit

The passionfruit, *Passiflora edulis* Sims (Figure 4.3) has a family to itself, the Passifloraceae, of which there are roughly 400 members. Few of them have edible fruits and the only other members of any real interest are the banana, passionfruit and the giant granadilla which are closely related.

The fruit is a native of South America and it became known to Europeans when it was discovered by the Spanish conquistadores. They were reminded of the pomegranate (Spanish *granada*) and they called it granadilla, meaning little pomegranate, a name still used in certain parts of the world, particularly South Africa. Another name for the passionfruit is the Portuguese or Brazilian maracuja, a name also used in Germany. There is a popular misconception that the English name refers to some interesting and unusual properties possessed by the fruit but, alas, this is not the case, and the name actually refers to the structure of the flower which takes the form of a cross.

The fruit was first exploited commercially in Australia, but today the main producing areas are in South America, particularly Brazil, Peru and Colombia. There is also significant production in Venezuela and Ecuador, Kenya, the Ivory Coast and South Africa, Hawaii, Australia and Fiji. There are no reliable production statistics because the quantities grown have tended to be most irregular. Passionfruit will yield fruit within a year of planting so that production varies widely according to demand. It is estimated, as an example, that in 1988 the Brazilian crop will yield about 130 000 tonnes of fresh fruit.

The fruit is small, about 5–8 cm in diameter and slightly ovoid in shape. A thin leathery skin which is either yellow or purple encloses numerous soft edible seeds each surrounded by a sac which also contains the yellow, slightly fleshy juice. It grows on a vine which in the wild climbs up amongst other vegetation, but which commercially is supported on wires or trellises.

Propagation is either by seed or by grafting but the plants are not always very hardy and have in the past been very vulnerable to viral diseases which has restricted production in many areas.

There are basically two distinct varieties, the purple and the yellow (*Passiflora edulis flavicarpa*). The latter grows under lowland tropical conditions whilst the purple type tends more to be cultivated in sub-tropical areas or at higher altitudes in the tropics.

The yellow variety is the more resistant to disease and has a higher yield averaging 20–40 tonnes per hectare. The purple type has a much finer quality of fruit, but yield is less than half that of the yellow. Much work has been and is still being done to produce hybrids and cultivars which combine the best characteristics of both types.

Passionfruit is grown both for fresh consumption and processing into juice. The fresh fruit travels well, owing to its tough skin but it wrinkles very quickly as the fruit loses moisture. Although this does not affect the eating quality or its flavour, it does give the fruit a rather odd leathery appearance which may not immediately commend itself to the uninitiated when first seen on a supermarket shelf.

The juice has a strongly intense and aromatic flavour with a soluble solids content of about 15%. The acidity is high with a Brix/acid ratio averaging 5 and a pH value of 2.6–3.2. Carotene and vitamins A and C are present in quite high quantities.

During processing, after the selection and washing of the fruit, various different methods are employed to extract the juice. One system involves a perforated centrifugal bowl into which sliced passionfruit pieces are placed. As the bowl spins, the pieces gradually mount the side and the pulp and seeds pass through perforations and are collected below. In another method the whole fruit is fed into a space between revolving cones which are attached to the ends of inclining shafts. The cones rotate the fruit, eventually causing it to burst and disgorge its contents. The Passypress, produced by Bertuzzi (Figure 4.4) achieves much the same result using rollers. All these methods yield a mixture of juice, pulp and seeds which is then passed through finishers and centrifuged. After de-aeration and pasteurisation, the juice is concentrated to about 50° Brix. Passionfruit juice has a high starch content which means that the pasteurisation has to be at a rather high temperature, around 120°C to be effective, as the granules can protect microorganisms, and also necessitates frequent cleaning of the evaporator.

Figure 4.4 Bertuzzi Passypress.

Table 4.3 Estimated production of passionfruit juice 1987–1988, 50° Brix concentrate equivalent

	Tonnes
Brazil	8500
Peru	3200
Colombia	1600
Ecuador	850
Kenya	350
Sri Lanka	350
Fiji	250
South Africa	250
Australia	150

The world production of passionfruit juice has increased significantly in recent years. Hawaii and Australia have traditionally been important producers, and Fiji, New Guinea and Kenya were developed as additional sources by the Australians.

More recently, the most important developments have been in South America which is now the largest supplier to Europe and there are extensive new plantations in Zimbabwe. Some production has also been established in Sri Lanka, Thailand and the Philippines. Production figures, when available, are difficult to interpret since often no distinction is made between single strength juice and concentrate. Table 4.3 gives a tentative estimate, based for convenience in terms of 50° Brix concentrate although the equivalent quantity of single strength juice is included.

It is interesting to trace the growth of the market for passionfruit juice over the last 50 years. In the late 1940s Harold Cottees, who had already established a successful business in the production of jams and preserves in Australia, was using locally produced passionfruit as one of his raw materials. He decided to launch a passionfruit-based soft drink which quickly became successful in Australia. In order to secure continuity of supplies he established plantations in Queensland, Western Australia, New Guinea and later, in Kenya.

In Britain, A. F. Cade, an associate of Cottees, set up Pashon Products Ltd. in Slough in the early 1950s and produced a passionfruit squash under the Pash brand (add a dash of Pash to life) using the 'Empire' passionfruit juice from Kenya. This company was taken over in 1954 by Cantrell and Cochrane and their Technical Director, Victor Walkley, developed a carbonated passionfruit drink which was introduced into the Club range of soft drinks.

Both these passionfruit drinks were reasonably successful but production was discontinued in about 1960 when supplies of raw materials were halted by the viral disease which was affecting the Kenya plantations, no acceptable alternative sources being available at that time.

At the same time Unilever had introduced an orange and passionfruit squash under their Treetop range which was also successful but again the problem was that of finding a reliable source of supply and eventually the product was dropped for the same reason.

Meanwhile, in 1959, Dr R. A. Barth of Rothwist in Switzerland, who had established the Rivella soft drinks company, was travelling in Australia to arrange local distributors for his brand and came across passionfruit for the first time. He immediately recognised its potential and made a point of meeting Cottees. The outcome was that they agreed to cooperate and Barth undertook to develop a market for Cottees passionfruit juice in Europe. In 1962, the Barth group launched in Switzerland a new soft drink, Passaia, a carbonated passionfruit drink packed in glass. Not long afterwards, the Cottees business was taken over by General Foods who took the view that the growing and processing of fruit was outside their proper sphere of activity, and as a consequence, in 1965, Barth bought the majority of the Kenya operation. Following this, Barth established a new Swiss subsidiary, Passi A.G., in order to develop sales of passionfruit juice to the European food industry. Subsequently Passi further increased its production with additional investments, mainly in Brazil, and went on to become the first major promoter of tropical fruit juices in Europe.

Today passionfruit juice is widely available mainly as frozen concentrate and single strength juice and although supply is bound to fluctuate from one year to another, depending on harvests, demand is such that growers feel secure in the knowledge that there is an established market for their produce. New, more reliable cultivars are available, reducing the risk of failure, and the potential to increase production is large enough to satisfy considerably increased demand.

4.5 Pineapple

The pineapple, *Ananas comosus* Merril (Figure 4.5), a member of the Bromeliaceae family, is thought to be a native of that part of South America which is now Paraguay. Certainly it became widely distributed in South America and the West Indies most probably by the Guarani Indians and first became known to Europeans when it was discovered by Columbus in 1493 and given the name Pina because of its resemblance to a fir cone. The name Ananas is derived from the original Indian name.

It is most successfully cultivated in those parts of the tropics which enjoy a mildish climate and unlike most tropical fruits no significant quantities grow in the wild. By the 17th century plantations had been established in India, Malaya and South Africa in addition to the existing ones in Brazil. Another 100 years showed further development in Australia, Singapore and particularly in Hawaii, until by the 1930s Hawaii was the largest producing area, accounting for over 50% of all the pineapple grown in the world. In 1938 the

Figure 4.5 Perola pineapple, Brazil.

total world production of the fresh fruit was 1.1 million tonnes and Hawaii was producing about 600 000 tonnes. By 1958 the total had increased to 1.8 million tonnes and the proportions remained much the same. By that time, and indeed until today, the world pineapple production was largely in the hands of two large American corporations, Castle and Cooke, who own Dole, and Del Monte, now owned by R. J. Reynolds.

In the meantime all was not well in Hawaii. By 1920 the cost of land and labour was increasing and in addition the Del Monte company Calpak revealed that they were having problems with plant diseases, largely insect borne. In 1926 Del Monte had set up Philpak in the Philippines and after 2 years of experimentation, their first commercial groves in that country were planted in 1928. During the 1930s they began to transfer their operations to that area although developments were interrupted by the Japanese invasion in 1941. In 1965 they bought the existing Kenya Canners and began to develop Kenya as their second source. Dole also faced similar problems and established pineapple groves in Thailand as well as in the Philippines. Del Monte closed their processing operation in Hawaii in 1982, although they have retained their groves. Dole still operate their plant there but the majority of pineapples grown there today find their way to the fresh fruit market.

Table 4.4 Fresh pineapple production in 1987

	Tonnes $\times 10^3$
Philippines	1670
Thailand	1510
Brazil[a]	785
Hawaii	628
Mexico	306
Ivory Coast	273
South Africa	267
Kenya	203
Taiwan	193
Malaysia	184
Australia	174

Source: Foreign Production Estimates Division, USDA, Washington, DC.
[a] Estimate.

These changes considerably reduced the importance of Hawaii as a source but resulted in a dramatic overall increase in world production to well over 6 million tonnes per annum. Table 4.4 shows the 1987 production figures for fresh pineapple.

The pineapple plant generally grows to a height of around 80 cm and consists of a rosette of sharp spiny leaves. Within about 15 months a flower spike appears from the centre of the plant and produces a hundred or more small flowers whose fruits coalesce together to produce the multiple fruit which we recognise as a pineapple. The plant continues to fruit but each successive crop is smaller in size and eventually becomes uneconomic. Propagation is normally achieved by planting the crown from the top of the fruit, or from the shoots and suckers that develop from the stem. Planting density can be from about 45 000 to 60 000 plants per hectare yielding around 100 tonnes of fruit per year.

The principal variety used for juice production is the Smooth Cayenne which is also grown widely for the canning industry. This has a fairly dark yellow flesh yielding a similar coloured juice and tends to have a higher acidity than most other pineapples.

Another variety growing in popularity in the juice industry is the Abacaxi type, also known as the Pemambuco or Perola. This is grown widely in Brazil and has a whitish-yellow flesh. This fruit is particularly juicy and is low in acidity.

Other varieties of note for juice production include the Queen, which has a deep yellow coloured flesh and is grown mainly in South Africa and Australia and the Spanish types, particularly the Singapore Spanish, another yellowish variety, which is popular in the West Indies and Malaysia.

The juice has a soluble solids content averaging around 12–15% and a high Brix/acid ratio which can vary between 14 and 35 although both these figures can be higher in juice produced from ripe fruit particularly of the Brazilian Perola variety. The pH tends to average 3.8–4.0 depending on origin, although the Brazilian Perola can be as high as 4.5.

The processing industry is dominated by the canneries and pineapple juice has traditionally been a by-product of this industry. During the preparation of the fruit for canning it is first trimmed into cylinders and then cut into the final segments, rings etc. The waste from this operation, including the core, is pressed and the resulting juice is added to the juice collected from the cutting table. This blend is then centrifuged to remove excess pulp, de-aerated, pasteurised and concentrated.

In 1979, encouraged by one or two of the most quality conscious European juice packers, there was a move in Brazil to start producing juice from the whole fruit which resulted in a product with better flavour and more body and stability than that previously available. This process involves crushing the whole fruit rather than those parts rejected during the canning process. Such was its success that purpose built extractors, such as the Indelicato 'Polypine', which cuts the fruit in half against a screen, soon became available. The juice is then passed through a normal finisher before de-aeration and concentration.

The resulting product is, of course, much more expensive but its quality is so superior that it has made an immediate impact on the market and now accounts for nearly 20% of the EEC consumption of pineapple juice. There is no doubt that this development has made an important contribution to the increased consumption which has taken place in recent years.

Today, the principal areas producing pineapple juice are the Philippines and Thailand. The juice is mainly available in the form of concentrate, at 60 and 72° Brix packed either aseptically or frozen. Recent production figures are indicated in Table 4.5 which, for the sake of convenience, includes both single strength and concentrates all converted to 60° Brix.

Table 4.5 Estimated world production of pineapple juice expressed as 60° Brix concentrate

	Tonnes × 10³	
	1980	1986
Philippines	23.0	57.2
Thailand	18.0	26.0
Brazil	8.0	16.6
Kenya	3.8	13.5
Australia	8.5	8.6
South Africa	2.5	6.9
Mexico	4.0	2.8
Ivory Coast	1.9	0.9
Malaysia	0.5	0.5
United States (Hawaii)	13.2	n.a.

The market for pineapple juice is well established and can be traced back to 1932 when Dole in Hawaii began to can the single strength juice (referred to as hot-pack, i.e. pasteurised and hot-filled). After a slow start this juice began to gain popularity in the United States and by the 1940s it was accounting for some 20% of the total hot-pack single strength fruit juice market. After the war its market share declined largely as a result of improved production techniques and marketing efforts for other juices, particularly orange, although at that time sales of the hot-packed juice were beginning to start in Europe and especially the United Kingdom with supplies coming mainly from South Africa and Australia. The Philippines were also becoming a source and smaller quantities were produced in Martinique and the Ivory Coast which were mainly exported to France. In 1957 Hawaii was producing around 150 000 tonnes of single strength juice which accounted for over 85% of total world production. This was mainly sold in America.

The total fruit juice market in Europe began to expand in the mid-1960s when, firstly, more sophisticated methods of concentration became available, which went hand in hand with the availability of bulk frozen concentrates, and, secondly, new and improved methods of packaging were developed. This meant that good quality juices could efficiently and economically be packed close to the consumer. This general expansion of the juice market also had its effect on pineapple juice and sales were further enhanced by the availability of improved qualities resulting from the new techniques of producing juice from the whole fruit as discussed above. The original sales volume, based largely on low prices has thus been increased and to a large extent replaced by higher quality products selling at premium prices.

The largest market for pineapple juice remains the United States, followed by the United Kingdom, France and Canada. It is also gaining popularity in other parts of Europe, and in Spain particularly consumption is increasing rapidly.

Traditionally the most important outlet in the United Kingdom was the 4 oz (114 ml) bottle sold through the licensed trade. This market has shown growth over the years and pineapple has lately increased its share of this trade to about 12%. Consumption increased rapidly with the advent of the Pure-pak juice carton in the 1960s and today throughout Europe the majority of pineapple juice is sold in aseptic Tetra Brik and Combibloc cartons. A recent development was the advent of the bag-in-box package and it looks as if sales in the catering industry using post-mix dispensers will further increase sales. Today, the EEC countries import about 40 000 tonnes of concentrated pineapple juice per annum of which 40% is consumed in the United Kingdom.

4.6 Other tropical fruits

There are several other tropical fruits which hold potential interest for the soft drinks industry. Most of these are not available commercially and a major difficulty in introducing them results from the problem of supply and demand. No soft drink manufacturer is likely to introduce a new drink unless he can be reasonably certain that supplies of the raw material are going to be available. At the same time he cannot guarantee a market until he has launched the product and ascertained the demand. On the other hand, the producer has little incentive to make the necessary investment to guarantee such supplies until he can be assured that there will be a demand.

Nevertheless, apart from the products mentioned so far there are a few other tropical fruit juices and purées which are becoming available because of their applications in other food industries, because they are consumed locally, or because some enterprising producer has felt it worthwhile trying to develop a market. These are beginning to be used in the manufacture of fruit drinks and brief details of the main ones follow.

4.6.1   Acerola

Also known as the acerola cherry or West Indian cherry, the acerola (*Malpighia punicifolia* L.) is cultivated in Puerto Rico and other parts of the West Indies and Central America. The fruit grows on small bushy trees which can reach a height of about 3 m. The trees have small dark green oval leaves and dense thorns which make hand picking unattractive to local labour. Acerolas are similar to cherries in appearance, about 2 cm in diameter, and can be converted into a palatable juice by crushing and screening.

The chief interest in acerola centres around its exceptionally high ascorbic acid (vitamin C) content, up to 3000–4000 mg/100 g in some cases. This was first noted in 1946 in Puerto Rico and was enthusiastically taken up by the nutritionists as a valuable source of vitamin C. Commercial development was promoted, largely by universities and agricultural research centres and a processing plant was established in Puerto Rico in the 1960s. Apart from juice, the plant also produced spray dried acerola powder which was exported as an enriching agent for mixing with fruit juices and drinks to increase their vitamin C content. This was largely unsuccessful, however, because there are many varieties of the fruit in a half wild form, some of which are less rich in ascorbic acid than others, and it became a relatively expensive source of vitamin C with the advent of synthetic ascorbic acid.

Today, the main interest for this product is in the health food area where it appeals to those who perceive advantages in natural over synthetic vitamin C. With the current interest in new types of juices it may again become more popular, particularly since supplies are still available; however, its rather bland taste ensures that this will be most likely confined to mixtures with other juices.

4.6.2   Banana

Grown in all tropical countries, the banana is the most widely cultivated tropical fruit and is very important to the world's economy. The total production of bananas and plantains combined is over 55 million tonnes and in terms of total fruit production is second only to grapes. The plant, which grows to a height of 3–10 m, is in fact a giant herb. The fruit grows on stems which emerge from the trunk, and one plant can sometimes yield over 200 bananas. Banana purée is mainly produced in Central America and in Brazil.

There are several different varieties of banana and the fruit most commonly processed is *Musa sapientum* L. After peeling it is mashed, deseeded, sterilised and then generally packed aseptically in drums as a purée although it is sometimes canned and for some applications it is spray dried. It is not available in concentrated form since the high pulp content makes it impossible to separate a free running liquid. The Brix averages 22° and as the acidity is so low the Brix/acid ratio, which averages 86, can go up to about 130. pH is 5.0.

Its principal uses are in the baby food, ice cream and bakery industries but it is also used in soft drinks, where the flavour blends well with other juices. It is principally used to lighten the colour and to improve the texture in mixed fruit drinks.

4.6.3 Kiwifruit

A native of China, the kiwifruit or Chinese gooseberry (*Actinidia chimensis* Planch) was introduced in 1906 to New Zealand where it thrives. It is not strictly speaking a tropical fruit, and is in fact grown widely in Europe and Russia as well as Chile, California and Australia. The plant is a vine and the fruit is green and oval, about 6–8 cm long. It has a hairy skin and in cross-section presents a decorative green pattern with small dark seeds which radiate from the centre. Its popularity as a fresh fruit relies largely on this decorative appearance.

The flesh is processed into a paste which can then, with pectolytic enzyme treatment, be transformed to a juice and concentrated, and is then either frozen or aseptically packed. The natural juice has a Brix of about 12° and an average Brix/acid ratio of 10. pH is 4.0. The flavour is very delicate and tends to be lost during processing, particularly when concentrated. It is used mainly as an ingredient in mixed fruit drinks.

4.6.4 Lulo

The Lulo (*Solanum quitoense* Lam), otherwise known as the Quito orange or naranjilla, is believed to have originated in Ecuador and grows in the Amazonian districts of that country as well as in Colombia and Peru. It is a member of the very large Solanaceae family which includes a host of poisonous and drug producing plants such as belladonna (deadly night-shade), tobacco and henbane, in addition to the more palatable potato, tomato, cucumber and pimento, etc.

The fruit grows on a thorny perennial shrub which reaches a height of 1–3 m and has very large leaves. It is a round, yellowish green fruit about 5 cm in diameter and has a leathery skin covered with short prickly hairs. The flesh is greenish and has a jelly-like consistency divided into compartments filled with a large number of seeds.

It is processed into a yellow juice which has a Brix of 8–9° and is quite acid with a Brix/acid ratio between 3 and 5, pH 3.6. Currently it is only available as a single strength juice packed either frozen or aseptically. The flavour is sharp and aromatic and it is used as an ingredient in mixed fruit drinks.

4.6.5 Papaya

Also known as pawpaw, the papaya (*Carica papaya* L.) is a native of Central America and is grown widely in all Central and South American countries as well as Hawaii, South Africa, India and other tropical countries. Economically it is quite important and world production exceeds 1.3 million tonnes.

Papaya trees which can grow up to 10 m high have no branches and the leaves grow out of the top of the softish, often hollow trunk. The fruit is borne in a cluster just under the leaves. The fruit, which is not unlike a rather smooth fleshed melon with a space inside containing many black seeds, has the shape of an avocado pear. When ripe it can vary in size from about 200 g up to 4–5 kg and has a bright orange flesh.

It is processed into a purée which has a Brix of about 9° and the acidity is low with Brix/acid ratios varying between 35 and 75, pH 5.0. As with mango, concentrates are beginning to be available with a Brix up to about 22°, usually frozen. Papaya purée is mainly produced in Brazil where it is used locally in soft drinks and ice cream. It is gaining popularity in Europe particularly in mixed drinks.

4.6.6 Soursop

Also known as the guanabana, the soursop (*Annona muricata* L.) is a member of the Annonaceae family which includes the fairly similar cherimoya and custard apple. A native of Central America and the West Indies it is grown commercially in many tropical regions where it is consumed as a fresh fruit. The fruit is soft and very easily damaged, and since the flavour deteriorates quickly after picking and refrigeration can cause discoloration, it is relatively unknown in Europe.

The tree is evergreen and grows from 5 to 8 m high. The fruit which, like pineapple, is a composite with a rather fibrous flesh has black seeds and is enclosed by a soft green skin. It is usually about 0.5 kg in weight and has a most refreshing sweet-sour flavour with an agreeable and distinctive aroma reminiscent of strawberries and pineapple.

After hand-peeling the fruit is pulped and screened to yield a purée with a Brix of around 16° and a Brix/acid ratio of between 25 and 30, pH 3.7. It is not usually available in a concentrated form, and at present the main commercial source of the purée is the Philippines, where it is available frozen or hot-packed in 5 kg cans.

So far, soursop purée has remained relatively unexploited by the European soft drinks industry, probably for cost reasons, although it is used in mixed fruit drinks, notably in Italy. It is popular in Central America where it is used in jellies and it also mixes well with milk and ice cream. Modern cultural methods have been introduced into some of the larger producing areas which should make the juice more readily available in future; it is felt that there is an interesting potential for soursop-based drinks.

4.6.7 Umbu

A member of the same family as the mango, the Anacardiaceae, the umbu (*Spondias tuberosa* Arruda), or imbu, grows wild in the tropical regions of Central and South America. It is rarely cultivated since the fruit is widely available in the wild, particularly in northern Brazil. The tree is distinguished by a low spreading crown which can achieve a diameter of 8–9 m and the fruit, which is soft and greenish yellow in colour, resembles a greengage plum. The yield is very prolific and if unpicked the ripe fruit falls to the ground and forms a yellow carpet around the trees.

The pulp is whitish in colour, succulent, and has a bittersweet flavour. It has a Brix of around 9–10° and it is fairly acid with a Brix/acid ratio which is generally around 7.

The fresh fruit is popular in Brazil where it is used to make various desserts and jellies. It is relatively unknown in Europe but the frozen single strength purée is beginning to be used as an ingredient in mixed drinks.

4.7 Tropical fruit juices in Europe today

With the exception of pineapple, total retail sales of single strength tropical fruit juices in Europe are very small. The market for pineapple juice is, of course, well established and in the United Kingdom, most of the major supermarket groups have their own label, often in a choice of more than one pack. The branded market in the United Kingdom is dominated by Del Monte who also produce a range of mixed juices under their Island Blends range and their more recent Fruit Bursts. Del Monte is also strong in Italy and sells pineapple in addition to other tropical juice and nectar blends.

The other principal markets for pineapple juice in Europe are France and Spain. The French market began to develop in the late 1970s and by 1987 pineapple juice accounted for 8.2% of all juice sales. Spanish interest in all fruit juices is more recent; the growth has, however, been very rapid. Recently pineapple has taken a very large share of this market, and by 1987 it accounted for 25% of all juice consumption when imports of pineapple concentrate increased by 67% over the previous year.

The other tropical juices, being largely unsuitable for direct consumption are sold either as nectars or juice drinks, or in tropical blends and fruit cocktails. The first of these was probably Trink 10 launched by Eckes in Germany in the late 1970s. This is a blend of juices with the addition of vitamins and first became a success in the health food outlets in Germany. This was quickly followed in the United Kingdom by Sunfruit which was launched by Marks & Spencer in 1980. A drink with a 25% juice content containing a mixture of ten different juices, Sunfruit has become the archetypal tropical blend. In the United Kingdom, most supermarket groups now sell their own brand of mixed tropical drink and similar products are sold throughout Europe. As with many other products, the UK supermarket trade has largely been dominated by own label, but one or two noteworthy brands have also emerged, including Five Alive and Libby's Um Bongo. Soft drink companies have also launched carbonated tropical drinks including Quattro, originally launched by the C & C group and St. Clements tropical, now produced by A. G. Barr. In the licensed trade, however, such drinks have made a less certain start.

Tropical blends are represented in Scandinavia with Pripps, Arla and Bob all having their own brands. The La Bamba range packed by Granini is well established in Germany and in Belgium Looza, Sunnyland and Chaudfontaine all have tropical blends in their ranges; Sunnyland is also exported, mainly to Italy. Top Tientje, a ten-fruit mixture packed by Reidel, is the main tropical blend in Holland but here tropical juices are also used in the whey- and yoghurt-based drinks produced by the dairies Campina and Melk-Unie. The other main consumers in Europe are in France and Italy where important brands include Oasis, Fruite, Pampryl and Oro Fruit in France and Santal and Brasil in Italy. In Switzerland, Migros have their own tropical blend.

In addition to these national brands, the multinationals have also introduced tropical mixtures. Notable are Coca Cola with their Five Alive and Minute Maid ranges across Europe; Procter & Gamble, mainly in Germany; Passaia in Switzerland; Del Monte and Sunkist in Scandinavia, Belgium and Germany; and also Nestlé and Schweppes who are both selling tropical blends in Spain.

Today's market in Europe owes much to the enterprise and commitment of the pioneers. Passi A.G. was the first company to make reliable supplies of tropical fruit juices of good flavour and stability available to the soft drinks industry. Another company to have had a profound affect on the market is Euro Citrus B.V. of Oosterhout in Holland. Originally, Euro Citrus blended single tropical juices from different origins in order to guarantee consistent quality. This is necessary because most tropical juices, including pineapple, can vary significantly not only by origin and variety but also by season or harvest. Euro Citrus also developed the concept of producing blends of mixed juices, again guaranteeing consistent quality, which has provided the foundation for the market as it is today, not only in Europe, but in many other parts of the world. Now that these two companies belong to the same group they are cooperating to improve existing sources of supply and to develop new sources of reliable raw materials.

4.8 The future

There can be no question that tropical juices are here to stay. From the marketing point of view they have enabled soft drinks manufacturers to open up new areas of the market and expand their business. Because tropical juices are increasingly used in mixtures with other, better known, fruit juices they encourage growth and diversification in the consumption of all fruit juices. As the soft drinks market increases, and the trend moves away from artificial flavours towards flavouring by higher juice content, so interest in tropical fruit juices will continue to grow.

Supply need not be a major problem. All tropical fruits can be produced in countries where suitable land is available and whose economies are desperate for sources of foreign income. Finance and technical assistance are available. Indeed one of today's problems is the enthusiasm with which many of these countries have grasped these new opportunities. This could easily lead to supplies outstripping demand to an unacceptable level, creating problems for some of the producers until the market grows enough to absorb this additional production. There is every reason to believe that this will happen in the years to come, and that tropical fruit juices will gain an increasing share of the fruit juice and fruit beverage market.

Acknowledgements

The author would like to thank the following for their help and advice in preparing this chapter: Alfa Laval Food & Dairy International AB., Lund, Sweden; APV International Ltd., Crawley, England; Castle & Cooke Food Sales Co., Henley, England; Del Monte Foods Europe, Staines, England; Euro Citrus B.V., Oosterhout, Holland; The Foodnews Company, Sidcup, England; The Golden Circle Cannery, Queensland, Australia; Industria Alimenticias Maguary S.A., São Paulo, Brazil; Jugos del Norte S.A., Chiclayo, Peru; ODNRI, Chatham, England; Passi A.G., Rothwist, Switzerland; Plantation Foods Iberica S.A., Madrid, Spain; Sunbase USA Inc., Ocala, FL., USA; USDA, Foreign Production Estimates Division, Washington, DC, USA; V. T. Walkley Esq., Holmfirth, Huddersfield, England.

Further reading

Colayco, M. T. (1987) *Crowning the Land*, Philippine Packing Corporation, Makati, Metro Manila, Philippines.
Desrosier, N. W. (1984) *Elements of Food Technology*, AVI, Westport, CT.

Gysin, H. R. (1984) *Tropenfruechte*, At Verlag Aaran, Stuttgart.

Harman, G. W. (1984) *The World Market for Canned Pineapple and Pineapple Juice*, Report No. G186, Tropical Development and Research Institute, London.

Hoest, O. (1978) *Fruechte und Gemuese aus Tropen und Mittelmeerraum*, Franckh, Stuttgart.

Mott, J. (1969) *The Market for Passionfruit Juice*, Report No. G38, Tropical Products Institute, London.

Nagy, S. and Shaw, P. E. (1980) *Tropical and Subtropical Fruits*, AVI, Westport, CT.

Pijpers, D., Constant, J. G. and Jansen, K. (1985) *The Complete Book of Fruit*, Co-operative Vereniging Pampus Associates v.a., Amsterdam.

Thoenges, H. and Rinder, H. (1984) Raw materials for the fruit juice industry. In: *Confructa Studien*, Vol. 28 No. 1/84, Fluessiges Obst GmbH, Schoenborn, West Germany.

Tropical Products Institute (1960) *The Production and Trade in Pineapple Juice*, Report No. 5, TPI, London.

5 Growing and marketing soft fruit for juices and beverages

M. F. MOULTON

5.1 Introduction

The growing and marketing of soft fruits (berries and currants) for the juice-processing industry is still being developed. Only in recent years have processors sought growers who were prepared to grow fruits specifically for the industry instead of depending on product that was surplus to the fresh fruit market. For the grower this has meant a change in growing and harvesting techniques, with the need to appreciate the juice industry requires standards parallel to those of the fresh fruit market. No longer is juice production considered the home for sub-standard reject fruit and the industry has had to accept that to produce the higher quality required, the grower deserves and must have, an adequate return for his labour. Despite this realisation, as demand for various soft fruits expands, the fruit grower is faced with an increasing number of options when he sets out to grow and sell to the juice-processing industry.

5.2 Selling the fruit crop: the options

The fruit grower has traditionally believed that his job was simply to grow a crop. Selling was another matter and one which for a long time received scant attention. Growers depend on fresh fruit markets, nightly sending their produce to a trader who sells it for the best price he can get, paying the grower for his produce after deducting the merchant's fee. In years of plenty, the processor would take the surplus fruit from the market, paying as little as possible, and occasionally leaving the grower without any return to cover his costs.

Whilst many growers are still very happy to sell to the fresh market, others choose to avoid the problems that result from inevitable swings in supply and demand for selling their crop in the open market, preferring instead to enter into a firm commitment and sell to a selected outlet. The need for the grower to be more aware of specific market requirements was increased by the development of pick-your-own (PYO) selling which brought him face to face with the end user. Historically juice processors purchased their fruit requirements annually but the development of the market for processed fruit and juice led to the supplier and the buyer establishing far closer links. Nowadays many large scale growers produce specific crops to clearly defined and agreed standards for national and international companies. Some growers and merchants are still happy to negotiate annually, the grower confident that in a short year buyers will compete with each other to give a good price and the buyer happy in the knowledge that in a heavy crop or low demand year he has no firm commitments and if he wishes, he can make good any past losses. This type of arrangement appeals to those who grow such crops as strawberries.

In recent years growers and processors have combined to put considerable effort, investment and research into fruit juice production. As a result of this investment both are keen to protect their interests by means of long term contracts. These allow growers to concentrate on growing and enable processors to plan the production and development of the market knowing that the supply of essential raw material is reasonably secure, subject to the usual hazards of weather.

5.2.1 The market place

Since writing this chapter for the first edition there have been many changes in the marketing pattern for soft fruit world-wide. Recession, turmoil and the emergence of developing sources as well as the gradual improved standard of living have all played their part.

In 1989/90 the market for raspberries was well supplied, prices were low

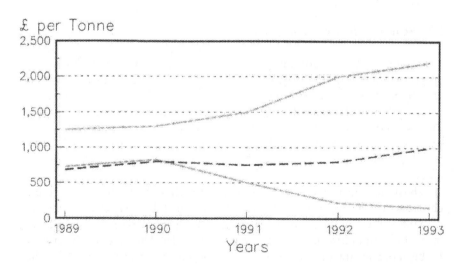

Figure 5.1 Chart showing soft fruit prices from 1989 to 1993. Dashed line, strawberries, 22/25 mm IQF (individually quick frozen); top shaded line, fresh blackcurrants; lower shaded line, raspberries IQF (individually quick frozen).

and buyers could obtain their requirements relatively easily. Since then there has been a dramatic fall in supplies, for example, no fruit emerging from Yugoslavia, which was one of the world's major producing areas and production cutbacks in Scotland due partly to low prices and high labour costs. As a result prices have risen substantially, with many buyers unable to cover their requirements. The chart (Figure 5.1) on soft fruit prices shows vividly what has happened in the market.

Following widespread frost damage across Europe in 1989/90 the total crop of blackcurrants was severely reduced and as a result prices were correspondingly very high. Blackcurrants were used to help sell other slower moving products. However to those who visited the growing areas a few months later it was very apparent that this shortage would not continue for long and that prices would plummet. Many growers and some processors were keen to get into blackcurrants at almost any price. The overall outlook was frightening but few people listened to the warnings that prices would crash in the near future. In 1991 prices were on the slide and by 1992 they were well below the cost of production; by 1993 the growers' main concern was to find a buyer who could be persuaded to make an offer. In some areas no buyers were evident and growers were advised to leave fruit on the bushes. A few growers put fruit into deep freeze on their own account and were still paying the storage account years later. Others sold their crop in the hope that it would be paid for at a later date (see chart on average returns 1970–1993 in Figure 5.2).

In recent years blackcurrant crops have become established in many

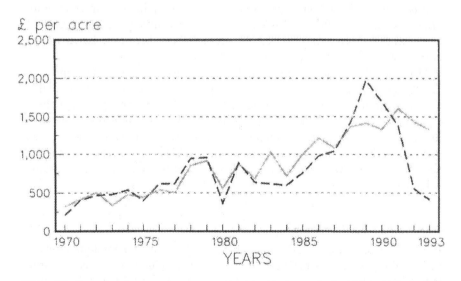

Figure 5.2 Chart showing average returns per acre for blackcurrants for 1970 to 1993. Dashed line, market prices; shaded line, contract prices.

regions of the world and these are now being sold at what appears to be uneconomical prices. Buyers and traders have found that they are being offered fruit as part of a barter deal which has often been put together by people with little or no knowledge of fruit or the fruit industry. The buyer has of course the opportunity to select the best deal for his product and knows that he can dictate the price, but as in all such deals the buyer needs carefully to check and confirm exactly what he is buying to avoid major problems.

5.2.2 Minimum import prices

Following the collapse of the market for blackcurrant crops in 1992 and in an attempt to bring some stability into the market place, the UK National Farmers Union, with the support of other farming groups in Europe persuaded the European Union (EU) to put in place a minimum import price aimed at preventing the dumping of cheap fruit from countries such as Poland into the EU. At the same time, it is hoped that by raising market prices the action would protect existing blackcurrant growers within the European Union.

This scheme was not fully effective in 1992 but following further lobbying additional safeguards were put in to the Regulations for 1993. These appear to have achieved their objective in reducing the flow of cheap fruit from Eastern Europe. However in December 1993 the EU acting against its advisors lifted the minimum import price structure. At the time of writing frozen fruit and juice concentrate is being offered around Europe at prices which are considered to be well below the accepted cost of production.

It is questionable if there is the political will to keep in place any minimum price structure in 1994 and 1995 and with the recent signing of the general agreement on tariffs and trade (GATT) accord it is fairly certain that any price barriers will be destroyed by 1996.

Most people would assume that following two years of very depressed prices many blackcurrant growers would pull up their un-economical plantations; at present we are not aware that this is happening and under these circumstances it is difficult to see any justification for a significant price rise in the next two years. In the medium term it is reasonably certain that given normal conditions, the price for raspberries will fall substantially within the next three to four years and at the same time that the price for blackcurrants will rise.

5.2.3 The long term contract

For some fruit growers, especially those in the United Kingdom growing perennial crops such as blackcurrants for juice production, the security of a long term contract is very appealing. One of the first processors to introduce this system was the British Company H. W. Carter back in the 1950s. Later when they were taken over by the Beecham Group, the contract system was retained and developed to include free technical advice and other benefits including interest-free loans for machinery, etc. The present SmithKline Beecham blackcurrant contract for growers runs for a period of seven cropping years and then, by mutual agreement, may be further extended for up to seven additional years. Many of the growers who now supply Smith-Kline Beecham have done so for over 30 years. During this time a strong bond of understanding and respect has been established between growers and the Company.

With any long term purchase contract, which may be tonnage or acreage based, the risk of glut or scarcity has to be carried. The SmithKline Beecham blackcurrant contract is acreage based and the company carries the stock as cover against crop shortage due to frost, etc. In some years this involves holding large quantities of juice concentrate in store, which is expensive. With the development of late flowering cultivars the risk of frost damage is lower and therefore there is less need to hold such large stocks (see chart (Figure 5.3) on seasonal variation in yield).

One of the major problems to resolve in any long term contract is the price. In an ideal world with zero inflation and regular crops there should be no problem. However, with blackcurrants marketed internationally, with an average yield varying from 1.5 to 8.5 tonnes/hectare and whose price may be influenced by political effects on currency, or the sudden acreage change

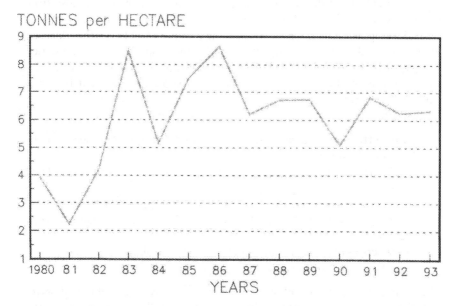

Figure 5.3 Chart showing seasonal variations in yield of blackcurrants in the UK between 1980 and 1993.

resulting from the individual decisions of thousands of growers, the determination of a realistic and fair price for a long term contract is extremely difficult.

To set a long term contract price that is based on 'costs plus' could in a year of large crops and low market prices find the buyer with excessive stocks of expensive raw materials and in a short year strain growers' loyalties. It is under these conditions that the relationship between the processor and its growers will be fully tested. Usually it is found that if you treat growers fairly then the majority will respect your actions and remain loyal. For many blackcurrant growers in the United Kingdom, the knowledge that they have an assured outlet for their fruit, immediate access to specialised technical advice and confidence in prompt payments through the strength of a buyer who has built up a world-wide reputation for a high quality blackcurrant product, far outweighs the annual concern of seeking the highest price with the possible risk of disputes over quality.

5.3 Producing the soft fruit crop

Soft fruits can be successfully grown over a wide area of Europe from Northern Sweden to the Mediterranean and along the coast into Asia. The last few years have seen the development of aggressive fruit production areas in many different parts of the world. The soft fruit buyer can now not only source supplies from the Northern and Southern Hemispheres but also from different areas within that hemisphere.

As developing countries seek to raise their standard of living their farmers are encouraged to produce for export high value crops such as strawberries. The growers have access to government loans and relatively cheap labour. They are thus well able to sell at prices well below those sought by many existing fruit producing areas.

Most fruit crops are usually associated with specific growing areas, e.g.:

—strawberries are grown widely throughout Europe, America and the Southern Hemisphere.
—blackberries (cultivated/wild) in Eastern Europe, Italy, New Zealand and Chile.
—raspberries in Scotland, America and Eastern Bloc countries.
—blackcurrants in the United Kingdom, Poland, New Zealand, Europe and China.
—gooseberries and red currants in Hungary and Poland.
—elderberries in Northern Europe, Spain and Chile.

To illustrate the important aspects of soft fruit growing, a detailed description is given of blackcurrant production. This is followed by notes on the other principal soft fruits.

5.4   Blackcurrants

5.4.1   General

The grower of bush and cane fruits normally identifies the ultimate purpose for which his fruit will be destined, whether canning, freezing, juice manufacture or the bakery trade, and organises his production methods to meet the quality requirements of the end user. With blackcurrants, considerations of flavour, natural colour intensity and vitamin C content are so important that the buyer may be willing to enter into a contract which will enable him to specify the varieties that he is prepared to purchase. This will give him a vital influence in the growing and harvesting factors, including specifying what treatments the grower may apply to the crop under contract. This is normally based on a restricted list of permitted crop protection agents of known safety and freedom from taints and off-flavours. In some cases the buyer will maintain a close liaison through the medium of field advisors who combine agronomic and quality control roles. Such an Agronomic Advisory Service also extends to counselling the grower on all aspects of the growing operation from choice of variety and initial selection of the planting site through every aspect of the establishment and routine management cycle. This assists the grower to maximise his yields and overall profitability whilst at the same time ensuring that the buyer receives the desired varieties at the correct stage of ripeness to optimise juice yield and quality. The fruit will be produced in a succession which will permit the optimum utilisation of processing capacity and duration of harvest season.

Among the key issues which need to be considered by any soft fruit grower are those discussed below.

5.4.2   Siting

Correct siting for bush and cane fruit crops calls for considerable experience as well as some localised knowledge of the growing areas and climatic conditions. In the United Kingdom, the preferred site typically would enjoy shelter either from rising ground or other features such as a belt of woodland, to the north, north-east and the north-west.

Elevated ground allowing drainage of frosty air to lower levels is desirable, provided this is compatible with the further requirement for a deep soil and freedom from undue exposure or elevation above sea level which will result in the risk of prevailing lower temperature and/or exposure to wind damage. If necessary, especially in exposed sites and large fields, the use of wind breaks can be of considerable advantage although these must not interfere with the natural drainage of frost.

Blackcurrants perform well on a wide range of soil types provided there is adequate depth, drainage and sub-soil aeration. Establishment tends to be more rapid on light sandy loams compared with heavier soils although the former will be more at risk from droughts. It would be inadvisable to consider establishing blackcurrants on soils with less than 60 cm of workable top soil otherwise the exploitable reserves of soil moisture would be insufficient to maintain growth during a prolonged dry spell. The availability of irrigation either as laterals with sprinklers or more usually as mobile rainers of the reel variety can be of enormous benefit in ensuring a satisfactory establishment of soft fruit. Alternatively, the laying of polythene mulch usually as a metre band, through which hardwood cuttings are inserted at the chosen planting distance, has proved to be extremely successful where there is no irrigation supply and where inherent weed problems are present, as the polythene provides valuable suppression of weed competition during the first two years of the plantation's life.

5.4.3   Manuring

The requirements for manuring are different when setting up a plantation or for established bushes. For pre-planting, most fruit crops will benefit from a generous application of a bulky organic manure. Well rotted farmyard manure spread at the rate of up to 40 tonnes/acre and then ploughed in will provide cuttings and young bushes with adequate nutrients during the establishment period, thus encouraging the development of a strong root system; it also helps retain soil moisture which will prove invaluable to young stock in times of drought. Before planting, the soil should be analysed and any deficiency corrected. Whilst blackcurrants can grow on a wide range of soil types and varying acidities the ideal range is a pH of 6.0–6.5 and it is far easier to apply lime to a bare field than in between rows of blackcurrants.

Once a plantation is established the use of organic manures is difficult and there is a risk that rotting manure may be picked up by the harvester and could contaminate the fruit. Most growers therefore depend on inorganic fertilisers once the bushes have been planted. These are relatively easy to apply and by careful timing they will be effectively utilised by the plant. Application rates vary according to soil type, varieties and rainfall or the use

Table 5.1 Application rates for inorganic fertilisers

Units/annum	Small bushes i.e. Baldwin type non-irrigated	Large bushes i.e. Ben types irrigated
Nitrogen*	60–80	30–40
Phosphorus	30	30
Potash	60	100

*Often applied as a split dressing in Spring according to crop prospects.

Table 5.2 New cultivars recently introduced from the Scottish Crop Research Institute

	Parents
Ben Lomond	(Consort × Magnus) × (Brodtorp × Janslunda)
Ben Nevis	(Consort × Magnus) × (Brodtorp × Janslunda)
Ben Alder	Ben More × Ben Lomond
Ben Tirran	Ben Lomond × (Baldwin × Ribesia o.p.)

of irrigation, the aim being to promote strong healthy growth, good fruit quality and high yields.

A rough guide is shown in Table 5.1.

Some growers find that foliar feeds are very beneficial to crops especially when the plants may be under stress, i.e. cold dry springs. These are relatively cheap and can be applied at the same time as other crop protection materials.

A recent trial in the UK has set out to monitor regularly the major elements such as nitrogen, potash phosphate as well as other minor ones to enable a graph of nutritional needs to be established for the whole season. It is hoped from this work that it will be possible to design manuring programmes which meet the plants' needs throughout the year and ultimately improve fruit quality and total yields.

5.4.4   Varieties

It is only in recent years that the breeding of cultivars aimed specifically at the juice market has recorded much progress. Following the severe crop losses from spring frosts in 1967 and again in 1977, great efforts have been made to develop cultivars that give good juice quality and are late flowering, thus offering some protection against spring frosts.

Many of the modern cultivars have Northern European parentage; recent introductions from the Scottish Crop Research Institute are shown in Table 5.2.

The introduction of Ben Lomond from 1972 (now 50% of UK acreage) has led to improved yields and, combined with frost protection measures, has given some stability to the growing of blackcurrants as can be seen in Figure 5.4. The variation in flowering time between varieties can be seen from the photographs of Ben Lomond and Ben Tirran taken on the same day in the same field (Figure 5.5).

5.4.5   Propagation

Historically, blackcurrants have been propagated by means of hardwood cutting taken from the ripened annual extension growth produced in specially grown stool-beds which are maintained only for the propagation of such

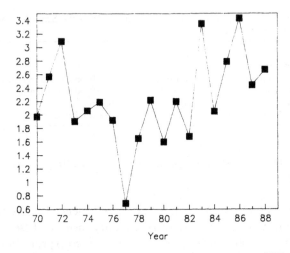

Figure 5.4 Blackcurrants. Average yield, tonnes per acre (UK).

shoots and can be intensively sprayed to ensure freedom from galled buds and reversion infection. Stool-beds are grown in complete isolation from fruiting plantations, requiring a distance of at least one mile to ensure freedom from disease.

Figure 5.5 (A) Ben Tirran; (B) Ben Lomond.

For the rapid propagation and multiplication of new cultivars where material is scarce, successful results have been obtained from the use of softwood tip cuttings rooted under mist propagation in peat modules or beds. This can take place during the growing season and by means of successive batches of cuttings a multiplication factor of 10–20 can be achieved under ideal conditions. It is however, expensive.

The use of short hardwood cuttings of 1–3 buds has given very variable results, unless under ideal conditions, dehydration being the main problem. Single bud hardwood cuttings placed vertically in pots of peat/sand mixture and rooted under glass is a possible alternative in times of stock scarcity. Rooting hormones should not be used on blackcurrant cuttings.

Specialised plant raisers have used micropropagation on blackcurrants but results have varied considerably with each cultivar. Once established the young plants will require very careful weaning before they are strong enough to be put out in a field plantation. Needless to say this is an expensive system and is only likely to be used when stock of new cultivars is being bulked up for small scale field trials.

5.4.6 Planting

To permit modern machine harvesting it is necessary to establish a hedgerow plantation which presents a continuous flow of fruiting branches to the harvester. Most blackcurrant plantations are set to give three metres between rows and an in-row spacing of 0.38–0.50 m. The use of rooted bushes has now virtually disappeared although some growers purchase a limited quantity of yearling bushes from a reliable source and then use the shoots further to extend the area being planted.

Ideally, any new plantation should be at least 100 m from an older plantation to minimise the risk of gall mite spreading reversion virus in to the young bushes. This also will help reduce the spread of leaf curling midge and mildew.

5.4.7 Weed control

The control of weeds is important in all fruit crops but especially in young blackcurrants in which excessive competition reduces growth and subsequent yields. The use of selected weed killers applied with a carefully shielded ground sprayer at low pressure and with large droplets can successfully control most weeds without damaging or contaminating the crop.

Several growers are now going back to the old system of establishing a green sward in the alleyways with the objective of increasing the numbers of predators in the plantation and hopefully reducing the need to use chemicals.

5.4.8   Frost protection

To encourage the development of frost protection schemes, the Beecham Group, in 1968 offered its contract growers interest free loans. These were taken up by many growers and in the east of England most of the black-currant acreage is now frost protected by means of permanent water sprinklers capable of applying 3–4 mm of water per hour for periods of up to 10 h per night. Most of these schemes were installed between 1970 and 1984. Since then costs have risen considerably and with the development of varieties offering some degree of frost tolerance few growers are now consider-ing putting in new frost protection schemes. For those growers who have established systems, these have proved to be very beneficial giving above average crops (on average about an extra tonne per acre) and a marked degree of stability of yields.

Blackcurrant flower tissue will withstand a limited severity and duration of radiation frost, however below −1.3°C (29.5°F) damage will occur. This increases with the duration of frost. The modern frost protection system should give protection down to about −5.6°C (22°F). A wind frost is far more damaging to bush fruits than radiation frost. It rapidly removes any minute pockets of warm air around the flower and additionally upsets the application of water from the frost protection sprinklers causing uneven protection. Unprotected and wet blossoms suffer a super-chill effect and suffer greater damage during wind frosts.

5.4.9   Harvesting/handling

Nearly all blackcurrants are now picked by machine. As the harvester passes down the row of blackcurrants the fruit is removed by vibrating fingers. The fruit then falls on to a conveyor which carries it to the back of the machine where it passes through fans which extract the leaves and other rubbish prior to collecting the fruit in trays. Trials have taken place to increase the size of the unit handled but it was found that the skin of the fruit is rather thin and tends to break under compression. As a result quality suffers as pectin is released causing the fruit to gel with the consequent reduction in juice yield.

Table 5.3 Guide to harvest date for three varieties of black-currant

Variety	Number of days from first flower to harvest	Variation
Baldwin	96	±4 days
Ben Lomond	90	±4 days
Ben Tirran	85	±2 days

Trials have taken place using pre-chilled/milled fruit pumped into a lagged bulk tanker for delivery to the processing plant for pressing.

In Sweden and Australia, fruit is put directly into polylined bulk bins holding up to a tonne, and in New Zealand blackcurrants have been washed and chilled prior to loading into the bins which are then stored frozen. Freezing time varies but with suitable chilling and good dry fruit it can be achieved within 7–10 days.

5.4.10   Control of fruit quality at harvesting

The first indication of the likely harvest date for blackcurrants is the date of general first flower. This varies between varieties and Table 5.3 should only be considered as a rough guide.

The optimum condition of blackcurrants for processing to juice has been accepted as being when 90% of the berries are black with the remaining 10% red/black. However, recent work has shown that as the fruit matures complex changes occur related to colour, vitamin C levels, pH, Brix and concentration of glucose.

If the fruit is underripe, then removal from the bush will be difficult, requiring extra shake on the picking fingers. This can cause bush damage and produce lower fruit yields, leading to slow juice extraction and lower yields.

Blackcurrants start to change colour some four weeks before they will be fully ripe. The actual timing depends on many factors including temperature, rainfall, crop size and variety. Cold weather, the use of irrigation and a heavy crop will delay fruit ripening.

The grower and the processor must ensure that the total crop of each variety is harvested at the optimum time in order that they may both achieve the best results in quality and yield of fruit and juice.

It is at harvest that the buyer/processor and the grower need to work particularly closely together to ensure that the total crop of each cultivar is collected at its optimum stage, in order that they both achieve the best results in terms of quality yield of fruit and juice. SmithKline Beecham depends on its team of skilled agronomists to make daily visits to every farm and oversee fruit quality.

5.4.11   Plantation life

The life of a blackcurrant plantation is extremely variable; the average is about 12 years, although some, well managed, can go on for 15–16 years. When yields start to fall or disease or weeds take over then the field should be grubbed, i.e. cleared. This is usually done immediately after harvest. The bushes are destroyed by rotavating the stools to destruction after pulverising the top growth. In this way there is no risk of carry over of reversion to a subsequent crop. In practice, however, the weed problem, coupled with the lack of good physical condition of the soil due to compaction means that a break of 1–2 years is really necessary.

5.5   Control of fruit pests and diseases

Fruit growers can either accept those pests and diseases that occur naturally under the conditions of 'organic' farming or can use manmade chemical control agents — pesticides. The majority of fruit producers choose the latter alternative and the reason for this will become clear from consideration of the problems of 'organic' production and the steps taken to ensure the effectiveness and safety of crop protective substances. In principle there is a third alternative method of protecting fruit crops from pests and diseases, that is to breed plants with natural resistance. Work on this is in progress in many research institutions world-wide. However, the breeding of new varieties is a lengthy process which can take at least 10–12 years. Experience shows that positive results are unlikely to be achieved quickly, and that even when progress is made, it is in small steps.

5.5.1   Organic production of fruit

Growers and processors would keenly welcome the facility to grow fruit without having to depend on the use of chemicals to ensure a crop. However, there is limited knowledge and experience of the intensive production of perennial fruit crops without the use of any crop protection agents. Yields from organic production may vary widely and as a result prices tend to be higher. In general the only place where juicing fruit has been farmed in this way is in previously neglected orchards and on a limited scale on specific farms. Some growers in Chile are now offering fruit which is stated to be organically grown. In some areas of Germany, many growers produce apples and pears without the use of any pesticides but since this fruit is processed for juice, appearance defects from pests and disease damage do not worry the end user.

In some areas of Eastern Europe crops are grown without the use of chemicals but this is believed to be due to the shortage of money rather than the deliberate attempt to produce organic crops.

To illustrate the potential severity of the problem, the principal pest and disease hazards affecting blackcurrants are reviewed below; other crops face different but similar threats. However, it is first appropriate to consider the care taken to ensure that safe and effective crop control agents are available as an alternative to organic farming.

5.5.2 *Selection of pesticides for crop protection*

To obtain reliable yield and maximise fruit quality, the majority of soft fruit growers throughout the world use a range of substances specially developed to control pests and diseases while being as safe as possible to the fruit plants and to higher animals and man. In most fruit growing countries, there are strict regulations on the use of crop protectants.

The system in the United Kingdom is illustrative. Once a chemical manufacturer has produced an experimental pesticide which it is believed may have an application in fruit he applies for a special trial licence from the Ministry of Agriculture, Fisheries and Food (MAFF). This licence is granted only after acceptance by independent experts of evidence of safety to workers and the environment. When issued, the licence allows the manufacturer to carry out small field trials to establish efficacy. Extensive testing is then carried out to confirm the acceptability of the application system, e.g. spraying, and the safety of any pesticide residues and metabolites remaining in the fruit.

All the produce from these trials will be destroyed apart from any samples required for analysis to meet statutory requirements. These trials will continue for a number of years and the results submitted to independent expert review before the pesticide is approved for wider use. In addition to establishing safety and efficacy, the pesticide manufacturer will also have to satisfy growers and processors that there is no risk of taint. For juice crops this usually involves up to seven tests over a three-year period. Only when everyone is fully satisfied will a new pesticide be placed on the market, almost certainly with strict regulations regarding its usage, including the number of applications, the usage rate and the pre-harvest interval.

Checking does not stop once the pesticide is released for sale. Growers are required to maintain full records of all pesticides used. Many major food processors specify the crop protectants they will accept and ensure compliance by checking growers' records. Fresh fruit grown under contract is subjected to detailed analysis at the time of harvesting and all major purchasers of fruit materials nowadays specify limits for pesticide content and routinely analyse to confirm that good agricultural practice is being followed.

In addition to these checks on the fruit further checks are made on a regular basis of the juice and concentrate produced.

5.6 Pests and diseases of blackcurrants

5.6.1 *Viruses*

Bush and cane fruits are subject to a wide range of virus infections. To control and eliminate these, growers should ensure that new planting material is free from all known viruses and must plant new plantations well away from any old or diseased fields.

To maintain clean plantations, it is vital that routine inspection is made during the growing season and roguing carried out, i.e. suspect material is removed and destroyed immediately. Viruses affecting bush and cane fruit can be transmitted by aphids, leaf hoppers and soil nematodes and usually result in mosaic (yellowing) symptoms as well as a reduction in yields and vigour.

Reversion.   In blackcurrants the most dangerous virus is reversion virus. It causes sterility of the flowers and eventually total suppression of yield. The only known vector of reversion is the blackcurrant gall mite (*Cecidophyopsis ribis* (Westwood)) also known as big bud mite on account of the highly characteristic enlarged winter bud form caused by the mites feeding in the bud tissue. The approved chemical treatment for gall mite is the non-persistent organochlorine pesticide Endosulfan which has been used widely since the 1960s.

More recently Meothrin has been approved in the UK for the control of gall mite. Meothrin has the added advantage of being active against gall mite at lower temperatures than Endosulfan; this has permitted growers to extend the total period of control against this pest. However Meothrin is lethal to many predatory insects, especially red spider mite predators (adults and eggs) requiring a twelve-week interval between spraying and the introduction of *Phytoseiulus* (see graph in Figure 5.6).

Emergence of mites from the galled buds starts in the spring when temperatures exceed 50°F (10°C) and continue for a period of over 100 days. Opportunities for applying Endosulfan are controlled by the regulatory

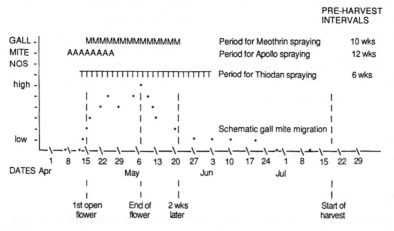

Figure 5.6 Strategies for the effective use of Meothrin. Graph showing relationship between the various growth stages and when Thiodan and/or Meothrin can be used to maximise control.

authority, which stipulates a six-week pre-harvest interval, and by the need to avoid spraying during flowering to minimise harmful effects to pollination and to bees.

In practice this is achieved by three sprays at medium to high volume, starting at first open flower with a second some three weeks later at the end of flowering and a final application 10–14 days later. This should be supplemented by careful roguing of the plantation twice every year, starting at the grape stage where the bright red reverted grapes are fairly easy to spot then again in July on the leaf symptoms.

Despite the absence of resistance to Endosulfan, changes in growing practices throughout the world are making it far more difficult to control reversion disease. These can be summarised as follows:

— release of new cultivars without adequate safeguards concerning virus freedom and inspection;
— increase in planting densities and larger units, increasing the risk of infection from older plantations;
— establishing plantations from cuttings near older plantations rather than raising stock in isolation;
— reduction in roguing by fruit growers;
— slow progress on the development of new cultivars that are resistant to reversion;
— mild damp springs have improved the survival rate of the mite in the United Kingdom.

5.6.2 Arthropods

Leaf curling midge. Blackcurrant leaf curling midge (*Dasyneura tetensii* Rubs) adults emerge from the soil in the spring to lay their eggs in the young shoot tops. There can be up to three generations in a season. If left uncontrolled, the shoot tips will be killed and extension growth prevented. This would be a serious problem in a young plantation and speedy control is necessary if strong healthy bushes are to be produced. Control is by regular application of systemic organophosphorus pesticides to kill the larvae. Many of the cultivars grown commercially are susceptible to attacks from leaf curling midge, however the plant breeders are making progress in achieving one of their objectives and hope to have resistance to this pest in new cultivars in the near future.

Vine weevil. The vine weevil (*Otiorhynchus sulatus* Fab) is a pest on a wide range of fruit and ornamental crops. Some perennial crops such as strawberries and blackcurrants appear especially at risk due to damage to roots resulting from the feeding of the larvae. Egg laying occurs in late summer just below soil level and adjacent to the crop. Vine weevils are all females and are capable of laying up to 400 eggs each per annum by parthenogenesis. As the larvae hatch out they commence feeding on the root system, gradually working their way down into the soil. In late spring the larvae cease to feed and return to within 2–5 cm from the surface to pupate in small earthen cells. When the adult weevils emerge they cause little damage to the crop but can be spotted by the characteristic notching of leaf margins. The adult weevil is very mobile and can move rapidly from one plantation to another and frequently will gather in the hedgerows. Control is very difficult and research is now being directed towards biological control by soil inhabiting nematodes and parasitic fungi.

Red spider mite.   The two-spotted red spider mite (*Tetranychus urticae* Koch) is frequently found on all soft fruit crops. In a hot summer, the rapid increase in numbers can very quickly stunt growth and reduce crop potential as the leaves bronze, then drop off as a result of the feeding of the mites. Control on bush fruits is difficult due to the degree of resistance to organophosphorus compounds and growers now have to depend on the suppressant properties of certain fungicides such as Morestan and Karamate and specific compounds such as Meothrin, Apollo and Childion.

Clearwing moth.   In the Southern Hemisphere the feeding of the currant clearwing moth (*Synanthedon tipuliformis* Clerek, the currant borer) on the pithy interior of the young shoots gives rise to severe checks to growth and branch breaking as well as failure of hardwood cuttings used for propagation.

5.6.3   Fungi

Diseases caused by fungal pathogens are of significance in blackcurrants.

Botrytis cineria.   This universally distributed parasite gains entry into fruit by invasion of the flower style or via surface wounds. In conditions of high humidity and temperature, infection of the fruit can take place by contact leading to epidemic outbreaks, notably in crops such as strawberry, raspberry and blackcurrant. The risk of attack is heightened in plantations where air circulation is restricted, crop hygiene is poor or there is injury from earlier wind or frost damage to flowers.

Spores may be released by *Botrytis* inoculum early in the flowering period and observations made by Scottish Crop Research Institute (S.C.R.I.) confirm that the *Botrytis* spores mimic the growth of the pollen tube down the style thus gaining entry into the ovary resulting in infertility and subsequent fruit losses.

The expenditure by fruit growers to control *Botrytis* is likely to exceed that of any other pest or disease and success is reflected in terms of yield and marketable quality.

The action of botryticides is mainly protective; these include Dichlofluanid which is free from resistance despite its widespread usage over many years.

The multi-site fungicide, Chlorothalonil, is also widely used, it is free from resistance problems but needs to be used in a sustained programme as it will not eradicate problems.

American gooseberry mildew. *Sphaerotheca mors-uvae* is a powdery mildew closely comparable with apple mildew. On *ribes* species the fungus over-winters as mycelium on the affected shoot tips in which may be embedded the black resting spores or perithecia. Primary infection is released from this mycelium as wind blown spores once the temperature rises to 15–20°C. During hot spells this disease spreads rapidly on the tender young shoots which quickly become covered with heavy overlay of white mould. A severe infection will lead to premature leaf fall, destruction of the flower initials and weak distorted shoot growth. Modern growing techniques have resulted in softer growth in blackcurrants which has favoured the development of diseases such as mildew. Successful control depends on accurate targeting and an early start to spraying before the fungus becomes established. The choice of mildewicide plays an important role because of the difficulty in adequately wetting the blackcurrant leaf, rapid growth of fungus on the bush and the risk of resistance to some of the triazoles.

Blackcurrant leaf spot. *Orepanopeziza ribis* Kleb spores are released early in the season from dead leaves. Normally infection begins at the bottom of the bush and is spread by rain splashes, rapidly infecting the older leaves and leading to premature leaf fall, fruit shrivel and, if left unchecked, to re-growth which decimates the following year's crop. Infection can take place at low temperatures (10–18°C). There is a wide range of fungicides available for the control of leaf spot including zinc and manganese dithiocarbamates, Chlorothalonil and Triforine. The benzimidazole fungicides Benomyl and Carbendazim can no longer be relied upon to control mildew successfully due to the development of resistance.

Rust. The fungal disease rust (*Cronartium ribicola*) can be a major problem in some areas, requiring specific control measures but produces only local difficulties in the Northern Hemisphere.

5.7 Other soft fruits

5.7.1 Strawberries

This fruit is widely grown throughout the world with the majority of crops being sold for consumption as fresh fruit. Processing, originally for jam then canning and later freezing, takes an increasing tonnage and now with the development of mixed fruit juices there is a growing demand for this fruit. For juicing, good quality evenly coloured firm, ripe fruit is required. The preferred varieties for juice include Senga Sengana and Elsanta which have red flesh. Most of the varieties grown in the United Kingdom have white flesh and are therefore unsuitable. To avoid degradation of colour and to retain flavour the strawberries are washed to remove sand, etc. and then frozen. The frozen fruit is made into juice to meet requirements.

5.7.2   Raspberries

Tradionally, raspberries are picked by hand especially if the outlet is for the fresh market or for individually quick frozen (IQF) processing. Several manufacturers have attempted to design suitable picking machines. Initial quality of fruit picked was only suitable for juice production but recent changes have been made to the harvesters resulting in greatly improved fruit quality with some machine picked fruit suitable for the fresh fruit market (see Figure 5.7).

Fruit destined for the juice market can be picked into buckets or pails as fruit damage is not so critical. This fruit is normally milled and then passed through fine sieves to remove all extraneous vegetable matter (EVM) and if required, seeds. It is then frequently frozen in the form of slabs weighing 10 or 20 kg each and measuring about $100 \times 30 \times 8$ cm or it can be aseptically packed. Varieties that ripen over a short period and produce fairly firm berries are best for machine harvesting.

Figure 5.7 Machine-picked raspberries suitable for the fresh fruit market.

Raspberry varieties suitable for juice production include Glen Moy and Glen Prosen. In Hungary, Yugoslavia and the United States one of the major varieties for juice production is Willimet.

5.7.3   Gooseberries

Gooseberries are still mainly picked by hand although some efforts have been made to mechanise the harvest. For processing, the fruit needs to be hard and firm. Once picked it can either be tipped into sacks or palletised containers prior to processing or freezing. Whilst the majority of gooseberries are sold for canning or freezing there is a limited market for fresh fruit and an expanding interest in using gooseberries in mixed fruit juices. However, the flavour of gooseberries is not universally liked and thus demand for this outlet is limited. The most commonly grown variety for processing is Careless.

5.7.4   Red currants

The majority of red currants are frozen after harvesting in order that they can be strigged and inspected prior to being sold for canning, juicing or for jelly manufacture. Red currants have a sharp tart flavour which is now being increasingly sought after partly to meet the aim of balancing the sweet flavour of other fruit products and partly as companies search to find new flavours to attract the customer.

5.7.5   Blackberries

The high cost of production of cultivated blackberries has resulted in a severe and dramatic drop in the total acreage and now very little is grown for processing. In addition, cultivated blackberries tend to lose their colour on freezing; tradionally this was overcome by the use of artificial colours but these are no longer acceptable in some countries and therefore the juice industry seeks its supplies from the wild blackberry which has a more stable colouring. These can be picked straight into polythene bags, frozen and then converted into juice to meet requirements. One of the major new fruit growing areas now supplying fresh and frozen blackberries is Chile.

5.8   Storing fruit for processing

Fruit for juicing may be handled in varying ways, all aimed at maintaining quality whilst at the same time affording the processor the maximum flexibility of usage.

(1) *Fresh.* For the majority of juice processors, the most cost effective way

of processing fruit is fresh direct from the field. However there can be major problems associated with processing fresh fruit, among which are continuity of supply, variation of quality and high capitalisation required in manpower and machinery to handle the large volume over a relatively short period of time. In addition, very few processors are fortunate to be established in the centre of a convenient fruit growing area. To augment these requirements processors can and do use chill storage.

(2) *Chill storage.* Most soft fruits can be held for 3–4 days in chill storage (2–4°C). To be successful, fruit will need to be firm, free from mould or *Botrytis* and dry. The field heat needs to be removed as quickly as possible if maximum benefit is to be gained. This method gives a short-term benefit and suits those growers who have suitable cold stores on their farms.

(3) *Gas storage.* As a variant on chill storage, gas storage of blackcurrants was widely used in the 1970s and involved placing blackcurrants in a gas tight apple store at 1–3°C in a mixture of 40% carbon dioxide/60% air for up to four weeks. This method has now been discontinued due to the adverse effect it had on juice yield and colour.

(4) *Sulphited pulp.* Where fruit is to be used for juicing or other manufacturing processes it can be stored as pulp with an added preservative. The use of sulphur dioxide as a preservative has been partly discontinued within the EU, due to statutory limits being imposed, but continues in some areas especially in Eastern Europe. Its decline is also due to the potential health hazard to operators in adding the sulphur dioxide and the adverse effect it has on the resultant juice quality and colour. Blackcurrants, gooseberries, strawberries and raspberries can be held in barrels with the addition of 1500–2000 ppm sulphur dioxide. The fruit may be either milled or placed whole in barrels and sulphur dioxide added as a soluble salt such as sodium metabisulphite. The barrels are normally stored at ambient temperature. They need turning at regular intervals to ensure the fruit does not compact and that no part becomes depleted in preservative. Sulphited pulp is a relatively cheap form of storage for mixed quality fruit.

(5) *Frozen fruit.* The method preferred by most processors for storing fruit to process after the season is freezing. With crops such as blackcurrants, which are all picked by machine, up to 78% of the crop can be harvested in 10 days whilst the total crop may be picked over 32 days. By freezing fruit that is surplus to the daily factory capacity, the processor can spread the harvesting peak.

Frozen fruit has the advantage of being in a very flexible form for marketing. For example, a processor can have his surplus fruit frozen and then graded to meet the specification of bakeries, canners or yogurt manufacturers. The more any fruit is processed the greater the limitations of outlets to other potential users. Some soft fruits, e.g. straw-berries and cherries, are mainly converted into juice to order. This maximises the shelf life of the juice especially in respect of its colour and flavour. The main disadvantages are the additional costs associated with freezing, i.e. the charge for freezing, weight loss in store (4–6%), storage and transport costs and the need to re-heat fruit for processing.

There are varying methods of freezing fruit for juice processing, including:

(i) *Store freezing.* Here full trays of fruit are placed in a deep-freeze chamber at about −22°C. Depending on such factors as the field heat, depth of fruit, air circulation, etc. fruit will be surface frozen within 24–36 h and should be solid in 3–4 days.

(ii) *Blast freezing.* Similar to store freezing in all respects except that a stream of cold air is directed through and around the stack of fresh fruit thus ensuring that it is frozen rapidly, generally in less than 20 h.

(iii) *Belt freezing.* In this process the fresh fruit is tipped onto a slow-moving perforated belt through which is forced air at sub-zero temperatures. The fruit takes between 20 and 30 minutes to pass through the freezer and is skin hardened by the time it emerges. It can then be re-packed into palletainers, bulk bins or paper sacks. The resultant product is a high quality IQF free flowing material which is suitable for specific market requirements.

(iv) *Nitrogen freezing.* This method is reserved for high-valued raw crops such as raspberries and is rarely used on fruit intended for juicing as it is expensive. Nitrogen freezing is gentle on the fruit causing little damage and produces clean IQF berries.

(v) *Irradation.* Although this method of food preservation has received approval in many countries there remains considerable public concern over its use and as a result there is considerable reluctance to use this form of preservation. In most countries there are strict laws on the labelling of irradiated food and at present there are only limited sites that can carry out this work.

5.9   The future—blackcurrant research and development (R & D)

In 1987 the UK Government announced cuts of about 60% in the funding of horticultural research. Cuts were mainly targeted at what was loosely defined as 'near market research'. In anticipation of this action, SmithKline Beecham and its growers set up their own Blackcurrant Research Fund aimed at meeting their defined specific requirements for the blackcurrant industry. The fund obtains the majority of its finance by way of a levy on the contract growers and by a contribution from the Company. The fund is managed by

Table 5.4 Trial site yields

	Cumulative yields (tonnes per hectare)	
	1991	5-year average
Ben Lomond	3.63	5.64
Baldwin	6.53	4.50
Ben Alder	6.27	7.66
Ben Tirran	4.82	6.26
C1/9/10	5.35	5.58
P7/7/19	0.00	2.20
P10/18/116	6.00	6.00

the Company free of any charge thus ensuring that all monies are applied to appropriate R & D projects.

One of the main objectives of the fund is the continuance of the blackcurrant breeding programme which is mainly aimed at developing new blackcurrant cultivars which give consistent yields of high quality fruit and juice, natural resistance to major pests and disease resistance.

This programme is backed up by two trial sites where the new cultivars are grown under commercial conditions and the harvested fruit is assessed for processing qualities. Those that pass the three-year testing programme are then released for growers to take up commercially (see Table 5.4).

Other programmes relate to the need to meet the day-to-day problems facing growers including the evaluation of suitable predators for use in the crop as well as new handling techniques. The industry needs this work to continue in order that growers can be in the forefront of development and well situated and prepared to meet the challenges of the future.

6 Apple juice

A. G. H. LEA

6.1 General background

Apples are amongst the most widely grown and widely consumed of temperate crops, taking second place only to grapes. The annual world apple crop is of the order of 40 million tons (Way and McLellan, 1989), of which at least five million tons is processed into juice (Possman, 1986). Fresh apple juice is a most unstable material both from the chemical and the microbiological point of view. Consequently, the distinct types of apple juice which are available on the market largely reflect the preservation technique used for their production.

Pure apple juice, which is unobtainable outside the laboratory, is a colourless and virtually odourless liquid. Within seconds of its expression from the fruit, however, it undergoes a sequence of enzymic changes to produce the colour and the aroma with which we are familiar. Such raw juice is occasionally found for sale on a farm-gate basis, particularly in North America where it is known as 'apple cider'. (It should be noted that the term 'cider' in Britain (and equivalent terms in France and Spain) refers to *fermented* apple juice which will not be discussed in this chapter. Excellent reviews of this subject are available elsewhere (Beech and Carr, 1977; Proulx and Nicholls, 1980; Durr, 1986; Lea, 1994.) The raw juice can be protected from microbiological degradation for a few days by storage in a refrigerator, or may be protected indefinitely by pasteurisation or by the use of permitted preservatives. Such juice is nearly always turbid, brown in colour, and tends to sediment on storage. A clear juice can be obtained by filtration, fining or the use of pectolytic enzymes before bottling.

An alternative product which has recently become popular on a commercial scale is prepared by flash heating or by the addition of ascorbic acid to the raw juice immediately after pressing, followed by pasteurisation or aseptic packaging. The effect of this is to produce a juice with an aroma much closer to that of the fresh fruit than in other types of apple juice. The turbidity is high, but the cloud is light in colour and is relatively stable to sedimentation. Such 'opalescence' is generally regarded as a positive quality factor.

The greatest volume of apple juice, however, is processed into 70° Brix apple concentrate before its eventual reconstitution. It may then be stored and shipped around the world as a relatively stable product occupying approximately one-sixth its original volume. The aroma which is inevitably removed during concentration is often recovered as an 'essence' and traded as a separate commodity. Peaks and troughs in fruit production may thus be evened out, and the juice packer can ensure reliable and consistent quality by blending together both concentrates and essences from different sources. Concentrate also lends itself to blending with other fruit juices or to the manufacture of carbonated apple beverages. The technology and chemistry involved in producing both opalescent juices and juices from concentrate will be considered in detail later in this chapter.

6.1.1 Juice extraction

Apple juice production begins with fruit harvesting, transport and washing facilities (see chapter 7). It is unusual for fruit to be grown specifically for juice production except in parts of Central Europe, and most juice is derived from second-grade or 'cull' fruit which is unsaleable on the fresh fruit market. In the United States, where there is a significant commercial apple sauce industry, much juice is produced from misshapen fruit which is too irregular to pass through the mechanical peelers and corers. Lower quality juice is also produced from the peels and cores themselves. However, all fruit must be sound and free from gross damage or contamination. In particular, it should be free from mould or rot which can lead to flavour taint, patulin contamination and microbiological instability. Certain types of mechanical harvesting equipment can lead to excessive bruising and skin penetration which cause off-flavours or even infection by pathogenic microorganisms. This is particularly the case where animal manure slurries have been used in the orchard, or where the fruit comes into contact with bare soil during harvesting. In such cases, it is good practice for fruit to be washed with clean water in the orchard before transport to the factory. Ideally, fruit for juice production should have been picked when fully ripe, with dark brown seeds and no residual starch as detected by an iodine test. Fruit which is picked slightly under-ripe for the dessert market or for controlled atmosphere storage should therefore be allowed to mature for a few days at ambient temperature before pressing (Figure 6.1).

As described in chapter 7, the fruit is washed, milled to a pulp, and then pressed using traditional pack presses or horizontal piston presses of the Bucher-Guyer type. In the USA, screw presses are still popular for apple juice production. Belt presses, both of the traditional Ensink type and of the serpentine Bellmer-Winkel design, have also established a considerable market niche for apple juicing in Europe. Diffuser extraction systems, using a current of hot water to extract the soluble solids from apple slices, were intensively investigated in Germany and Switzerland in recent years (Schobinger, 1987). However, they have proved unpopular in practice except in South Africa where the very firm 'Granny Smith' apple is well suited to them.

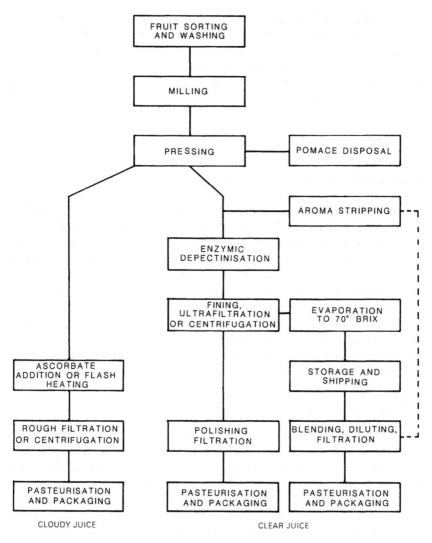

Figure 6.1 Flow chart for the processing of apples into clear or cloudy juice.

Freshly pressed juice may be packed as 'opalescent' or 'natural style', as described later, or treated to produce a clear single strength product. Juice for concentrate production is generally centrifuged or filtered, aroma stripped, depectinised, fined and finally evaporated. (Enzymic depectinisation is generally delayed until after aroma stripping. This makes more economical use of heat and provides a superior quality of essence.) Within about two minutes, therefore, it is possible for a clarified and depectinised apple juice at 12° Brix to be converted into a sixfold concentrate plus an aroma essence of one hundred-fold. Providing that the concentrate is immediately cooled below 20°C and is maintained below 5°C if it is to be stored for any length of time, the rediluted concentrate is virtually indistinguishable from fresh clarified juice. Most of the adverse changes associated with concentrates arise from the way in which they are stored and handled and not from the evaporation step itself.

6.1.2 Pomace disposal

Disposal of pomace is a vexing problem for all juice factories. If not dried or removed within a few hours, it soon turns 'acetic' and becomes a breeding ground for fruit flies and for spoilage microorganisms. If the pomace is already fairly dry after pressing, and has not been treated with pectolytic enzymes, it may be further dried below 10% moisture for storage and ultimate use as a feedstock for pectin production. Typically, hot acidified water is used to extract the pectin, which is then precipitated with isopropanol and dried to a powder. However, the economics of this are generally unattractive except to supply existing pectin production business. Alternative uses of pomace are as cattle feed (fresh, dried or after ensilage), as a 'high-fibre' food additive or in the past as a source of apple pips for raising seedling rootstocks. The seeds will germinate readily because their hard coats have been weakened by the milling and pressing operations. Most apple pomace is simply dumped on fields or in middens because it is uneconomic to dispose of it in any other way.

6.1.3 Blending and packaging

Fresh apples vary greatly in their sweetness, acidity and other flavour characteristics, although the consumer expects any given brand of apple juice to be a relatively consistent product. It is usually impracticable to blend fresh apple juices from different sources because of the limited time span permissible between pressing and packaging, and a more usual route to achieve the sugar/acid ratio required is to blend the stored fruit itself before pressing. Up to the 1960s the Boehi system of storing fresh juice at 10°C under CO_2 top-pressure was popular in Europe, but this is now of historical interest only. Mould and yeast growth was inhibited by this technique, but lactic acid bacteria were still able to grow, causing acid loss and flavour changes (Pollard and Timberlake, 1971). It is also possible to store fresh juice in bulk for months in the presence of 1000–2000 ppm SO_2, but special de-sulphurisation equipment is then required to reduce the SO_2 level below 10 ppm before use. This system is not commonly used for apple juice intended for direct consumption.

Since apple juice concentrate is now so widely available, it is relatively easy to set specifications for a juice in terms of its sugar and acid levels and to ensure consistency by blending different concentrates with water or fresh juice to the required specification. However, it is less easy to know exactly what levels are optimum for consumer acceptance (see later). Similarly, there is little published data on acceptable levels of aroma essence in juices blended from concentrate. Many juices at the cheaper end of the market have no essence added at all, while some incorporate a fixed amount. Some blenders vary the addition of essence according to its strength as assessed by sensory or instrumental analysis.

The blending and dilution of concentrates must be carried out *before* final filtration or the juice may become turbid on storage. The aroma essence, however, should be dosed into the juice line *after* the final mixing and filtration stages, to avoid losses by evaporation and adsorption by filter aids which have a high affinity for many of the small polar molecules which are responsible for volatile flavour.

The diversity of packaging for apple juices is largely reflective of national or regional markets. In North America, for example, 64 oz. cans, glass jars or their plastic equivalents are popular for domestic use. In the UK and Europe, glass bottles of 1 litre or smaller sizes have been more usual. The juice is sterilised in a flow-through heat-exchanger by high-temperature short-time (HTST) treatment, e.g. 10–30 s at 95°C or 15–120 s at 85°C and hot-filled directly into the pre-heated bottles. Immersion sterilisation of pre-filled bottles (e.g. held at 75°C for 30 min) is only usual these days in small-scale 'farmhouse' operations, since it is wasteful of heat and may lead to the generation of typical 'cooked' flavours. Continuous or 'tunnel' sterilisation of the filled bottles by a spray of hot water is a more modern equivalent but still tends to suffer by the generation of characteristic flavours. The response of microorganisms to increasing temperature is much steeper than is the chemical generation of cooked flavours by the Maillard reaction, so that it is usually beneficial to choose the highest practicable temperature regime and to sterilise for the minimum time required for microbial stability.

Over the past few years, new forms of laminated packaging technology such as 'Tetra Pak' and 'Combibloc' have been enthusiastically adopted, particularly in Europe. In the UK the majority of apple juice is now HTST treated and aseptically packed for supermarket sale in these 'long-life' bricks or cartons which are rapidly replacing glass except in small-scale operations (see chapter 10). Although the capital cost of aseptic packaging systems is high, many small-scale apple juice producers in the UK and Europe are also able to take advantage of the new technology by the use of sterile road tankers or demountable containers to deliver their product to contract packing houses.

Although it is both practical and legal in many countries to preserve apple juice for retail sale by chemical means, such as the addition of benzoic acid, sorbic acid or by a combination of these with sodium metabisulphite (Moyer and Aitken, 1971), such practices have become virtually obsolete in most countries. Apple juices are increasingly associated in the public mind with a 'healthy' image and the use of any preservative, however innocuous, could be commercial suicide.

6.2   Natural style and opalescent juices

Although the major market for apple juice is probably directed towards a clarified product, it has to be admitted that this is not particularly reminiscent of the fresh fruit even when well made, on account of the rapid enzymic and chemical changes which take place during processing. The mere removal of particulate and colloidal material also has a considerable effect on flavour, both by removing 'body' and by removing adsorbed volatiles. From time to time attempts have been made to produce a more natural style of juice which retains the body and flavour of the freshly pressed material. Conventional in-bottle pasteurisation of a turbid juice almost inevitably leads to failure, since the long heating time gives cooked flavours accompanied by polyphenol oxidation, pectin degradation and subsequent flocculation. The heavily sedimented brown product which results has little consumer appeal.

However, there is an increasing market for a premium product which retains more of the fresh fruit flavour and texture in an appetising form. One approach to this problem is to flash heat the juice at 95–110°C immediately after pressing, then centrifuging or coarsely filtering before final sterilisation and packaging as quickly as possible. This approach minimises browning and flocculation since the native enzymes of the fruit are quickly destroyed and the natural pectin/protein cloud is established. (This cloud is not primarily due to cellular debris as is often believed. The initial action of the natural pectin methyl esterase in the apple is sufficient to cause a pectin cloud with native apple protein, which is then heat-set by the sterilisation step.) The rapid heating and cooling also minimises cooked flavours so that high quality natural style juices can be produced in this way. However, even the few minutes which elapse between pressing and initial heat treatment may permit oxidation and flocculation to take place to the detriment of both colour and flavour.

One way of overcoming this problem to produce an 'opalescent' juice was suggested by Pedersen at the Geneva Experiment Station, New York, as long ago as 1947 (Pedersen, 1947), but never became commercially popular in the USA although it was adopted by some processors in British Columbia (Bauernfeind, 1958; Moyer and Aitken, 1971). It was introduced into the UK by the Long Ashton Research Station during the 1960s, and has since been adopted enthusiastically by small-scale British producers since it is ideally suited to growers or processors of high-quality fresh fruit. More recently it has become popular in Europe as well. The method relies on adding large amounts of ascorbic acid (about 500 ppm) to the fruit while it is being milled, or to the juice as soon as it has been pressed, followed by sterilisation as quickly as possible thereafter. This prevents the polyphenol oxidation which is a major source of browning and a contributory cause of flocculation. It also appears to modify the activity of the natural pectolytic enzymes. Although the chemistry of the flavour changes has not been studied in detail, the ascorbic acid also has the effect of retaining a fresh fruit aroma, probably by preventing the oxidation of volatile aldehydes (McKenzie and Beveridge, 1988; Stähle-Hamatscheck, 1989).

Refinements of this method consist in chilling the fruit to 4°C before pressing, maintaining the press under nitrogen (only possible with the enclosed Bucher-Guyer type of press), and transporting the juice in refrigerated tankers under an inert atmosphere if it is to be contract packed at a distant location. One drawback is that juice blending is impossible, and so the flavour will vary during the season even if fruit can be blended beforehand. However, this is not necessarily a disadvantage for the intended market which places a premium on the 'fresh' and 'natural' qualities of such juice. In the UK, cull dessert apples such as Discovery or Worcester (early), Cox (mid-season or cold-stored) and Crispin or Idared (late season or cold-stored) are blended with the acid cooking apple Bramley (available from September to August from cold store) to achieve a balanced sugar/acid ratio. This enables production of opalescent juice for nearly 12 months of the year from home grown or EU fruit. Juices pressed after Easter from cold-stored fruit have generally poor flavour characteristics, however, compared to those pressed in peak season.

It is important to use sufficient ascorbic acid during processing, and the recommended level of 500 ppm must be adhered to, since a minimum of 250 ppm free ascorbic must be present in the packaged juice to maintain its colour and stability. If it is not, the initially colourless oxidation products of ascorbate will be further oxidised and rapidly take part in Maillard or other non-enzymic browning reactions which will enhance the discoloration of the juice rather than prevent it. It is important to understand that the enzymic oxidation of polyphenols and loss of ascorbate still continues rapidly after pressing until the enzymes are inactivated during sterilisation, even though the initial oxidation products are colourless in the presence of excess ascorbate and so cannot be seen. For this reason it is beneficial to keep both fruit and juice at temperatures as low as possible, to minimise the activity of polyphenol and ascorbate oxidases.

Although these opalescent juices are normally packed in a sterilised form as previously described, there has been a recent demand in the UK for a lightly pasteurised product (e.g. 75°C for 20 s) which is packed into plastic or gable-top containers and kept in supermarket chill cabinets with a two week shelf life. There is even a small market for completely unpasteurised juice kept under the same conditions, which is very demanding on juice processors since the initial microbial load must be kept below 5×10^4 cells/ml (see section 6.5.9) by scrupulous sanitisation of equipment. Such juices are sold at a premium price to capitalise on the (completely unjustified) consumer belief that they must be of higher quality than their long-life equivalents since they have received less processing. In practice, the reverse is often the case, since short-life products are subject to continued microbial and enzymic degradation which has an adverse effect on the hedonic qualities of flavour, colour and texture after only a few days' storage.

6.3 Clarified juice and concentrate

6.3.1 Enzyming

It was long ago observed that the development of mould growth on a cloudy apple juice may lead to flocculation and clarification of the juice, due to the secretion of pectolytic enzymes by the microorganisms. Since the 1930s this effect has been put to good use by the commercial production of enzyme preparations with defined pectolytic and other activities for use in fruit juice production. The chemistry and biochemistry of their action is much more complex than can be covered here; detailed reviews are given by Kilara (1982), Pilnik and Rombouts (1981), Janda (1983) and Lea (1991). Two major subdivisions of enzyme usage are for the clarification of juices on the one hand and for the treatment of apple pulp for yield improvement on the other. Both techniques are most extensively used in Germany, Switzerland and in the Pacific North West of North America.

The principles of apple juice clarification by pectolytic enzymes have been intensively studied (Endo, 1965; Yamasaki et al., 1967; Kilara and Van Buren, 1989). Apple juice pectin is highly methoxylated and forms a relatively stable colloidal suspension. The action of a pectin methyl esterase (PME) will partly demethoxylate the pectin giving some free galacturonic acid groups which are negatively charged. These may combine with strong complexing cations such as added calcium to form an immediate floc which soon sediments, or with weak cations such as proteins to form a more stable hydrated cloud which initially remains suspended in the viscous juice. If polygalacturonase (PG) activity is also present in the system, however, the long pectin chains will be broken down and the viscosity of the system will diminish markedly. The PG activity affects the charge distribution of the pectin–protein complexes too, causing them to aggregate into much larger particles. Thus any macroscopic debris or pectin–protein complexes which have already formed will drop through the juice as sediment. Due to the considerable viscosity reduction the juice becomes much easier to filter and a clarified juice may easily be obtained. Both PG and PME activities are required because PG will not attack native methoxylated pectin.

For the production of a clarified single strength juice the use of pectolytic enzymes is helpful but not absolutely essential, particularly when working with firm fruit in good condition, since much removal of debris may be achieved by fining alone (see later). For the production of 70° Brix concentrate, however, the use of pectolytic enzymes on the juice is mandatory, since the pectin will otherwise form a gel in the presence of the high levels of acid and sugar once the solids content exceeds 60° Brix or so. Although full pectin 'half-concentrates' at 55° Brix can be made and were at one time popular, they are not microbiologically stable unless refrigerated and have virtually disappeared from world trade.

It is important to have the right balance of PME and PG activities in the enzyme preparation and that they should be active at juice pH. Other pectolytic activities such as pectin lyase, which acts directly to break the polygalacturonate chains of methoxylated pectin, may also be incorporated. Appropriate enzyme mixtures are supplied adsorbed onto a powder, or in an inert liquid carrier, by specialist manufacturers who have wide experience in their formulation. A very useful additional activity in enzymes for apple juice work is that of an amylase, which helps to break down the starch often found as granules in early season fruit. If these granules cannot be removed by filtration or centrifugation beforehand, it is important that they are fully dissolved by heating to 60°C before enzyme treatment, because commercial amylases are unable to degrade the native granular starch.

Enzyme treatment is not instantaneous so it is necessary to provide a juice holding tank after the press and before filtration, which can be maintained at temperatures between 15 and 55°C. Typical treatment times for juice depectinisation are 8 h at 15–20°C or 1 h at 45–55°C. Intermediate temperatures are not recommended because contaminating yeast growth is at a maximum between 20 and 40°C. It is often advantageous to enzyme the warm juice which has already been aroma stripped at 70°C, since the heat treatment provides a near sterile medium in which oxidising enzyme activity is also greatly reduced. Complete depectinisation is judged by the alcohol precipitation test (see section 6.6).

6.3.2 Pulp enzyming

The use of enzymes to clarify apple juice has now become commonplace, but the use of enzymes on apple pulp before pressing to improve yield is not yet so widespread. In the USA, for instance, press aids such as rice hulls are widely available and used to improve the yields in screw press systems. Pulp enzyming has been developed most extensively in central Europe where the Bucher-Guyer HP press is virtually standard equipment and the enzyme and press manufacturers have worked closely together to provide a complete system. Pectolytic enzymes are dosed into the pulp and allowed to stand for up to 24 h at 5°C or 1 h at 15–30°C before or during pressing. This helps to reduce the sliminess of the fruit so that the juice flows more freely and press throughput is increased. This is particularly valuable for softer cultivars such as cold-stored Golden Delicious. The HP press is well adapted to this system, because initial enzyming can take place for about 30 min inside the machine before pressing itself begins (Baumann, 1981).

There are certain disadvantages attached to this procedure, however. For instance, the necessity to hold and/or to heat the pulp is tedious and costly. The partial breakdown of structural pectin can lead to the formation of arabinans which cause haze formation later. The oxidation of polyphenols in certain varieties can lead to tanning which may inhibit the required pectolytic activity. This can be overcome if the pulp is pre-oxidised or is heated to 85°C first to inactivate the oxidases, but this tends to be detrimental to the flavour. There can be a considerable increase in juice methanol content to around 400 ppm compared to a level in normally pressed juice of about 50 ppm, raising problems for aroma essence production. Nevertheless, the use of pulp enzyming is growing in popularity, particularly as a means of handling difficult fruit from cold stores late in the season.

The concept can be taken one stage further, however, by incorporating cellulase and hemi-cellulase activities into the enzyme preparations, which degrade the fruit structure and which lead ultimately to a total liquefaction of the fruit. The advocates of this method regard the press as little more than a drainage basket since the pulp can be so completely broken down that the juice flows out by itself! Yield increases to 96% and threefold improvements in throughput have been claimed (Dörreich, 1986).

Although this idea is appealing in principle, it has the same drawbacks associated with conventional pulp enzyming. In addition, the much greater degradation of the fruit leads to an increase in non-sugar solids and an increase in free galacturonic acid and its breakdown products, which appear to have undesirable effects on flavour and colour (Voragen et al., 1986). It also calls into question the whole definition of 'pure' apple juice since these degradation products are not normally present in the fruit or the juice when traditionally extracted. Although the yields are certainly increased, there are some indications that this is at the expense of juice quality although this remains a very subjective area (Schork, 1986).

For a variety of reasons, therefore, total liquefaction is not yet standard procedure and it is not clear whether it will ever become so. For the present, pulp enzyming using conventional preparations is more commonplace, though in many factories it tends to be restricted to occasional use on specific batches of fruit which are difficult to handle in any other way, rather than being universally applied.

6.3.3 Fining

The traditional fining agent for apple juice is gelatin, a protein which carries a positive charge at the pH of apple juice (about 3.5). When added in solution it forms an insoluble floc by electrostatic neutralisation of other juice debris, which mostly carries a negative charge. The floc slowly sediments and the juice is thereby clarified. It can then be racked, filtered or centrifuged to remove it from the debris. If the juice contains substantial quantities of tannin-like polyphenols (more properly 'procyanidins') or added tannic acid, then hydrogen bonded interactions between gelatin and tannin produce a much denser floc and improve the fining efficiency (Zitko and Rosik, 1962).

Typical levels of gelatin addition are 50–500 g per 1000 l. However, it is very unwise to rely on the standard addition of a fixed amount of gelatin to all juices, because excess residual gelatin can remain in solution and cause further precipitation as post-bottling haze (PBH) when denatured after pasteurisation. Worse still, the addition of excess gelatin during fining can lead to 'charge-reversal' which stabilises the gelatin–tannin complex as a positively charged colloid which will never flocculate. This condition is known as 'over-fining' (Krug, 1969). It is therefore important to carry out a small-scale 'test-fining' on any particular batch of juice, to establish the minimum level of gelatin required for clarification. Such a procedure is given in section 6.6.

Not all food grade gelatins are suitable for fining purposes and it is important to select one that it specifically designed for clarification. The 'Bloom number', which is simply a measure of gelling power, is not a helpful guide for beverage use. More important criteria are the lack of low molecular weight peptide fragments (which can cause post-bottling haze) and the residual surface charge (Bannach, 1984). Gelatins are made by hydrolysis of collagen from animal hide and bone, and are classified as type A (acid hydrolysates) or type B (basic hydrolysates). Since it is important that the gelatin should be positively charged at pH 3.5, its isolelectric point should be high and a type A gelatin is therefore preferable. Special 'non-gelling' gelatins, made by enzymic proteolysis of collagen, may also be suitable as fining agents. Gelatins must be brought into solution slowly and dosed carefully into the juice. A procedure recommended by Wucherpfennig et al. (1972) is to dissolve the gelatin as a 5% solution in demineralised water at 40°C, allow to age for 4–5 h and then dose slowly into the juice with gentle stirring before allowing to stand at 15°C.

The action of gelatin in clarifying apple juice is slow and it can take many hours for the floc to form fully. Such a delay may be unacceptable from the standpoint of flavour quality and microbiological spoilage. The addition of food grade tannic acid at 50–100 g per 1000 l will speed up the process by promoting the formation of hydrogen bonded complexes, and will also tend to minimise the danger from over-fining, although a test-fining is still advisable. It is also possible to use gelatin in conjunction with bentonite, a naturally occurring clay which is negatively charged and therefore neutralises the added protein, forming a floc which sediments and brings down juice debris from suspension. It is typically used at 500–1000 g per 1000 l. As an insoluble mineral with a huge surface area (750 m^2/g), bentonite has the advantage that it will remove some suspended material by simple adsorption and entrapment too. For this reason, bentonite is sometimes used on its own, particularly after pectolytic treatment when it helps to remove the added enzyme protein. As with gelatin, there are many forms of bentonite, and a grade specifically recommended for fining must be used. Other grades, used in applications from toothpastes to pond-liners, may be unsuitable. The sodium form is generally preferred to the calcium form, due to its greater swelling power. As a clay, bentonite must be hydrated into a 5% slurry very carefully by steaming or with warm water, or it may set into hard and unworkable lumps.

The most elegant fining adjunct to be employed with gelatin is undoubtedly kieselsol. This is a synthetic silica sol (*not* a conventional silica gel) with a particle size of around $5 \times 10^{-3}\,\mu m$, sold under trade names such as 'Baykisol', 'Klarsol' and 'Syton'. When properly dosed into apple juice either before or after the addition of gelatin, it forms compact flocs within minutes. Kieselsol/gelatin fining is now used extensively in Germany and Switzerland, a typical dosage being 500 ml of 30% silica sol per 1000 l of juice. The corresponding gelatin dosage is less than it would be in the absence of kieselsol, typically 20–200 g gelatin per 1000 l of juice. Although it is generally used in a batch process as with other fining agents, it can work so quickly that continuous 'in-line' clarification becomes a possibility (Gierschner *et al.*, 1982).

There are few polyelectrolytes available to substitute for gelatin. Isinglass, a fish protein, generally shows no advantages over gelatin and suffers from the need to be dissolved in acid solution before use. Casein has been reported as effective, but is rarely used (Kilara and Van Buren, 1989). A recent introduction is chitosan, which is a chemically de-acetylated chitin derived from crab-shell waste, sold in solution under trade names such as 'Cel-Flock' and 'Vinocell' (Soto-Peralta *et al.*, 1989). Because of its glucosamine structure, it contains both positively and negatively charged groupings which make it more effective than gelatin under certain circumstances, for instance, in the clarification of intractable hazes formed on re-dilution of cloudy concentrates. Typical dosage rates are 1–10 l of solution per 1000 l of juice. However, because it can be up to ten times more costly than gelatin, it is rarely used routinely but only to solve specific problems.

It is often sensible to carry out the enzyming and fining operations simultaneously before concentration or bottling of single strength juice. Enzyming makes the fining more efficient by reducing juice viscosity, while decantation or centrifugation after fining reduces the load on the subsequent filters and reduces the likelihood of post-bottling haze by removing soluble polyphenols. Although the temperature requirements for the two processes are somewhat conflicting, since fining with gelatin alone is usually more efficient at lower temperatures, the use of adjuncts such as kieselsol and bentonite in combination make it perfectly practicable to complete both enzyming and fining within say 3 h at 50°C, rather than 2 h at 50°C followed by 4 h at 15°C (Grampp, 1977).

New types of fining system present themselves from time to time. For instance, the use of honey as a fining agent has recently been investigated (McLellan et al., 1985; Wakayama and Lee, 1987). In Canada, the 'Clarifruit' system has been developed to remove apple juice flocs in under 2 h by flotation with nitrogen (Anon, 1984; Otto et al., 1985).

Ultrafiltration (UF) seems to offer many advantages to the apple juice processor, since it would appear to combine both fining and filtration steps in one operation (Möslang, 1984). When first introduced on a pilot scale, however, many problems were encountered in practice. Low flux rates, high rates of membrane fouling, microbiological contamination in the recirculated feedstock and development of severe post-bottling and post-concentration hazes were amongst the drawbacks. However, later work has shown that ultrafiltration can be very successful provided that attention is paid to certain details. The juice must be adequately depectinised and starch-free before UF to ensure reasonable flux rates and to prevent the rapid clogging of membranes. Even so, operation at 55°C is necessary to reduce the juice viscosity. Turbulent or aerated flow of the feedstock juice must be avoided, because this induces polyphenol oxidation which contributes to membrane fouling and later to haze formation in the permeate. For this reason, flat or tubular cross-flow membranes are preferable to the 'spaghetti bundle' type in which the juice velocities are very high. Membrane cleaning and sterilisation at regular intervals is also important because of the long residence (incubation) time of the feedstock juice; it is important to avoid re-infection of fresh juice by microorganisms which are retained on the membrane from batch to batch. Modern membranes are generally made of polysulphone materials which are bonded to a mechanically strong support and will last for at least two seasons of continuous use. This is important because frequent membrane replacement is a tedious and costly business, particularly if a membrane leaks without warning. Ceramic membranes (e.g. zirconia) are also available; they are mechanically stronger and more resistant to aggressive cleaning agents. Membranes made from sintered stainless steel can withstand very high pressures and if fed directly with apple pulp could perhaps replace both press and filter in one unit; such a technique remains at the experimental stage (Barefoot et al., 1989).

Ultrafiltration may now be competitive in cost with conventional fining and filtration, particularly if new capital equipment must be installed anyway. Both techniques reduce the level of insoluble solids, but one thing that UF cannot do in contrast to gelatin fining is to reduce the level of soluble haze precursors, principally polyphenols. The UF user, therefore, must be ever watchful that the processed juice is given no chance to develop post-bottling or post-concentration haze by subsequent aeration or temperature cycling, for example (Nagel and Schobinger, 1985). It is perhaps advantageous to reduce the polyphenol levels by encouraging their oxidation onto the pulp before pressing, as described later, if juice is to be ultrafiltered without fining. An alternative strategy is to hyperoxidise the juice under controlled conditions in the presence of an added laccase, to flocculate the potentially unstable procyanidins before UF (Dietrich et al., 1990).

6.3.4   Concentrates

The principles of concentration and of juice treatment prior to concentration have been described above. Although apple juice concentrate (AJC) is a relatively stable commodity and can be kept wholesome almost indefinitely, it can nevertheless lose quality markedly if it is poorly stored and handled. One consideration is that of clarity. A juice which has been clarified before concentration will generally give a clear juice on re-dilution and requires only a final polishing before bottling. A concentrate made from cloudy juice, however, particularly if stored for a long time, may be extremely tedious to clarify on re-dilution; fining may be ineffective and filter sheets may blind up quickly. It is nearly always better practice to store AJC clarified rather than cloudy (Grampp, 1981).

A further consideration is that of storage temperature. At 5°C or less, concentrates and aroma essences will remain essentially unchanged for at least six months and probably longer. At 15, 20 or 30°C, however, the re-diluted juice becomes noticeably lower in quality, the effects being proportional to storage time and temperature; 2 weeks at 20°C can produce detectable changes (Weiss et al., 1973). Caramelised flavours develop and browning is increased because of the Maillard reaction between reducing sugars and amino acids at high concentration (Dryden et al., 1955; Kern, 1962). There is a loss in apparent polyphenols and in titratable acid too (Grob, 1958). Attempts have been made to define a single index to measure the overall quality loss, for instance by monitoring the increase in HMF (5-hydroxymethyl furfural). This results from the Maillard degradation of fructose and is easily measured by high performance liquid chromatography (HPLC) or by colorimetric procedures. It is undetectable in fresh juices and present below 1 ppm in fresh AJC made in modern flash evaporators, although Weiss and Sämann (1985) found it at 17 ppm in AJC made in an open pan. When concentrate is stored, however, HMF levels may rise as high as 700 ppm although levels of 100 ppm are more commonly encountered. HMF is a generally useful guide to deterioration of any given juice but results between juices are not easily comparable and do not correlate directly with sensory scores (Poll and Flink, 1983). Some concentrate buyers have tried to set HMF levels, e.g. < 20 ppm in their purchasing specifications to ensure products of the highest quality.

If AJC is not cooled to 5°C on exit from the evaporator, loss of flavour quality can begin immediately. Because its viscosity slows convective cooling,

AJC piped into a large uninsulated unstirred tank (say 50 000 l) at 50°C may take literally weeks to cool to ambient temperature in the centre, during which time Maillard reactions can progress significantly. Although oxygen itself does not take part in the Maillard reaction (Kern, 1962), AJC should be handled and stored in an inert atmosphere to prevent oxidative changes which could otherwise occur. Concentrate stored at ambient temperature may slowly produce CO_2 which must be vented. This can arise both from chemical decarboxylation of Maillard products and from the action of osmotolerant yeasts.

It is possible to inhibit AJC deterioration by the use of SO_2 at levels of about 250 ppm (Dryden *et al.*, 1955; Jacquin, 1960). The early Maillard products, including HMF, are characterised by carbonyl groupings to which the bisulphite will bind and so prevent further reaction. However, on dilution the juice will contain 30–40 ppm SO_2 which is nowadays unacceptable to most juice processors. Ascorbic acid is not an alternative because its anti-oxidant properties are not appropriate in this case and it does not bind to carbonyls as SO_2 does. Indeed it has been shown that the addition of ascorbic acid to AJC actually enhances the rate of browning because its breakdown products are themselves carbonyls and so they participate in the non-enzymic brownng reaction too (Pilnik and Piek-Faddegan, 1970).

6.3.5 Hazes and deposits

Post-bottling hazes (PBH) in clear juices have already been described in passing and represent a chronic problem for all clarified apple juice producers (Heatherbell, 1976; Görtges, 1982; Van Buren, 1989). Pectin and starch as sources of haze should be eliminated by the proper use of enzymes, although it is often difficult to know whether sufficient enzyme action has taken place. Test methods for this are given in section 6.6. Residual pectin is usually detected by precipitation with alcohol which is a fairly reliable test. Residual starch is tested by discoloration with an iodine reagent. However, it has been shown that partly degraded starch (10 glucose units or less) does not react to this test, and that such oligomers can re-form and gelatinise during heat treatment later. These re-formed polymers cannot then be further broken down by added amylase. It is therefore recommended that, in any fruit where starch is known to be a problem, amylase treatment should be continued well beyond the point at which 'detectable' starch disappears.

A non-traditional haze is that due to arabinans, which tends to manifest itself as small bodies which are superficially similar to yeasts on microscopic examination. These are due to the action of prolonged heat or added pectolytic enzymes on apple pulp, which split soluble arabinose oligomers away from the pectin backbone. During juice concentration, these oligomers condense and form characteristic 'diploid cocci', often only visible after re-dilution with water (Schmitt, 1985). Once isolated, arabinans are easily detected by microchemical hydrolysis to arabinose or detected by techniques such as thin layer chromatography (TLC) or nuclear magnetic resonance (NMR) (Churms et al., 1983). It is more difficult to predict their formation in advance although tests have been proposed (Endress et al., 1986; Ducroo, 1987). The major enzyme manufacturers are well aware of the problem, however, and arabinase activities are now incorporated in all their formulations intended for apple pulp treatment.

The major cause of post-bottling or post-concentration haze, however, is the polymerisation of polyphenols (Johnson et al., 1968, 1969; Van Buren and Way, 1978). Only the procyanidins or 'tannins' are involved in this process; the simpler phenolics such as chlorogenic acid play no part in haze formation. The simplest procyanidin unit in apples (B2) is a dimer with a molecular weight of 580. Other procyanidins with a molecular weight up to about 2500 are also soluble in the juice (Lea, 1984). Under the influence of low pH, aldehydes or oxygen, these oligomeric procyanidins can polymerise by various routes to form insoluble materials which are displayed as haze. In extreme cases they will fall out of solution as sediments to leave a bright juice, but more often they remain in colloidal suspension. This is particularly the case if any protein also remains in solution (e.g. from previous fining), because the colloids are then stabilised as hydrophilic complexes. Indeed most polyphenolic hazes contain significant amounts of protein too. Microscopic examination with phase-contrast illumination shows them as roughly circular particles with a dark rim but a light centre, helping to distinguish them from bacteria which may be much the same size (Tanner and Brunner, 1987; Beveridge and Tait, 1993) (Figure 6.2).

To a certain extent such hazes may be temperature-dependent ('chill hazes') and may be solubilised on warming to 60°C but re-appear on cooling to 4°C. Indeed it is often thermal shock or cycling which causes them to appear in the first place. For instance, hot bottles from a pasteuriser may display a chill haze if rapidly cooled in an unheated warehouse in winter but may remain stable if cooled much more slowly in summer. Thus bottles on the outside edge of a pallet may show more PBH than those on the inside (Fischer, 1981). Such an effect is the basis for one of the predictive tests for in-bottle stability; heat to 80°C and cool immediately to 4°C in a refrigerator. If no haze appears under these conditions, or on cycling again, the product is regarded as stable.

It is obviously not possible to remove the major haze precursors, the procyanidins, by filtration since they are far too small. Fining with gelatin is helpful because it will preferentially remove the larger oligomers which are likely to become insoluble first (Lea, 1984; Van Buren, 1989). Treatment with PVPP (polyvinylpolypyrrolidone), which is an insoluble contact stabilising agent widely used to protect against chill haze in brewing by removing polyphenols, has much the same effect. However, excessive reduction of polyphenols is undesirable, since the procyanidins contribute to juice 'mouth-feel' and the product becomes insipid if they are reduced too far. Nor is it

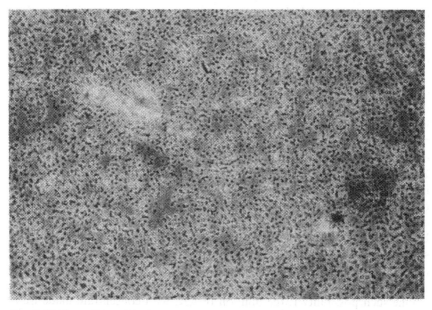

Figure 6.2 Photomicrograph of a procyanidin-protein haze isolated from apple juice by high-speed centrifugation (magnification about 1000 ×). When viewed by phase contrast optics, such hazes superficially resemble coccoid bacteria about 1–0.5 μm in diameter. However, they may be distinguished by their more irregular shape, and by the bright centre of each particle as it is racked in and out of focus.

possible to set analytical 'pass/fail' levels for procyanidins. The analyses are complex and in any case the stimulus for haze formation is not so much the absolute levels of procyanidins as their processing history and the level of potentiating factors. It is therefore best to minimise these potentiating factors as far as possible. In particular this means that all possibilities for oxidation during and after filtration must be avoided; contact with air should be minimised, turbulence and foaming in fillers and pipework should be eliminated. Transition metals such as iron and copper catalyse the oxidation and must be kept as low as possible; Fe < 10 ppm; Cu < 1 ppm. Even 0.5 ppm copper is enough to catalyse apple juice oxidation (Pollard and Timberlake, 1971). The presence of aldehydes such as HMF will also polymerise procyanidins by cross-linking them (the 'Bakelite' reaction) (Delcour *et al.*, 1982). Although this is not an oxidative reaction, poorly stored concentrates are more likely to show PBH on re-dilution for this reason. Removal of excess protein with bentonite may help to delay the onset of PBH.

Other metals may also contribute to haze formation. If hard water is used to dilute a concentrate containing demethoxylated pectin fragments, calcium pectate gels or deposits may form. Likewise, it is known that contamination with aluminium or tin from processing equipment can form a pectate cloud (Krug, 1969).

6.4 Authentication and adulteration

The producer who processes exclusively from fresh fruit has no need to worry that his juice might not be authentic. However, since most apple juice is traded commercially as concentrate, and many packers never handle the fruit itself, there is an ever-present possibility for the perpetration of fraud by the unscrupulous. This came to public attention in 1988 with the conviction and imprisonment in the United States of those involved in the so-called 'Beech-Nut Scandal', whereby totally artificial mixtures of sugar, acid, water, colourings and flavourings were sold as apple concentrate (Associated Press, 1988). However, most frauds are considerably less blatant and involve the 'stretching' of concentrate rather than wholesale substitution with synthetic materials. The detection of such fraud is an interesting challenge to the analytical chemist, because of the several possibilities available to the fraudster and because of the inherent natural variation in the genuine product itself. The subject has been reviewed in depth by Lee and Wrolstad (1988a), Mattick (1988), Elkins et al. (1988) and in chapter 1 of this book in relation to orange juice.

One such possibility is the dilution of AJC with concentrate made from other fruits. Since apple is generally a fairly cheap fruit, the most likely adulterants are other inexpensive concentrates light in flavour and colour, such as pear and grape. Pear is distinctive for its relatively high levels of citric acid and proline compared to apple. Grape also contains relatively high levels of proline, but in addition is unique amongst common fruits for its content of tartaric acid. More subtle is its phenolic acid pattern, which is based on tartrate esters of caffeic acid (caftaric acids) rather than the more common quinic acid esters (chlorogenic acids). Detailed analysis of organic and amino acids, therefore, is a fairly reliable method of detecting the addition of non-apple concentrate to AJC.

A more insidious type of fraud, however, is the addition of water to AJC. The resultant loss of gravity is compensated for by the addition of sugar from cane, corn or beet. If acid levels should thereby fall below the normally expected levels, the fraudster can add small quantities of synthetic food grade acids to make good the deficit. The addition of citric acid is readily detected by chromatography, since genuine apple juice contains virtually no citrate, but the addition of malic acid is less easy to recognise. However, chemically synthesised malic acid is a racemic mixture of the d and l optical isomers whereas the naturally occurring acid in apples is exclusively the l form. Since the d and l forms can be specifically determined by enxymic assay or chiral HPLC, this form of adulteration is much less common than hitherto. Biologically produced l malic acid is now available in commercial quantities, which is more difficult to detect, although the handling and drying of bulk malic acid always produces small quantities of fumaric acid as a contaminant which can easily be detected by HPLC using a UV detector.

Unfortunately fumaric acid may also arise naturally in apple juice concentrate during heating. Junge and Spädinger (1982) suggested that this figure would not exceed 3 ppm; Evans *et al.* (1983) reported that none was produced in AJC held at 60°C for 3 days but as much as 10 ppm could be generated in AJC held at 100°C for 3 h. Typically, levels > 5 ppm give rise to suspicion of adulteration by added malic acid.

When analysing malic acid, it must be remembered that up to 20% of the natural malic acid in AJC may be esterified with sugars or in the form of a dimeric lactide (Gierschner, 1979). This leads to a depression of all the malic acid figures unless an alkaline hydrolysis is first employed (Burroughs, 1984). This problem is only apparent in juices made from concentrate, not in fresh or 'natural' juices.

A further problem is presented by the addition of sugar. The sugars of apple are principally fructose and glucose in a 3:1 ratio, with varying amounts of sucrose depending on fruit maturity and age of concentrate. However, this ratio varies widely in nature and so the fraudulent addition of a certain amount of invert sugar with a 1:1 fructose/glucose (F/G) ratio may not necessarily be detectable. The further addition of isomerised high fructose corn syrup may then be tailored to give the 'correct' ratio, presenting the analyst with an insuperable task by normal chemical means. However, the cheapest and commonest sources of sugar outside the EU are maize hydrolysate ('corn syrup') or sugar cane. Both of these derive from tropical grasses which fix their carbon by the C4 photosynthetic route rather than the C3 pathway adopted by most temperate dicotyledenous plants. Although the sugars produced by these pathways are chemically indistinguishable, they contain different ratios of the stable carbon isotopes ^{13}C and ^{12}C, usually expressed as a value 'δ' (Krueger, 1988). Typically, maize and cane sugar give a δ of 8–11 parts per thousand, while apple sugar gives a δ of 22–30 parts per thousand (Doner, 1988; Lee and Wrolstad, 1988b). Stable carbon isotope ratio analysis (CSIRA) is therefore a method for detecting such adulteration, and is offered by a number of specialist laboratories at relatively reasonable cost (typically £40 per sample in 1994).

In Europe the major commercial source of invert sugar is the root of sugar beet, which is a temperate plant whose δ $^{13}C/^{12}C$ ratio is indistinguishable from that of apple. The addition of beet sugar to AJC is best detected by examining the deuterium/hydrogen (D/H) ratios after fermentation to ethanol (details of the technique are given in chapter 1). The natural D/H ratio of apple sugar is typically 98 ppm whereas that for beet sugar is typically 90 ppm. Low levels of adulteration (15% or less) are not reliably detectable by this technique, and skilful 'tailoring' of AJC with corn and beet syrups of known D/H ratios can circumvent it entirely. It is also a very expensive analysis to run (about £500 per analysis in 1994).

The HPLC analysis of trace 'oligosaccharides' produced chemically during the manufacture of invert syrups may also be used for the detection of added sugar (Low and Swallow, 1991). Unfortunately AJC contains confusing amounts of naturally occurring material from pectin breakdown which interferes with this analysis and makes it less reliable than for citrus juices.

There is therefore no single analytical technique which will ensure the authenticity of apple juice and concentrate. Most laboratories in this field therefore offer a 'multicomponent' or 'matrix' approach, based not only on the techniques described above but on a range of conventional chemical analyses too. By carefully examining the individual data and, most importantly, the interrelationships between the data, it is easier to judge whether adulteration has taken place or not.

A prerequisite of any analytical method which seeks to monitor adulteration is a wide ranging database of genuine values encountered in practice, and a number of compilations of analytical data towards this end have been published in recent years (Burroughs, 1984; Mattick and Moyer, 1983; Evans et al., 1983; Brause and Raterman, 1982). The German Fruit Juice Association (1987) has taken this concept furthest through the publication of its 'RSK values' (see chapter 1) which are based on many hundreds of authentic samples (Koch, 1984). The RSK tables, which have been produced for many fruit juices in addition to apples, contain the median values, the range and the 'guide value' for those analytes which are judged to be most relevant (see Table 6.1). It must be stressed that quite authentic juices may, on occasion, fall outside the given range for one or two analytes, and that the main purpose of these tables is to provide a framework from which a competent analyst can decide whether or not there are grounds for suspicion about a particular sample which may then need more detailed investigation. Conversely, an adulterated sample may display all its data within the RSK ranges and yet the 'internal relationships' between the values may indicate that the juice is not authentic. The RSK values are not immutable and may be revised with time; for instance, authentic samples of Dutch and German apple juice from orchards close to the sea show particularly high levels of sodium and the original RSK values have been adjusted to take account of this. Some analysts also find it useful to supplement the RSK data with other analytes such as chlorogenic acid.

Attempts have been made to codify the interpretation of authenticity data using 'expert systems' or by the use of multivariate statistical techniques such as principal components analysis. This latter procedure enables the creation of a multidimensional sample space or map from the analytical database, so that the relationship of the unknown sample to a group of authentic samples can be more easily perceived. A particular application of multivariate techniques for authenticity studies is exemplified by the technique of pyrolysis-mass spectroscopy (Py-MS), which produces data that can only be handled in this way. Small samples of juice are pyrolysed in a vacuum and a mass spectral 'fingerprint' is obtained of the gaseous products, taking only a few seconds to analyse each sample. A study in our own laboratory on 125

Table 6.1 RSK values for apple juice

		Guide value		Range		Median value
				From	To	
Relative density 20°/20°C		min	1.0450	1.0450	1.0570	1.0488
Brix, ref. corr.		min	11.18	11.18	14.01	12.08
Soluble solids	g/l	min	116.8	116.8	148.1	126.7
Titratable acids (pH 7.0)						
as tartaric acid	g/l	min	5.0	5.0	8.5	6.5
as mequiv./l		min	66.7	66.7	113.3	86.7
Ethanol	g/l	max	3.0	–	–	–
Volatile acids						
as acetic acid	g/l	max	0.4	–	–	–
Sulphur dioxide, total	mg/l	max	10	–	–	–
Lactic acid	g/l	max	0.5	–	–	–
D-Malic acid	g/l		n.d.			
Citric acid	mg/l		–	50	200	100
Tartaric acid	g/l		n.d.			
Glucose	g/l		–	18	35	26
Fructose	g/l		–	55	80	65
Glucose/fructose ratio		max	0.5	0.3	0.5	0.40
Sucrose	g/l		–	5.0	30.0	15.0
D-Sorbitol	g/l	min	2.5	2	7	4
Reduction-free extract	g/l	min	18	18	29	22
Ash	g/l	min	2.1	1.9	3.5	2.55
Alkalinity number		min	11	11	14	13
Potassium (K)	mg/l	min	1000	900	1500	1200
Sodium (Na)	mg/l	max	30	–	–	–
Magnesium (Mg)	mg/l		–	40	70	52
Calcium (Ca)	mg/l		–	30	120	59
Chloride (Cl)	mg/l	max	50	–	–	–
Nitrate (NO$_3$)	mg/l	max	10	–	–	–
Phosphate (PO$_4$)	mg/l	min	150	130	300	220
Sulphate (SO$_4$)	mg/l	max	150	–	–	–
Formol number						
(ml 0.1 mol NaOH/100 ml)			–	2.5	10	4.5
Proline	mg/l	max	15	–	–	8

samples of AJC from 13 different countries showed that some discrimination between geographical origin may be obtained, and that concentrates were differentiated partly on the basis of carbohydrate pattern, degree of pectin methoxylation and presence of HMF (Hammond, 1994). Although Py-MS is not an 'absolute' analytical technique, it may have value for checking quickly large numbers of concentrates for consistency and alerting the analyst to suspect or outlying samples which merit further detailed investigation.

The authenticity of the aroma or essence fraction of apple juice can be determined by chiral gas chromatography of 2-methylbutyric acid after hydrolysis of its esters. Natural essences are predominantly in the S form while synthetic essences are a racemic mixture of R and S forms (Rettinger et al., 1990).

6.5 Composition

The composition of apple juice necessarily reflects the composition of the fruit from which it is made. Hence varietal characteristics, climate and cultural practices play a major role in determining juice composition. The yearly variation in juice composition due to weather is often unrecognised. In an unpublished study of cider apple juices made over a ten-year period from the same individual trees in the same orchard under the same cultural conditions, the levels of juice acid and sugar showed a range of ±20% around the mean values, with a relative standard deviation of 12%.

A further source of variation in apple juice is processing itself. Even while fruit is being stored prior to juicing, its content of acids, sugars and pectins will slowly change due to normal respiratory action (Hulme and Rhodes, 1971). On milling, however, many fruit enzymes and substrates are brought into contact and rapid changes can occur in certain juice components, of which the most susceptible are pectin, polyphenols and volatile flavour compounds. Some aspects of apple juice, such as colour, are entirely a consequence of processing. Different mill types, different press-hall temperatures and different shapes of mash tanks will all play a part in determining the final colour of fresh juice.

Quantitative figures for juice composition, therefore, can only be taken as a general guide. A useful starting point are the values given in Table 6.1; further tabulations of quantitative data are to be found in Herrmann (1987), Mattick and Moyer (1983), Burroughs (1984) and Lee and Mattick (1989). A detailed general description of apple juice composition is also included in Lee and Wrolstad's (1988a) discussion of apple juice authenticity. In the following discussion, the individual chemical classes of juice constituents will be considered.

6.5.1 Sugars and sorbitol

Sugars are the major soluble constituents of apples, comprising from 7 to 14% of the fruit on a fresh weight basis. The specific gravity or the °Brix of apple juice is therefore very closely related to sugar content. The sugars present in the juice are almost entirely fructose, glucose and sucrose, with only traces of others. Xylose was reported at 0.05% in apple juices by Whiting (1960). Fructose always exceeds glucose in a ratio of 2–3 times, while the content of sucrose is usually similar to that of glucose. Fructose and sucrose concentrations increase markedly during fruit ripening while glucose tends to fall (Beruter, 1985). Free sugars continue to increase during storage as starch is broken down (Hulme, 1958) (Figure 6.3).

After pressing, sucrose under acid conditions may tend to hydrolyse to fructose and glucose ('inversion'). This is particularly noticeable in juices from concentrates and thus affects the glucose to fructose ratio (Mattick and

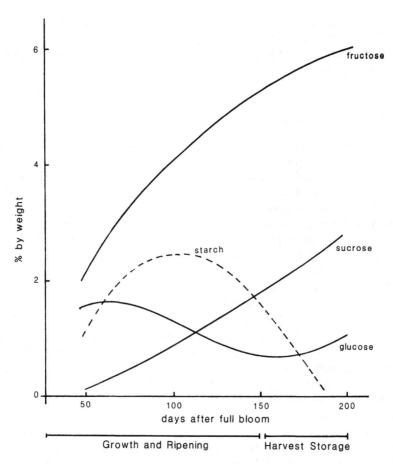

Figure 6.3 Typical changes in levels of fructose, glucose and starch during apple ripening and storage. Data taken from Berüter (1985) and Hulme (1958).

Moyer, 1983). Grob (1958) showed that sucrose could drop to one-third of its original value in concentrate stored at 20°C for three years. Babsky *et al.* (1986) showed a loss of more than 50% in concentrate stored at 37°C for 3.5 months. In both cases the reducing sugars increased in proportion to the sucrose loss, and there was no evidence for any significant consumption of sugars by the Maillard reaction although this reaction was evidently taking place as judged by browning and HMF production.

Nearly all the sugars in apples and pears are synthesised in the leaves, and then transported to the fruit in the form of sorbitol, a sugar alcohol. (This is peculiar to the *Rosaceae*; in most plants sucrose, not sorbitol, is the transport sugar.) Most of the sorbitol is then converted into fructose, glucose and sucrose, but a certain amount of free sorbitol remains and appears in the juice. It is easily analysed, together with sugars, by HPLC using refractive index detection. Mean levels of sorbitol in pear juice (20 g/l) are much higher than in apple (4 g/l) (Tanner and Duperrex, 1968), and this has been suggested as a test for adulteration, although Mattick and Moyer (1983) reported sorbitol levels in authentic American apple juices as high as 12 g/l.

6.5.2 Starch and pectin

Starch does not appear in solution in fresh pressed apple juice, but takes the form of insoluble granules about 1–16 μm in diameter which derive from storage vacuoles in the fruit and can constitute as much as 2% of the fruit on a fresh weight basis. The starch consists of glucose polymers, α-1,4 and α-1,6 linked and about 30 units in length. Although the granules tend only to be present in early season or under-ripe fruit, their small size means that they may escape filtration procedures. When juice is heated above 60°C, the granules are gelatinised and may cause haze problems as previously described unless they are fully degraded by added amylase.

Pectin, which is released from the middle lamella of apple cell walls by the mechanical action of milling and pressing, is present in all fresh apple juices to a varying extent. In early season fruit, amounts of soluble pectin may be low, in the order of 0.1% by weight of juice. In later fruit, or in fruit from cold store, pectin may rise as high as 1–2.5% in the juice. Juice prepared in traditional pack presses often contains far less pectin than juice from horizontal Bucher-Guyer type presses. Krug (1968) found values of 0.37% and 1.5% respectively when working with replicate batches of the same fruit pressed by the two different methods.

Native juice pectin consists very largely of polymerised galacturonic acid chains which are methoxylated up to a level of 95%. These chains form the so-called 'smooth' regions of the pectin molecule. At intervals along the chain, the so-called 'hairy' regions are incorporated, containing a variety of polymers based on neutral sugars such as arabinose, rhamnose and galactose (de Vries et al., 1986). Apples themselves contain varying amounts of pectin methyl esterase (PME). Thus a fresh pressed apple juice, kept cool to prevent fermentation, may spontaneously clarify by the action of native PME, which demethoxylates the pectin and leads to flocculation. This is known as 'défécation' (French) or 'keeving' (English) and was once an important step in the production of traditional fermented cider (Beech and Carr, 1977; Lea, 1991). In modern juices, the pectin is largely broken down by the action of added enzymes, yielding shorter galacturonic acid residues which are partly demethoxylated. From the technological viewpoint, pectin has generally been regarded as a nuisance. However, it contributes a major part of the texture and mouthfeel of opalescent or 'natural' juices, and also represents a 'non-dietary fibre' element in the juice, which may lead to some re-evaluation of its role.

6.5.3 Organic acids

The major organic acid of apple juice is *l*-malic, which is named from the very fruit itself (Latin: *Malus*). It is typically present at levels of 0.5%, with a range from 0.18 to 1.4% depending upon variety and season, and generally constitutes at least four-fifths of the total acid of the juice. The balance is largely accounted for by quinic acid at levels between 0.04 and 0.46%. Citric acid is usually detectable in traces of about 0.01%. Citramalic and shikimic acids have been reported at levels up to 0.05 and 0.015%, respectively (Beech and Carr, 1977). Galacturonic acid arises from pectin breakdown, and in juices prepared from concentrate may be about 0.2%. Lactic acid is detectable only as a result of bacterial spoilage. Tartaric and oxalic acids do not occur in apple juice.

Although the malic acid is partly neutralised at apple juice pH, it is not firmly bound to any other constituents in fresh juices. In concentrates, however, progressive malic acid losses of up to 20% during storage have been recorded due to ester or lactide formation (see section 6.4). It is not known what effect such complexes have on the perceived acidity of juice in flavour terms.

Of further interest is ascorbic acid. Levels in the fresh fruit generally do not exceed 100 ppm (Bauernfeind, 1958) although figures as high as 318 ppm have been reported (Herrmann, 1987). However, ascorbic acid levels in apple juice are generally regarded as vanishingly small due to oxidative losses during processing. It is probable that the ascorbic acid is destroyed by coupled oxidation with polyphenols via the polyphenoloxidase (PPO) enzyme, rather than by the action of a true ascorbic oxidase (Pollard and Timberlake, 1971). At any rate the ascorbate levels in processed apple juice are unimportant either nutritionally or technologically by comparison with the fresh fruit, except where ascorbic acid has been added for technological reasons (see section 6.2). In this case the vitamic C levels are as high as in most citrus juices (about 300 ppm).

6.5.4 Protein and amino acids

The true soluble protein level in apple juices, even when unclarified, is extremely small and generally does not exceed 100 ppm. In unpublished work, Lea found between 10 and 250 ppm in a range of natural style juices. Even this small amount, however, can be sufficient to play a major role in the formation of PBH and deposits when complexed with polyphenols (Hsu *et al.*, 1989). Such hazes, when analysed, show up to 50% each of protein and polyphenol on a dry weight basis. The complexes themselves may form only 1–10 ppm w/v of the juice. In natural style or opalescent juices, the combination of protein with pectin and polyphenols is largely responsible for the desirable cloud, which constitutes between 120 and 500 ppm w/v of the juice.

Table 6.2 Mean values for amino acids in
natural apple juices

	mg/l
Asparagine	893
Aspartic acid	107
γ-Amino butyric	11.3
α-Alanine	8.8
Serine	10.0
Valine	5.1
Phenylalanine	3.2
Threonine	1.8
Glycine	1.6
Isoleucine	1.8
Leucine	0.7
Proline	1.7
Methionine	0.9
Histidine	3.1
Lysine	2.6
Arginine	2.6

Burroughs (1984) showed that 89% of the total soluble nitrogen in apple juice is attributable to free amino acids and that 79% is accounted for by asparagine alone. His figures for amino acids detected in a range of natural apple juices are given in Table 6.2. Tyrosine, tryptophan and cysteine were not detectable, while glutamic acid, glutamine and 4-hydroxyproline were totally obscured by the huge asparagine peak in the analytical method used. Babsky et al. (1986) found glutamic acid levels in fresh concentrate were similar to those of aspartic acid. For analytical purposes, free amino acids as a group are often measured indirectly by titration with alkali after reaction with formaldehyde, the so-called 'formol index' (IFJU, 1985).

There is some evidence for the belief that dessert apples contain more soluble nitrogen than cider apples or the European 'most-apfeln' which are specifically grown for juice (Fischer, 1981), and that young trees produce juice with higher nitrogen levels than do older trees (Burroughs, 1974).

Lüthi (1958) showed that amino acids declined markedly during concentrate storage, glutamic acid declining most rapidly. Babsky et al. (1986) demonstrated that after 111 days at 37°C, only 13% of the original amino acids remained when analysed chromatographically, and that the rate of amino acid loss correlated very closely with the measured rate of non-enzymic browning. The formol titration remained at 82% of its original value, however, implying that the intermediate Amadori compounds (early Maillard products) reacted with the formaldehyde. This casts some doubt on the validity of the formol method. There can be no doubt that amino acids in apple juice are critically implicated in the deterioration of concentrate by the Maillard reaction. In the case of both apple and pear, it has been shown that removal of amino acids from concentrate markedly decreases the rate of non-enzymic browning (Cornwell and Wrolstad, 1981; Dryden *et al.*, 1955).

6.5.5 Polyphenols and colour

The polyphenols of apples fall into six groups. Of these, the anthocyanins (principally cyanidin 3-galactoside) and the flavonol glycosides (principally quercetin 3-glucoside, galactoside and arabinoside) are found almost exclusively in the skin and are not extracted with the juice. Hence apples with a red skin colour, which is due to anthocyanins, do not produce a red juice when normally processed. Typical members of the other four classes are shown in Figure 6.4 (Lea, 1974, 1978). Quantitative amounts present in two extreme cultivars, extracted under varying oxidation conditions, are shown in Table 6.3 (Bramley is a cooking apple; Dabinett is a cider apple. All normal dessert and juicing apples lie somewhere between the two.)

Some natural variations in phenolic pattern are known. For instance, although the principal phenolic acid in most cultivars is chlorogenic acid (5-caffeoyl quinic acid), certain cultivars contain major amounts of other acids such as 4-*p*-coumaroyl quinic acid instead or in addition (Whiting and

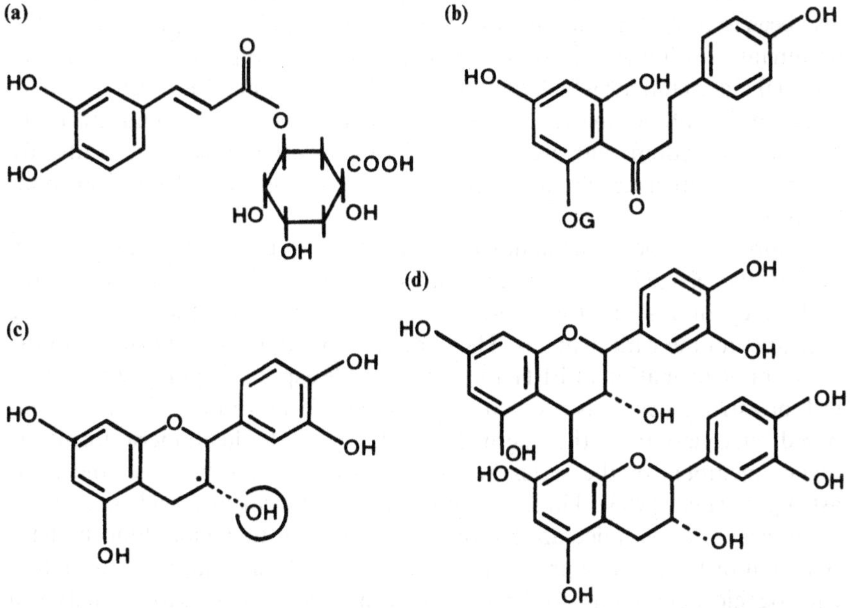

Figure 6.4 The four major classes of polyphenols found in apple juice. (a) Phenolic acids: chlorogenic acid (*p*-coumaroyl quinic acid has a similar structure but lacks a phenolic hydroxyl group). (b) Dihydrochalcones: phloridzin. (c) Catechins: (−) epicatechin ((+) catechin has the alternative stereochemistry at the hydroxyl group indicated). (d) Procyanidins: procyanidin dimer B2.

Table 6.3 Polyphenol levels in freshly pressed and oxidised apple juices

Component	Freshly pressed juice (15 min oxidation)	6 h oxidation after pressing	6 h oxidation on pulp before pressing
Bramley (mg/l)			
Phenolic acids	373	253	196
Epicatechin	27	7	2
Procyanidin B2	14	0	0
Phloridzin	39	32	30
Oxidised procyanidins	309	165	17
Total	762	457	245
Dabinett (mg/l)			
Phenolic acids	686	564	109
Epicatechin	308	165	7
Procyanidin B2	306	163	8
Phloridzin	195	183	2
Oxidised procyanidins	788	1185	93
Total	2283	2260	219

Coggins, 1975), and the ratio between the two may be cultivar specific (Coggins and Whiting, 1974). Phloridzin (phloretin glucoside) is the major dihydrochalcone but analogues with additional sugars such as xylose have also been found (Johnson et al., 1968; Timberlake, 1972). (−)Epicatechin is the dominant catechin derivative reported in all apples except 'Medaille d'Or', an unusual cider cultivar which contains equal amounts of (+) catechin. It may be noted that phloretin derivatives are virtually unique to the genus *Malus* in the plant kingdom and are not found in any other rosaceous fruits. By contrast, all other apple phenolics are widespread in other fruits and food-plants too.

Chlorogenic acid is the major phenolic of apple juice and as such is often incorrectly referred to as 'tannin'. Only the procyanidins, previously known as 'leucocyanidins' or 'anthocyanogens', are true tannins in the sense of being able to combine with proteins. They are therefore of great importance in the undesirable formation of PBH and the desirable aspect of apple juice 'mouth-feel' or astringency. It is now known that they form a range of epicatechin-based oligomers from the major dimer B2 up to the heptameric form, the higher members of the series being the most unstable but also the most astringent (Lea, 1983). Thus, measurements of chlorogenic acid or of 'total phenolics' in apple juice are relatively unhelpful in understanding the haze potential or the sensory properties of juice, except that fruit which is high in chlorogenic acid tends to be high in other polyphenols too. Analytical methods which distinguish between 'flavonoid' and 'non-flavonoid' phenols are more useful for this purpose since the procyanidins are flavonoids while the phenolic acids are not. Thus the analysis of juice phenolics by the Folin-Ciocalteau reagent, before and after the precipitation of flavonoids by formaldehyde, yields helpful results (Singleton, 1974). The controlled hydrolysis of procyanidins to the red pigment cyanidin, after selective adsorption onto a column of polyamide, is a potentially useful method too (Wucherpfennig and Millies, 1973), although it is difficult to make it truly quantitative. The best methods for analysing apple phenolics are those based on HPLC whereby the individual compounds can be separated and quantified (Wilson, 1981a; Lea, 1982; Spanos et al., 1990).

Apple juice phenolics change enormously on processing, principally due to oxidation during milling and pressing which is mediated by an active polyphenoloxidase (PPO) system. The procyanidins can be lost almost entirely by tanning onto the pulp and to some extent by precipitation from the juice, due to their facile polymerisation in the active quinoidal form which is produced by oxidation. Losses of chlorogenic acid are generally less, but final levels as low as 10 ppm chlorogenic in authentic apple juice have been reported (Van Buren et al., 1976). Such extreme losses are associated with depectinisation by added enzymes, containing side activities which cleave the chlorogenic acid into caffeic and quinic acids. Fining and concentration are additional sources of loss. Table 6.3 shows typical losses due to oxidation both in the presence and absence of pulp, while Table 6.4 shows the losses due to fining. Oxidative losses can be entirely prevented by the addition of an SO_2 solution during milling to give 100–250 ppm in the juice, but this is of no practical interest due to the sulphite residues remaining in the juice and the fact that enzymic generation of juice aroma (see below) is totally inhibited.

Hot water leaching (diffuser extraction at 50°C) will readily double the extraction of phloridzin and procyanidins from apple pulp (Lea and Timberlake, 1978; Haug and Gierschner, 1979) but the juice flavour quality is often poor and the oxidised procyanidins are potent PBH precursors. During concentrate storage for three years at 20°C, Grob (1958) showed losses of procyanidins (as formaldehyde precipitable polyphenols) of up to 25%. No such losses occurred at 0°C.

The effect of polyphenols on juice colour is interesting. Lea (1984) showed that the total colour of fresh apple juice is due to the enzymic oxidation products generated from phloridzin (25%), epicatechin (25%) and procyanidins (50%). The first two classes are chemically well-defined as orange yellow

Table 6.4 Typical polyphenol losses during gelatin fining of apple juice

Component	Loss (%)
Chlorogenic acid	8
Epicatechin	9
Phloridzin	12
Oligomeric procyanidins	17
Polymeric/oxidised procyanidins	31

pigments with λ_{max} at 420 nm in the visible spectrum (Goodenough *et al.*, 1983). The procyanidin oxidation products, however, are poorly defined with no definite absorbance maximum in the visible region and so have browner hues. Surprisingly, chlorogenic acid is not directly involved in colour formation although its quinone plays a critical role as a coupled oxidant in the oxidation of procyanidins, which are not themselves substrates for the polyphenoloxidase enzyme (Lea, 1991).

The final colour of a juice, however, is determined by a large number of different factors (Lea, 1984; Cummings *et al.*, 1986; Oszmianski and Sozynski, 1986). If juice is allowed to oxidise in the presence of pulp, the coloured compounds will be adsorbed, and successively pressed juices will become lighter after an initial peak of soluble colour. The total polyphenols will also diminish markedly. If bulk juice is oxidised with the minimum of solids, however, colour will increase progressively until the enzyme becomes inactivated by phenolic oxidation products after about 12 h (see Figure 6.5), although in this case the total polyphenol level remains largely unchanged. This effect is critically dependent on the amount of soluble enzyme present in the juice, since apple PPO tends to be tightly membrane bound although it becomes more soluble as the fruit matures on storage. Temperature is also critical for enzyme activity, reaching an optimum at around 40°C. Cold fruit pressed in late autumn at temperatures near freezing will often hardly brown at all. Acid juices, at pH 3, may also be well below the pH optimum for maximum PPO activity, usually pH > 4.5. Apple pulp stored in a tall hopper will often brown considerably on the top surface but will be quite unoxidised a few inches down where oxygen cannot penetrate. The presence of native ascorbic acid may also inhibit browning for a while. Addition of sulphur dioxide (> 150 ppm) during the first few minutes reduces juice colour absolutely, both by inhibiting PPO activity and by chemically reducing the chromophores of the phenolic pigments. As polymerisation and oxidation proceed, however, the oxidised procyanidin chromophore becomes progressively more resistant to SO_2 bleaching. Hence the addition of SO_2 to juice several hours after pressing will reduce the colour only by 50% or so.

Although it is easy to catalogue the factors which affect the colour of apple juice, it is much more difficult to predict accurately the colour which will be produced in any given case. When concentration is involved, the effects of non-enzymic (Maillard) browning during storage are overlaid on those of initial polyphenol oxidation. Maillard pigments cannot be significantly bleached by the addition of SO_2, although the development of further browning can be prevented. Ascorbic acid does not reduce the Maillard pigments at all and will actually stimulate the non-enzymic browning reaction. Non-enzymic polyphenol browning due to oxidation during storage may also be catalysed by traces of iron and copper.

Fining browned juices will usually reduce the colour somewhat, usually at the expense of some non-volatile flavour, by removing oxidised polyphenols.

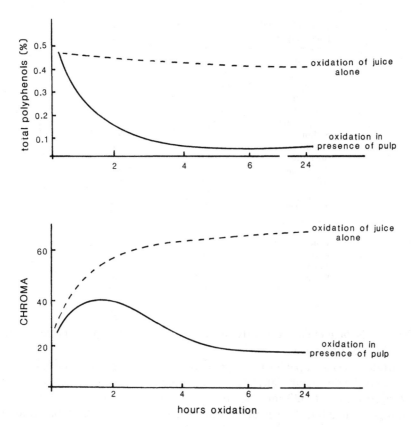

Figure 6.5 Colour development in Dabinett apple juices during oxidation in the presence and absence of pulp. Changes in total polyphenols for the same juices are also shown at the top of the diagram. Data from Lea (unpublished).

If the browning is mostly due to Maillard pigments, however, as in poorly stored concentrate, then fining will have very little effect and more drastic treatment is necessary with adsorbents such as activated charcoal. In such circumstances, considerable loss of both volatile and non-volatile flavour can be expected.

6.5.6 Minerals

Typical mineral levels found in a range of English juices, compared with levels found in clear commercial juices made from imported concentrate, are shown in Table 6.5 (Burroughs, 1984). Potassium and calcium are the dominant cations in both sources, although calcium and sodium are higher in rediluted concentrates due to the water used for their preparation. Iron and nickel are also higher in these samples, indicating metal pick-up during processing.

Table 6.5 Metal levels in natural and commercial apple juices (mg/l)

Element	Natural (mean)	Commercial (mean)
Potassium	1020	927
Calcium	95	131
Magnesium	45	50
Sodium	16	39
Barium	0.16	0.07
Strontium	0.17	0.32
Aluminium	1.05	1.6
Iron	1.53	4.7
Copper	0.22	0.27
Manganese	0.32	0.41
Nickel	0.26	0.81
Chromium	0.03	0.08
Lead	0.04	0.04
Zinc	0.40	0.22
Boron	2.5	1.6
Silicon	0.64	0.64

High levels of metals are nearly always indicative of contamination from packaging or processing.

Nitrate levels in apple juices are generally very low. Fraudulent addition of nitrate (as the potassium salt) has been described as a means of increasing the UV absorption at 320 nm, thereby defeating the rapid spectral test for chlorogenic acid but the elevated nitrate levels are readily detectable (Brause et al., 1986).

6.5.7 Volatile components

Several hundred components have been identified in flavour extracts of apples and apple juices (Dimick and Hoskin, 1981; Acree and McLellan, 1989; Paillard, 1990). Four major classes of compound found in apple juice may be distinguished, as listed in Table 6.6. The aroma of fresh apples is usually dominated by esters, which slowly increase during ripening and sometimes during storage. When the fruit is disrupted during milling, a series of very rapid enzymic changes leads to the generation of C6 aldehydes and alcohols (both saturated and unsaturated) by lipoxygenase action on saturated fatty acid precursors such as linoleic and linolenic acid. Drawert et al. (1986) have shown that these volatiles are generally at a maximum within 10–30 min of crushing. Simultaneously, esterases cleave some of the original fruit esters to their parent alcohols and acids. Terpene-like products such as 6-methyl 5-heptene-2-ol, octanediol and β-damascenone are also generated by hydrolysis of non-volatile precursors in the fruit, while benzaldehyde arises

Table 6.6 Typical volatiles of apple juice

Component	ppb in juice[a]
Esters	
Ethyl 2-methyl butyrate	40
Ethyl butyrate	300
Isopentyl acetate	50
Hexyl acetate	300
Hexyl 2-methyl butyrate	40
Aldehydes	
trans-2-Hexenal	1 500
Hexanal	1 500
Benzaldehyde	100
Alcohols	
Ethanol	30 000
Butanol	14 000
Isoamyl alcohol	1 500
Hexanol	3 000
trans-2-Hexenol	300
Process volatiles and miscellaneous	
Furfural	20
HMF	5 000
6-methyl 5-hepten-2-ol	30
β-Damascenone	20

[a] Quantitative data drawn from Poll (1985), Williams *et al.* (1980) and Lea and Ford (unpublished).

from amygdalin in the seeds (Schwab and Schreier, 1990). The effect of pasteurisation and other thermal processes then creates an additional range of 'process' volatiles such as furfural and HMF from sugar degradation. The final aroma of apple juice is therefore exceedingly complex and not necessarily characteristic of the fresh and intact fruit. Opalescent and natural style juices tend to be dominated by the 'green' note of unsaturated alcohols and aldehydes. Juices made from very aromatic fruits, e.g. Red Delicious, tend to retain high levels of esters. Juices made entirely from concentrates (without essence addition) are low in total volatiles and dominated by saturated alcohols and 'process volatiles'.

None of these compounds is specific to apple juice, however, and to some extent the characteristic aroma and flavour of any apple juice must rely on the blend of volatiles present rather than the presence of a particular 'impact' compound. Durr and Röthlin (1981), for instance, showed that a mixture of nine easily available compounds in water could be balanced to give an apple-like essence which was not recognised as synthetic when profiled together with natural essences by a sensory assessment panel. Odour-port gas chromatography by Flath *et al.* (1967) showed that hexanal, *trans*-2-hexenal and ethyl 2-methyl butyrate were scored as 'apple-like' and that the ester in particular was regarded as the character impact compound of 'Delicious' (i.e. Red Delicious) apples and essence. (It is unfortunate that this information has been misinterpreted by very reputable European workers (Drawert and Christoph, 1984; Dürr, 1981) and attributed to 'Golden Delicious', a totally different apple. In fact the level of ethyl 2-methyl butyrate in 'Golden Delicious' essence is particularly low (Wiley, 1985; Paillard, 1990).) The odour threshold of ethyl 2-methyl butyrate was found by Flath *et al.* to be 0.1 ppb, which is 10–100 times lower than the other esters and aldehydes found in the same essence and is 5000 times lower than butanol and hexanol. Lea and Ford (1991) used 'odour port dilution analysis' on commercial apple juices to show that ethyl 2-methyl butyrate was 30 times more important to juice aroma than hexanol, despite its 30-fold lower concentration in the headspace.

Damascenone, with an odour threshold in the parts per trillion range, was suggested as being important to the aroma of cooked 'Bramley's Seedling' by Nursten and Woolfe (1972), recently confirmed by Acree's group (Zhou *et al.*, 1993). Some apple cultivars (e.g. 'Ellison's Orange') are characterised by a 'spicy' or 'aniseed' aroma, which was identified as 4-methoxy allyl benzene (anethole) by Williams *et al.* (1977).

Several authors have tried to correlate specific volatiles or groups of volatiles with the hedonic flavour characteristics of juices. Jepsen (1978) showed that high juice flavour quality correlated positively with ethyl 2-methyl butyrate, *trans*-2-hexenal and butanol, but negatively with ethanol and ethyl acetate. Odour intensity was positively correlated with butanol, *trans*-2-hexenal and hexanol. Wiley (1985) considered that high levels of ethyl 2-methyl butyrate, hexanal, *trans*-2-hexenal, butanol and propyl butyrate should be sought when selecting the best apple cultivars for juicing. Poll (1985), working with stored single strength McIntosh juices, showed that a decrease in total aldehydes and esters correlated with a decrease in organoleptic quality. Such changes were less in juices stored at 3°C than at 20°C. The highest concentration of all volatiles was given by fruit pressed at the 'ripe for eating' stage. Immature fruit, or fruit stored for 5 months at 3°C, gave juices with generally fewer volatiles and lower sensory scores. Ethanol, however, increased more than 200-fold during fruit storage presumably due to anaerobic respiration of the fruit or perhaps due to incipient microbiological spoilage. In short, it would seem that the best quality juices are made from fruit which is also in peak condition for eating.

6.5.8 *Other flavour aspects*

The processor who packs from concentrate and essence is able to tailor his product to specific requirements in terms of aroma, acid, °Brix and tannin levels. Processors of single strength or 'natural style' juice are more limited in scope but can exert some influence on flavour specification by their choice of fruit varieties before juicing. It may be difficult, however, to decide exactly what that flavour specification ought to be. It is generally understood that consumers in the northern states of Europe or North America prefer a lower sugar/acid ratio than do consumers in the south. To some extent, therefore, the ideal balance has to follow these regional variations, which are presumably based historically on the composition of the fruit actually grown in these climates. Even within a small country, however, preferences may vary. Williams *et al.* (1983) for instance, surveyed 135 visitors to a UK agricultural show by presenting them with seven blended Cox/Bramley juices with different sweetness/acidity ratios. He showed that two approximately equal sub-populations emerged, one preferring a specific gravity of 1.042 and an acid of 0.55–0.68% malic, the other preferring a sweeter and less acid juice of specific gravity 1.045 and an acidity of 0.48%. These represent sugar/acid ratios of about 17 and 23, respectively. In New York State and Pennsylvania, typical ratios for blended juices produced commercially lie between 32 and 41, the acidity being somewhat lower than in the UK at about 0.3–0.4%. La Belle (1978) found sugar/acid ratios of 24–62 in single cultivar New York State juices, with a ratio of 35 being regarded as most desirable. Typical ratios set by British supermarkets and contract packers for cloudy 'natural style' juice lie between 14 and 28, with the lower end of the range being the commonest specification. On this basis it would seem that the British preference is for a markedly more acid juice than the American.

Schobinger and Müller (1975) assessed single strength juices from 15 different cultivars in Switzerland and showed that a ratio of 15–18 was regarded as optimum by their panel. Poll (1981), in a similar study of 18 Danish cultivars, found an optimum ratio of 15–16. Both groups also examined the optimum level of total polyphenols, which lay between 300 and 750 ppm in the Swiss study, and 200 and 900 ppm in the Danish work. Juices with lower or higher levels were assessed as too insipid or too astringent, respectively. Schobinger and Müller then plotted their optimum levels on a map shown in Figure 6.6. To make use of this figure, a line is drawn between the sugar content on the top axis and the acid content on the bottom. The polyphenol level is then projected horizontally to give a point of intersection. 'Harmonious' juices intersect within the circle whereas unbalanced juices fall outside it.

In Germany and Switzerland, formal schemes for the sensory assessment of fruit juices have been laid down by the Deutsche Landwirtschaft-Gesellschaft and the Schweiz. Obstverband, respectively. The Swiss scoring sheet is shown in Table 6.7. Each system results in a total score for overall quality. Schobinger and Müller used data from the Swiss scoring scheme to show that juices high in overall quality nearly always fell within the 'harmonious' circle of Figure 6.6, and that low scoring juices nearly always fell outside it. As might be expected, this confirms that both volatile and non-volatile flavour attributes play a part in determining overall juice quality. Unfortunately, there is no single analytical measurement which will acceptably represent the

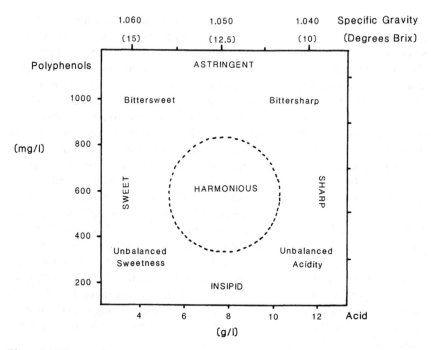

Figure 6.6 Diagram showing the optimum levels of sugar, acid and polyphenols in apple juices (adapted from Schobinger and Müller, 1985). The usage of this diagram is explained in the text.

volatile flavour in the same way that sugar, acid and polyphenols do for the non-volatiles.

It may be objected that the application of such scoring techniques is too mechanical or too simplistic to be of real value. Certainly it does not approach the much greater depth of a full-scale sensory profile analysis (Williams and Carter, 1977; Dürr, 1979). On the other hand, it is perhaps surprising that the apple juice industry in Britain and North America often does very little formal sensory testing at all, other than a quick check for gross taints and other major flavour defects.

6.5.9   Microbiology

The natural microflora of apple juice is dominated by yeasts (Beech, 1993). It is not generally appreciated that a variety of weakly fermenting yeasts (e.g. *Kloeckera apiculata*) is actually present inside the apple flesh and may rise as high as 4.5×10^4 cells/g in stored fruit. Under certain conditions, similar numbers of acid tolerant bacteria may also be found on the fruit surface. However, milling and pressing equipment that is not cleaned daily is the major source of inoculation, so that freshly pressed juices may contain 10^6 yeast cells/ml (Beech and Carr, 1977). Fermentation will therefore begin

Table 6.7 Sensory scoring scheme of the Swiss Fruit Association

	Points[a,b]
A. Clarity and colour	
Clarity: Clear, optically bright	1½
Slight sheen, light haze	1
Hazy	½
Colour: Normal	1½
Unnaturally dark or light	1
Totally uncharacteristic	½
B. Aroma	
Clean, fruity, typical, high-quality	5
Clean, full	4
Clean, weak, not typical	3
Unclean	2
Tainted, unusual, intrusive	1
C. Flavour	
Clean, harmonious, full-bodied, typical	5
Clean, fruity	4
Clean, not balanced, lacking in finish	3
Unclean	2
Defective, tainted	1
D. Overall impression	
Excellent	5
Adequate	4
Room for improvement	3
Inadequate	2
Unusable	1

[a] Half points can be given.
[b] Classification (based on total score):

Excellent	18–16½
Very good commercial quality	16–15
Good commercial quality	14½–13½
Satisfactory	13–12
Unsatisfactory	11½–10½
Unusable	10–9
Offensive, tainted	Below 9.

within a few hours if juices are not kept cool or promptly processed. With normal pasteurisation procedures the yeasts are destroyed and fermentation is prevented. Mould spores may be present up to 10^5/ml, however, and are often more heat resistant than yeasts (Dittrich, 1987). Occasional mould growth is found in juices which have not been adequately heated, and usually takes the form of a cotton-wool like colony floating in the juice. *Byssochlamys fulva* is the most heat-resistant mould to be encountered, whose ascospores will reportedly survive in apple juice for 180 min at 85°C (Swanson, 1989). Bacterial spoilage in processed juices is uncommon; it is generally due to acetic or lactic acid bacteria and is detectable by the off-flavours generated by

Table 6.8 Typical microorganisms associated with apple juice

Yeasts	Moulds	Bacteria
Candida spp.	Penicillium spp.	Lactobacillus spp.
Kloeckera spp.	Aspergillus spp.	Leoconostoc spp.
Saccharomyces spp.	Paecilomyces spp.	Acetobacter spp.
Torulopsis spp.	Byssochlamys spp.	Gluconobacter spp.
Rhodotorula spp.		
Zygosaccharomyces spp.		

these organisms (Carr, 1975). Most spoilage organisms can be adequately controlled by sterilisation. Heat-resistant moulds, yeasts and bacteria were found at typically less than 1 per 10 g in a survey of apple fruit, pulp and juice in New York State (Swanson et al., 1985). Microbial spoilage is very much increased as the acidity falls and apple juices should always be brought below pH 3.8 before bottling (Table 6.8).

Apple juice concentrates, although nominally stable because of their low water activity, nevertheless have a specialised microflora of their own. Osmophilic yeasts such as Zygosaccharomyces rouxii and Z. bailii will slowly ferment in stored concentrates. Z. bailli will grow even in the presence of 2000 ppm SO_2 and > 600 ppm benzoate (Davenport, 1981). Such organisms are found particularly on the surface of concentrates where condensed water vapour tends to cause a slight dilution. Not only are they a problem in their own right but they can also act as a preservative-resistant (though not heat-resistant) inoculum for diluted beverages if contamination occurs in downstream processing.

The low pH of apple juice generally means that pathogenic microorganisms, if present, cannot grow. Even so, occasional outbreaks of Salmonella-based food poisoning have been associated with unpasteurised apple juice in the USA (Swanson, 1989). Such outbreaks are rare and are caused by unwashed and damaged fruit in contact with animal manures from the orchard. A survey of UK cider apple juices by Goverd et al. (1979) showed more potential pathogens and soil-borne bacteria than expected, possibly due to the use of manure slurries and mechanical harvesting in such orchards, especially from bare soil. For juice production, therefore, the highest standards of orchard management and hygiene during fruit handling are required.

A more frequent public health concern has been the presence of mycotoxins, principally patulin (see chapter 13). This is produced in apples by the growth of secondary rotting fungi such as Penicillium, Aspergillus and Byssochlamys (Doores, 1983). In fully infected fruit, patulin levels of 200 ppm can be recorded, but in well-made apple juices patulin levels are in the low parts per billion range (ppb). Although patulin can be mutagenic to hamster cells in culture and toxic to animals at high levels, there is no good evidence for its carcinogenicity. Applying the usual 100-fold safety factor from animal studies, the WHO/FAO JECFA committee (1990) concluded that a provisional Tolerable Weekly Intake would be 7 µg/kg body weight. In 1993 an advisory limit of 50 ppb patulin in apple products was set by MAFF in the UK, similar to that in the USA and in most other countries where limits exist. Since patulin levels cannot be reduced significantly by pasteurisation or by preservative treatment (Wheeler *et al.*, 1987), this level must be achieved by proper fruit storage, followed by stringent grading and sorting of apples prior to juicing. Detailed guidelines are given in a Code of Practice issued by the British Soft Drinks Association (1994). In the UK, at least, all commercial apple juice must now be analysed to ensure that patulin levels are below 50 ppb before 'positive release' to trade.

6.6 Juice tests

Test for the presence of pectin in clarified juice. One part of single strength juice is mixed at 5–25°C with two parts of 96% ethanol containing 1% conc. HCl. If flocculation has not occurred within 15 min, then depectinisation is sufficient.

Test for the presence of starch. 10 ml juice are heated to 70°C to dissolve any starch granules present, and then cooled. A few drops of iodine solution are then added (the solution containing 1% iodine and 10% potassium iodide in water). Blue, violet or red coloration demonstrates that starch is still present. Enzyme treatment should be continued beyond the point where starch is apparently absent.

Test fining with gelatin. A 1% stock solution of fining grade gelatin should be made up by swelling 1 g in 5 ml cold water and then adding 95 ml of water at 50–80°C. The gelatin should be stirred to dissolve, cooled to room temperature and aged for a further 4 h. Do not keep the solution longer than 24 h or it may degrade.

Set up a number of identical flasks each containing 250 ml juice. Add a range of volumes of gelatin solution (e.g. 0.6, 1.25, 2.5, 5.0, 10.0, 20.0 ml) and mix well. Leave to stand at 15°C or whatever the fining temperature will be. After several hours, select whichever flask gives the greatest clartiy with the least amount of gelatin, and scale up for the factory in direct proportion (e.g. 25, 50, 100, 200, 400, 800 g gelatin per 1000 l of juice). Remember that the gelatin must be properly dissolved at 5% in demineralised water before adding to the bulk juice.

Test fining with gelatin/kieselsol. Make up a gelatin stock solution as described above, and a kieselsol stock solution by diluting commercial kieselsol (30% solution) 1:10 with water. Set up 250 ml flasks of juice as shown

Table 6.9 Test fining with gelatin/kieselsol

Diluted kieselsol (ml)	1% Gelatin (ml)	Undiluted kieselsol (ml per 1000 l)	Gelatin (g per 1000 l)
0.63	0.3	250	12.5
1.25	0.63	500	25
1.25	1.25	500	50
2.5	2.5	1000	100
2.5	5.0	1000	200

in the left-hand columns of Table 6.9, adding the kieselsol, mixing well, adding gelatin solution and mixing again. Then leave at fining temperature. After 3 h, assess for maximum clarity and scale up using the right-hand columns. (If necessary, the test can be repeated using a wider range of kieselsol concentrations at the optimum gelatin level). Remember the kieselsol and gelatin must be added sequentially, never together. The order of addition is generally kieselsol followed by gelatin, although if the purpose of fining is mainly to reduce procyanidins then some authorities claim that the gelatin should be added first (Table 6.9).

Similar tests may be set up for combination fining with bentonite, using levels of 125 and 500 mg in the test sample which correspond to 500 and 1000 g per 1000 l of juice respectively. Further details of fining procedures are tabulated in Lehmann (1983).

Test for overfining. A diluted kieselsol solution (as above) may be used to test for excess soluble gelatin, by adding two or three drops of solution to 5 ml of juice. The appearance of a haze or precipitate during the next 20 min implies the presence of excess gelatin.

6.7   Note

Advice on apple juice production is available in Germany, Switzerland and Austria from the research stations at Geisenheim, Wädenswil and Klosterneuberg, respectively. In France the appropriate centre is the INRA Station de Recherches Cidricoles at Le Rheu, near Rennes. In New Zealand, the DSIR Horticultural Processing Section in Auckland should be consulted. In the USA, advice is available from the Geneva Experiment Station, N.Y. and Oregon State University, Corvallis. In Canada the appropriate sources are the Vineland Research Station, Ontario, and the Summerland Research Station, BC. There is no publicly funded source of advice in the UK, although Reading Scientific Services offers consultancy and analysis relating to apple juice production, as do other independent consulting scientists and research organisations.

References

Acree, T. E. and McLellan, M. R. (1989) In: *Processed Apple Products*, D. L. Downing (ed.), AVI Van Nostrand, New York, pp. 323–342.

Anon (1984) *Food Engineering International*, September, 69.

Associated Press, 16th June, 1988.

Babsky, N. E., Toribio, J. L. and Lozano, J. E. (1986) *J. Food Sci.* 51, 564–567.

Bannach, M. (1984) *Confructa*, May/June, 198–206.

Barefoot, S. F., Tai, H. Y., Brandon, S. C. and Thomas, R. L. (1989) *J. Food Sci.* 54, 408–411.

Bauernfeind, J. (1958) *Report of the International Fruit Juice Union Conference*, Bristol, pp. 159–186.

Baumann, J. W. (1981) In: *Enzymes and Food Processing*, G. G. Birch, N. Blakeborough and K. J. Parker (eds.), Elsevier Applied Science, London, pp. 129–147.

Beech, F. W. (1993) In: *The Yeasts*, Vol 5, *Yeast Technology*, 2nd edn., A. H. Rose and J. S. Harrison (eds.), Academic Press, London, pp. 169–213.

Beech, F. W. and Carr, J. G. (1977) In: *Economic Microbiology*, Vol 1, *Alcoholic Beverages*, A. H. Rose (ed.), Academic Press, London, pp. 139–313.

Beruter, J. (1985) *Flüss. Obst.* 52, 454–459.

Beveridge, T. and Tait, V. (1993) *Food Structure* 12, 195–198.

Brause, A. R., Raterman, J. M., Doner, L. W. and Hill, E. C. (1986) *Flüss. Obst.* 53, 15–22.

Brause, A. R. and Raterman, J. M. (1982) *J. Assoc. Off. Anal. Chem.* 65, 846–849.

British Soft Drinks Association (1994) *A Code of Practice for the Production of Apple Juice*, British Soft Drinks Association, London.

Burroughs, L. F. (1974) *Report of Long Ashton Research Station*, Bristol, pp. 151–153.

Burroughs, L. F. (1984) *Report of the International Fruit Juice Union Conference*, Tel Aviv, pp. 131–138; also in *Flüss. Obst.* (1984) 51, 370–393.

Carr, J. G. (1975) *Rev. Microbiol.* (Sao Paulo) 6, 18–56.

Churms, S. C. *et al.* (1983) *Carbohydr. Res.* 113, 339–344.

Coggins, R. A. and Whiting, G. C. (1974) *Report of Long Ashton Research Station*, Bristol, pp. 146–147.

Cornwell, C. J. and Wrolstad, R. E. (1981) *J. Food Sci.* 46, 515–520.

Cummings, D. B., Beveridge, H. J. T. and Gayton, R. (1986) *Can. Inst. Food Sci. Technol. J.* 19, 223–226.

Davenport, R. R. (1981) *Report of Long Ashton Research Station*, Bristol, p. 170.

Delcour, J. A., Dondeyne, P., Trousdale, E. and Singleton, V. L. (1982) *J. Inst. Brew.* 88, 234–243.

De Vries, J. A., Voragen, A. G. J., Rombouts, E. M. and Pilnik, W. (1986) In: *Chemistry and Function of Pectins*, American Chemical Society, Washington DC, pp. 38–48.

Dietrich, H., Wucherpfennig, K. and Maier, G. (1990) *Flüss. Obst.* 57(2), 68–73.

Dimick, P. S. and Hoskin, J. C. (1981) *CRC Crit. Rev. Food Sci. Nutr.* 18, 387–409.

Dittrich, H. (1987) In: *Fruchte und Gemusesäfte*, 2nd edn., U. Schobinger (ed.), Ulmer Verlag, Stuttgart, pp. 470–518.

Doner, L. W. (1988) In: *Adulteration of Fruit Juice Beverages*, S. Nagy, J. A. Attaway and M. E. Rhodes, (eds.), Marcel Dekker, New York, pp. 125–138.

Doores, S. (1983) *CRC Crit. Rev. Food Sci. Nutrition* 19, 133–149.

Dörreich, K. (1986) *Report of the International Fruit Juice Union Conference*, The Hague, pp. 183–197.

Drawert, F. and Christoph, N. (1984) In: *Analysis of Volatiles*, P. Schreier (ed.), Walter de Gruyter, Berlin, pp. 269–291.

Drawert, F., Kler, A. and Berger, R. C. (1986) *Lebensm.-Wiss. Technol.* 19, 426–431.

Dryden, E. C., Buch, M. L. and Hills, C. H. (1955) *Food Technol.* 9, 264–268.

Ducroo, P. (1987) *Flüss. Obst.* 54, 265–269.

Dürr, P. (1979) *Lebensm.-Wiss. Technol.* 12, 23–26.

Dürr, P. (1981) In: *Criteria of Food Acceptance*, J. Solm and R. L. Hall (eds.), Forster Verlag, Zurich.

Dürr, P. (1986) In *Food Flavour B.—The Flavour of Beverages*, I. D. Morton and A. G. Macleod (eds.), Elsevier, Amsterdam, pp. 85–97.

Dürr, P. and Rothlin, M. (1981) *Lebensm.-Wiss. Technol.* 14, 313–314.

Elkins, E., Heuser, J. R. and Chin, H. (1988) In: *Adulteration of Fruit Juice Beverages*, S. Nagy,

J. A. Attaway and M. E. Rhodes (eds.), Marcel Dekker, New York, pp. 317–341.

Endo, A. (1965) *Agric. Biol. Chem.* 29, 129–136, 137–143, 222–228, 229–238.

Endress, H. U., Tableros, M., Omran, H. and Gierschner, K. (1986) *Report International Fruit Juice Union Conference*, The Hague, pp. 293–306.

Evans, R. H., van Soestbergen, A. W. and Ristow, K. A. (1983) *J. Assoc. Off. Anal. Chem.* 66(6), 1517–1520.

Fischer, G. (1981) *Flüss. Obst.* 48(11), 516–517, 526–530.

Flath, R. A., Black, D. R., Guadagni, D. G., McFadden, W. H. and Schultz, T. H. (1967) *J. Agric. Food Chem.* 15(1), 29–35.

German Fruit Juice Association (Verband der deutschen Fruchtsaftindustrie) (1987) *RSK Values—The Complete Manual*, Flüssiges Obst. Verlag, Schönborn, Germany.

Gierschner, K. (1979) *Flüss. Obst.* 46(8), 292–298.

Gierschner, K., Valet, R. and Endress, H. (1982) *Flüss. Obst.* 49, 574–582.

Goodenough, P. W., Kessel, S., Lea, A. G. H. and Loeffler, T. (1983) *Phytochemistry* 22, 359–363.

Görtges, S. (1982) *Flüss. Obst.* 49, 582–587.

Goverd, K. A., Beech, F. W., Hobbs, R. P. and Shannon, R. (1979) *J. Appl. Bacteriol.* 46, 521–530.

Grampp, E. (1977) *Food Technol.* (11), 38–41.

Grampp, E. (1981) *Process Biochem.* 16(3), 24–30.

Grob, A. (1958) *Report International Fruit Juice Union Conference*, Bristol, pp. 229–244.

Hammond, D. A. (1994) In: *Modern Methods for Detection of Fruit Juice Adulteration*, S. Nagy and R. Wade (eds.), AgScience, Auburndale, Florida.

Haug, M. and Gierschner, K. (1979) *Dtsch. Lebensm.-Rundsch.* 75, 248–253, 274–276.

Heatherbell, D. A. (1976) *Alimenta* 15, 151–154.

Herrmann, K. (1987) In: *Frucht und Gemüsesäfte*, 2nd edn., U. Schobinger (ed.), Ulmer Verlag, Stuttgart, pp. 39–88.

Hsu, J. C., Heatherbell, D. A. and Yorgey, B. M. (1989) *J. Food Sci.* 54, 660–662.

Hulme, A. C. (1958) *Adv. Food Res.* 8, 297–413.

Hulme, A. C. and Rhodes, M. J. C. (1971) In: *Biochemistry of Fruits and their Products*, Vol. 2, A. C. Hulme (ed.), Academic Press, London, pp. 333–373.

IFJU (1985) *Methods of Analysis Handbook*, International Fruit Juice Union, Paris.

Jacquin, F. (1960) *Industr. Aliment. Agric.* 77, 663–681.

Janda, W. (1983) In: *Industrial Enzymology*, Macmillan, London, pp. 315–320.

JECFA (1990) *Report of 35th Meeting—Food Additive Series no. 26*, Joint Expert Committee on Food Additives, WHO, Geneva, pp. 143–165.

Jepsen, O. (1978) *Report International Fruit Juice Union Conference*, Bern, pp. 349–362.

Johnson, G. (1969) *Food Technol.* 23, 1312–1316.

Johnson, G., Donnelly, B. J., Johnson, D. K. (1968) *J. Food Sci.* 33, 254–257.

Junge, C. and Spädinger, C. (1982) *Flüss. Obst.* 49, 57–62.

Kern, A. (1962) *Schweiz Z. Obst.-Weinbau* 71, 190–194.

Kilara, A. (1982) *Process Biochem.* July/Aug, 35–41.

Kilara, A. and Van Buren, J. P. (1989) In: *Processed Apple Products*, D. L. Downing (ed.), AVI Van Nostrand, New York, pp. 83–96.

Koch, J. (1984) *Confructa* 28, 55–62, 63–73 (see also *Flüss. Obst.* (1984), 510–518, 545–547).

Krueger, D. A. (1988) In: *Adulteration of Fruit Juice Beverages*, S. Nagy, J. A. Attaway and M. E. Rhodes (eds.), Marcel Dekker, New York, pp. 109–124.

Krug, K. (1968) *Flüss. Obst.* 35, 322–328.

Krug, K. (1969) *Flüss. Obst.* 36, 277–283, 333–343.

La Belle, R. (1978) Quoted by Acree and McLellan (1989), Way and McLellan (1989).

Lea, A. G. H. (1974) *J. Sci. Food Agric.* 25, 403–410.

Lea, A. G. H. (1978) *J. Sci. Food Agric.* 29, 471–477.

Lea, A. G. H. (1982) *J. Chromatogr.* 238, 253–257.

Lea, A. G. H. (1983) In: *Sensory Quality in Foods and Beverages*, A. A. Williams and R. K. Atkin (eds.), Ellis Horwood, Chichester, pp. 203–213.

Lea, A. G. H. (1984) *Flüss. Obst.* 51, 356–361.

Lea, A. G. H. (1991) In: *Enzymes in Food Processing*, G. A. Tucker and F. J. Woods (eds.), Blackie, Glasgow, pp. 194–220.

Lea, A. G. H. (1994) Cidermaking. In: *Fermented Beverage Production*, A. G. H. Lea and J. R. Piggott (eds.), Blackie, Glasgow, pp. 66–96.

Lea, A. G. H. and Ford, G. D. (1991) *Fruit Processing* 1(3), 29–32 (supplement to *Flüss. Obst.* 58(3)).

Lea, A. G. H. and Timberlake, C. F. (1978) *J. Sci. Food Agric.* 29, 484–492.

Lee, C. Y. and Mattick, L. R. (1989) In: *Processed Apple Products*, D. L. Downing, AVI Van Nostrand, New York, pp. 303–322.

Lee, H. S. and Wrolstad, R. E. (1988a) In: *Adulteration of Fruit Juice Beverages*, S. Nagy, J. A. Attaway and M. E. Rhodes (eds.), Marcel Dekker, New York, pp. 343–376; see also *J. Assoc. Off. Anal. Chem.* 71, 789–794.

Lee, H. S. and Wrolstad, R. E. (1988b) *J. Assoc. Off. Anal. Chem.* 71, 795–797.

Lehmann, H. (1983) *Fruchtsaftklärung*, VEB Fachbuchverlag, Leipzig.

Low, N. and Swallow, A. (1991) *Flüss. Obst.* 58, 2–5.

Lüthi, H. (1958) *Report International Fruit Juice Union Conference*, Bristol, pp. 383–390.

McKenzie, D. L. and Beveridge, T. (1988) *Food Microstructure* 7, 195–203.

McLellan, M. R., Kime, R. W. and Lind, L. R. (1985) *J. Food Sci.* 50, 206–208.

Mattick, L. R. (1988) In: *Adulteration of Fruit Juice Beverages*, S. Nagy, J. A. Attaway and M. A. Rhodes (eds.), Marcel Dekker, New York, pp. 175–194.

Mattick, L. R. and Moyer, J. C. (1983) *J. Assoc. Off. Anal. Chem.* 66, 1251–1255.

Möslang, H. (1984) *Confructa* 28(3), 219–224.

Moyer, J. C. and Aitken, H. C. (1971) In: *Fruit and Vegetable Juice Processing Technology*, D. K. Tressler and M. A. Joslyn (eds.), AVI, New York, pp. 186–233.

Nagel, C. W. and Schobinger, U. (1985) *Confructa* 29, 16–22.

Nursten, H. E. and Woolfe, M. L. (1972) *J. Sci. Food Agric.* 23, 803–822.

Oszmianski, J. and Sozynski, J. (1986) *Acta Aliment. Pol.* 12, 11–20.

Otto, K., Görtges, S. and Jost, V. (1985) *Flüss. Obst.* 52, 471–477.

Paillard, N. M. (1990) In: *Food Flavours C–The Flavour of Fruits*, I. D. Morton and A. G. Macleod (eds.), Elsevier, Amsterdam, pp. 1–41.

Pederson, C. S. (1947) *Fruit Prod. J.* 26, 294–313.

Pilnik, W. and Piek-Faddegan, M. (1970) *Schweiz. Z. Obst.-Weinbau.* 106(6), 133–137.

Pilnik, W. and Rombouts, F. M. (1981) In: *Enzymes in Food Processing*, G. G. Birch, N. Blakeborough and K. J. Parker (eds.), Elsevier Applied Science, London, pp. 105–128.

Poll, L. (1981) *J. Sci. Food Agric.* 32, 1081–1090.

Poll, L. (1985) *Lebensm.-Wiss. Technol.* 18, 205–211.

Poll, L. and Flink, J. (1983) *Lebensm.-Wiss. Technol.* 16, 215–219.

Pollard, A. and Timberlake, C. F. (1971) In: *Biochemistry of Fruits and their Products*, Vol 2, A. C. Hulme (ed.), Academic Press, London, pp. 573–621.

Possmann, P. (1986) *Report International Fruit Juice Union Conference*, The Hague, pp. 1–7.

Proulx, A. and Nichols, L. (1980) *Sweet and Hard Cider*, Garden Way Publishing, Pownal, Vermont, USA.

Rettinger, K., Weber, B. and Mosandl, A. (1990) *Z. Lebensm. Unters. Forsch.* 191, 265–268.

Schmitt, R. (1985) *Confructa* 29, 22–26.

Schobinger, U. (1987) *Frucht und Gemüsesäfte*, 2nd edn., Ulmer Verlag, Stuttgart.

Schobinger, U. and Müller, W. (1975) *Flüss. Obst.* 42, 414–419.

Schork, K. (1986) *Report International Fruit Juice Union Conference*, The Hague, pp. 177–182.

Schwab, W. and Schreier, P. (1990) *J. Agric. Food Chem.* 38, 757–763 (see also 36, 1238–1242).

Singleton, V. L. (1974) In: *Chemistry of Winemaking*, A. D. Webb (ed.), American Chemical Society, Washington DC, pp. 184–211.

Soto-Peralta, N., Müller, H. and Knorr, D. (1989) *J. Food Sci.* 54, 495–496.

Spanos, G. A., Wrolstad, R. E. and Heatherball, D. A. (1990) *J. Agric. Food Chem.* 1572–1579.

Stähle-Hamatschek, S. (1989) *Flüss. Obst.* 56, 543–558.

Swanson, K. M. J. (1989) In: *Processed Apple Products*, D. L. Downing (ed.), AVI Van Nostrand, New York, pp. 343–364.

Swanson, K. M. J., Leasor, S. B. and Downing, D. L. (1985) *J. Food Sci.* 50, 336–339.

Tanner, H. and Duperrex, M. (1968) *Fruchtsaft Ind.* 13, 98–114.

Tanner, H. and Brunner, H. R. (1987) *Getränke Analytik*, 2nd edn., Heller Verlag, Schwäbisch Hall.

Timberlake, C. F. (1972) *Report Long Ashton Research Station*, Bristol, pp. 145–146.

Van Buren, J. P. (1989) In: *Processed Apple Products*, D. L. Downing (ed.), AVI Van Nostrand, New York, pp. 97–120.

Van Buren, J. P. and Way, R. D. (1978) *J. Food Sci.* 43, 1235–1237.
Van Buren, J. P., de Vos, L. and Pilnik, W. (1976) *J. Agric. Food Chem.* 24, 448–451.
Voragen, A. G. J., Wolters, H., Verdonschot-Kref, T., Rombouts, F. M. and Pilnik, W. (1986) *Report International Fruit Juice Union Conference*, The Hague, pp. 453–462.
Wakayama, T. and Lee, C. Y. (1987) *Food Chem.* 25, 111–116.
Way, R. D. and McLellan, M. R. (1989) In: *Processed Apple Products*, D. L. Downing (ed.), AVI Van Nostrand, New York, pp. 1–30.
Weiss, J. and Sämann, H. (1985) *Mitt. Klosterneuberg* 35, 199–202.
Weiss, J., Sämann, H. and Jasanek, R. (1973) *Mitt. Klosterneuberg* 23, 367–378.
Wheeler, J. L., Harrison, M. A. and Koehle, P. E. (1987) *J. Food Sci.* 52, 479–480.
Whiting, G. C. (1960) *Report Long Ashton Research Station*, Bristol, p. 135.
Whiting, G. C. and Coggins, R. A. (1975) *Phytochemistry* 14, 593.
Wiley, R. C. (1985) *Confructa* 29, 112–119.
Williams, A. A. and Carter, C. S. (1977) *J. Sci. Food Agric.* 28, 1090–1104.
Williams, A. A., Tucknott, O. G. and Lewis, M. J. (1977) *J. Sci. Food Agric.* 28, 185–190.
Williams, A. A., Lewis, M. J. and Tucknott, O. G. (1980) *Food Chem.* 6, 139–151.
Williams, A. A., Langron, S. P. and Arnold, G. M. (1983) In: *Sensory Quality in Foods and Beverages*, A. A. Williams and R. K. Atkin (eds.), Ellis Horwood, Chichester, pp. 310–323.
Wilson, E. L. (1981) *J. Sci. Food Agric.* 32, 257–264.
Wucherpfennig, K. and Millies, K. (1973) *Flüss. Obst.* 40, 48–52.
Wucherpfennig, K., Possmann, P. and Kettern, W. (1972) *Flüss. Obst.* 39, 388–406.
Yamasaki, M., Kato, A., Chu, S. Y. and Arima, K. (1967) *Agric. Biol. Chem.* 31, 552–560.
Zhou, P. G., Cox, J. A., Roberts, D. D. and Acree, T. E. (1993) In: *Progress in Flavour Precursor Studies*, P. Schreier and P. Winterhalter (eds.), Allured, Illinois, pp. 261–273.
Zitko, J. and Rosik, J. (1962) *Die Nahrung* 6, 561–573.

7 Equipment for extraction and processing of soft and pome fruit juices

J. W. DOWNES

7.1 Introduction: modern juice processing methods

Fruit can be delivered fresh, frozen or from chilled storage to the fruit juice processor. The method of handling is dictated by the form in which it is received. Fresh fruit such as apples and pears are easily handled by the use of water or conveyor belts but soft fruit such as raspberries and strawberries are often received frozen requiring a defrost process prior to enzyming and pressing.

Prior to juice extraction the fruit has to be pulped to release the trapped juices and in the case of apples and pears needs to be pressed at fairly high pressure to force the juice to flow through the cell structure. In the case of berry fruits, cells are softer and such pressing inhibits juice flow so enzyme treatment is needed to break down the cellular structure to create juice flow channels. In certain fruits, press aids such as rice husks or Silva Cel tree bark have been found to create juice channels when added to the pulp.

Hand picked fruit is generally of good quality and needs little pre-press treatment, but the growth of automatic harvesting, particularly with apples, creates many new problems of debris and dirt such as leaves, stones and twigs, all of which need to be removed prior to pulping.

Fruit pressing can be split into two main areas of batch or continuous pressing. With increasing labour costs, continuous pressing with automatic control of fruit feed rate, milling and residue disposal is the obvious goal but this must be matched up to juice yield, plant capital cost, running costs and other costs such as effluent disposal.

Once the juice has been extracted and placed in storage, it will need considerable treatment before being acceptable to the customer. Insoluble solid contents of pressed juices are often high, up to 5%, and most customers need this figure reduced. In addition, clarity and stability in the finished product is often needed and a whole range of clarification and filtration methods can be used to create this state. Pasteurisation is also often required to stabilise the product by removing microorganisms that could produce fermentation and/or spoilage, affecting clarity, taste and shelf life.

Fruit, by its very nature, is seasonal but demand nowadays is fairly constant. Storage of fruit is often not practicable as its quality deteriorates while storage of natural strength juice is expensive and has potential quality problems. Methods of concentrating the juice up to 8–10 times its original strength by evaporation have become quite sophisticated as has the removal of aromatic substances for separate storage. This is beneficial for cost savings in storage and transport and for stability reasons.

Many available pressing, clarification and concentration methods are described in this chapter. The choice of which is most suitable is dependent on the raw material provided and the customer's requirements. Because fruit is subject to natural forces such as soil quality, rainfall, sun, etc. modifications to the methods must be continually assessed to ensure that optimum yield and quality are maintained.

7.2 Juice extraction systems

Juice extraction requires equipment for handling and cleaning fruit, for reducing it to pulp (mash) and for pressing the pulp to expel the juice. A typical modern installation for soft fruit processing is shown in Figure 7.1. The specific equipment used will depend on the fruit being processed. In the following description, apples are taken as an example and reference is made to essential variants for other fruit.

7.2.1 Fruit storage and handling

Equipment required for apple juice production begins with fruit harvesting, transport and washing facilities. It is unusual for fruit to be grown specifically for juice production except in Central Europe, and most juice is derived from second-grade fruit which, though entirely wholesome, is unsaleable on the dessert market because of factors such as size or shape. In the United States, where there is a significant commercial apple sauce industry, such juice is produced from misshapen fruit which is too irregular to pass through the mechanical peelers and corers. Lower quality juice is also produced from the peels and cores themselves. All fruit used for juicing must be sound and free from gross damage or contamination. In particular, it should be free from mould or rot which can lead to tainted juices. Certain types of mechanical harvesting equipment can lead to bruising and skin penetration which, without precautions, could cause off-flavours or even the growth of pathogenic microorganisms. This is particularly the case where animal manure slurries have been used in the orchard, or where the fruit comes into contact with bare soil during harvesting. In such cases, it is standard practice for fruit to be washed with clean water in the orchard before transport to the factory.

On receipt at the factory, apples are usually stored on concrete pads or pits before conveying to the mills. Since apples float in water, whereas orchard debris tends to sink, the transport of fruit to the mills is usually achieved at or below ground level via water-filled canals or flumes which

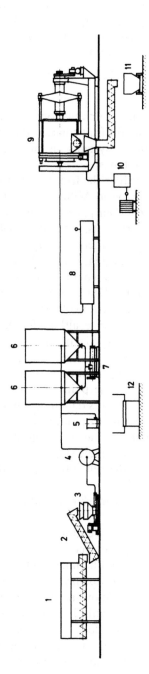

Figure 7.1 Berry and stone fruit processing for the production of juice. For black and red currants, strawberries, raspberries, bilberries, black berries, cranberries, cherries, prunes, plums and apricots. (1) Raw fruit silo; (2) feeding screw; (3) crusher; (4) heater; (5) enzyme dosing; (6) fermenting tank; (7) filling pump; (8) heater; (9) universal press; (10) settling tank; (11) pomace discharge; (12) control panel (Bucher-Guyer AG, Niederweningen/Zurich, Switzerland).

double as fruit washers. The transport water is often recycled many times through a simple screen, however, and so the fruit should always receive a final rinse in clean water before it enters the mills. Endless belt conveyors usually feed the fruit into the mills from above.

7.2.2   Milling

Modern systems have replaced the horse driven stone wheel in a trough but the principle remains the same in that the fruit concerned is crushed to release the juice. However, more advanced systems of milling the fruit have dominated the preparation of the fruit mash.

The function of fruit mills is to reduce the fruit to a pulp from which the juice can then be pressed in maximum yield. Milling is essential for fruits such as apples which are too hard to yield their juice directly. This contrasts with the situation for grapes, citrus or berry fruits, which need only light crushing. Experiments with hot-water diffuser extraction of apple slices in modified sugar-beet extractors were carried out extensively in the 1960s and 1970s, in the hope that both mills and presses could eventually be abandoned (Gierschner et al., 1978; Schobinger and Dürr, 1977; Schobinger et al., 1978). Problems with fruit texture and with extraction of unstable polyphenols made the concept generally unattractive in practice, although it has been successful in South Africa where very firm fruit low in polyphenols such as Granny Smiths are used.

Debris such as stalks and leaves is removed by passing the fruit through a rotating perforated drum. This allows the pulpy fruit to pass through the perforations but the harder stalks and leaves remain within the drum and can be collected by setting the drum angle at an optimum level. Leaves and rotted fruit can also be removed by feeding round fruit onto a rubber faced belt running up an incline. The fruit rolls down the slope to be milled whereas the leaves, sticks and rotten fruit stick to the belt and are propelled upwards to a collection point.

Descriptions of the various systems of mash preparation are as follows.

Hammer mill.   This was a fairly early invention and as the name infers it beats the fruit into a pulp with moving metal components. Generally used for apples and pears, it consists of a series of hammers actuated by mechanical cams which cause the hammer head to rise and fall on top of a continuous supply of fruit which is normally delivered by gravity. The number and size of the hammers determines the rate of the mill. It is claimed in some quarters that this type of pulp breakdown allows for a higher juice extraction than other methods. This method is commonly used in North America but can cause flavour problems with some fruits if too many seeds are broken up.

Fixed knife mill.   This is one of the most common systems available where the apples or other fruit are forced against fixed knives which shred the fruit. Normally the design consists of a circular chamber with a large top opening to allow the whole fruit to fall by gravity into this circular chamber. Inside the chamber is a three-armed 'spider' which rotates at high speed forcing the whole fruit onto the fixed knives under which are smaller holes through which the resultant mash falls to be conveyed to the pressing equipment.

Stoned fruit mill. Stoned fruit such as plums, damsons and apricots may need to be crushed without damage to the stone as this often contains substances which would affect the juice taste and/or storage stability. In this instance hard rubber lobed wheels rotate together forcing the fruit down and stripping most of the flesh but not affecting the stone.

Grape mill. A de-stalker usually removes the undesirable stalks from the grapes prior to crushing. The crushing of the grape is effected by a series of intermeshed arms mounted on two circular cylinders which shred the grape as it passes through, thus exposing the fleshy interior to the juice extraction process.

Mash transport. The normal means of transporting the pulp from fruit which has been milled or crushed is by stainless steel pipe using a positive displacement pump consisting of a rubber stator and stainless steel rotor. These are used primarily for apples, pears and other soft fruit.

Grape mash has a different consistency and, in Europe, grape mash pumps are often vertical piston pumps made of cast iron with a bronze sleeve.

7.2.3   Pressing

There are many and varied types of press covering the fruit market, some more suitable for one type of fruit than another. Nearly all the presses can be used for any fruit with varying degrees of success. Press efficiency is defined by yield of total juice from original fruit and this is described in the following paragraphs in more detail.

Pack press. The traditional presses which were originally made of wood have subsequently been replaced by steel. This type of press is widely used by the small cider makers in the United Kingdom and also by French cider makers. However larger producers of apple juice both in the United Kingdom and in the rest of the world use more sophisticated and less labour intensive systems. Having said that, it is still recognised that pack presses give a very high juice yield albeit with a high labour content and very manual operations. In recent times attempts have been made to automate the pack press principle but none of these has been particularly successful.

The modern pack press consists of a bed made of stainless steel which is attached to a hydraulic ram moving vertically. The press operators, normally two, create a 'cheese' of apple mash (so-called because a similar process used in loading curd into a cheese press results in an individual cheese after removal of whey), which is built up between slatted 'lathes'. The procedure for making this cheese has become quite sophisticated but in essence is still manual. The first action is to place a 'lath' on the bed and then place the cheese frame on this. This frame is just a little smaller than the bed and consists of four pieces of wood about 5 cm high joined in a rectangle. The pressing cloth, which nowadays is made of a nylon material with a fairly coarse weave, is placed over this frame but is almost twice the size of the frame. Apple or fruit mash is normally stored in a hopper above this position or a suitable mill is situated above and a predetermined amount of mash is dropped onto the cloth within the frame.

Next, using hand-held levelling trowels, the operators, working from either side of the press bed, spread out the mash within the frame so that it is level. The cloth is then folded over the mash to give a type of sandwich and the frame is removed. Another lath is then placed on top of this cheese and the whole procedure starts again until a pile of up to 20 cheeses is created. The hydraulic system is then activated pushing the bottom bed up towards a similar top plate and the cheeses are compressed, thus forcing the juice to flow out. The cloth acts as a filter as well as a container for the mash and the juice is collected in a suitable vessel. The skill in this system is getting each cheese as flat as possible because any angle will allow the whole pile of mash to slip and eventually slide out under pressure.

Variations upon this theme have been developed over many years. The original presses built by such companies as Beare's of Newton Abbott were normally water hydraulic with a high pressure reciprocating pump providing the pressure; many of these still operate. However, more modern units have oil hydraulics with very large rams up to 18 inch in diameter where very large loads such as 40 tons can be exerted with an attendant yield increase. Variations on the static hydraulic press are mainly related to the handling of the mash/cheese. For instance it will typically only take two operators 10 min to build up a stack of cheeses but it may take 30 min of pressing time to achieve optimum yield. In this case, a single cheese making station is matched to three hydraulic presses with a rail system to move the cheese. Some years ago a 'clover leaf' press was developed (Bucher Guyer, 1963) which rotated to each of three stations during making of the cheeses, before pressing and finally emptying the residue or 'pomace'.

It has been common practice in the United Kingdom to take the first pressed pomace and add water and then press it a second time. This increases yield but dilutes the extracted juice and is restricted to juice for beverage manufacture or concentrate.

Horizontal rotary presses.   About 30 years ago a Swiss company making agricultural and grape pressing equipment—Bucher Guyer AG of Niederwenigen—developed a new concept of fruit pressing known as horizontal piston (HP) press. The principle is still hydraulic pressing of the mash but the method of juice extraction is more sophisticated. The press is constructed mainly in three parts; first there is a vertical circular end plate which is hollow and has several hundred threaded holes passing through the plate to the hollow section. Secondly there is a circular body which moves in a horizontal mode up to, and away from the end plate. Finally there is a piston with a similar construction to the end plate which is also circular and moves under hydraulic pressure towards and away from the end plate, always remaining parallel to the end plate and within the circular body (Figure 7.2). Through the space enclosed by the cylinder run several dozen flexible rubber juice channels, each of which is ribbed (star shaped in cross-section) and covered by a coarsely woven nylon sleeve.

Mash is fed by a pump into the press until the space is partially filled. The piston's forward movement under hydraulic pressure expels the juice which runs through the filter sock and down the rubber cores, through the specially designed end plates to be collected in a juice tank. The main advantage of this

Figure 7.2 Universal fruit press HP5000 (from Bucher Guyer).

system, apart from its ability to extract more juice from the fruit being processed, is its ability to be operated by one operator or even automatically. The sequence of operation for a normal fresh apple pressing is as follows:

(1) The fruit is mashed as described earlier in this chapter and it is either pumped direct to the press or stored in an intermediate tank. The press outer basket is brought back until it is sealed against the end plate and the piston is brought forward towards the end plate with the rubber cores hanging between the two plates. Mash is then introduced into the space between the end plate and the piston at quite a rapid rate (perhaps 0.5 tonne/min) for a short period (2 or 3 min). This forces the piston back and a certain amount of juice flows via the rubber cores to the juice tank, known as 'free flowing juice'. The hydraulic piston is then brought forward to press the mash and release juice trapped in the mash which then also flows down the rubber cores to the juice tank. The tank includes a level probe, normally a float valve, connected to a juice pump which at a pre-set level is automatically started to pump the juice to an intermediate tank where further processing takes place as described later in this chapter.

(2) After this pressing has taken place, the cylinder is withdrawn allowing the mash to be released and simultaneously the drum is rotated, redistributing the remaining mash to allow a further pressing cycle to take place. Depending on the type of fruit, after several of these pressings more mash is introduced and further pressing cycles occur, until a full load has been received by the press and the maximum juice has been extracted.

(3) Following the complete cycle, the pomace residue is discharged. This is done by rotating the whole press with the piston fully retracted and opening the circular body from the end plate allowing the dry pomace to fall. This is normally taken by screw or belt elevator to a storage hopper or collection lorry and can be used for animal feed, pectin extraction or other industrial purposes.

Automation in the form of computer control systems has recently been introduced to control the optimum perfomance of such presses. This takes the form of following a typical graph of pressure versus time for a given type of fruit and allows the unit to fill, press and discharge at the optimum rate to give the best yield possible for a given fruit. The totally enclosed nature of the HP press is very suitable for operations such as pressing under a nitrogen blanket or the use of pulp-degrading enzymes under temperature-controlled conditions. However, due to its complexity this type of press has a relatively high capital cost.

Belt presses. Using two belts and squeezing a sludge between them has long been used as a 'de-watering' system for removing liquids from a solid mass. The most common use of this system is in sewage de-watering. Some 10 years ago several companies manufacturing this type of equipment—Belmer, Klein and Ensink—developed units where the juice extracted was the required product and the solid residue the by-product. A whole family of presses has since been developed to process fruit and obtain juice. The three basic configurations available are described; all the other systems are variants of these.

(1) The most common style of belt press, introduced notably by Belmer and Klein, is one where the two belts run either horizontally or vertically and the pumped mash is spread out evenly across the belt width using a type of 'doctor blade'. The two belts then converge pressing the mash between them. The belts are made of a porous material, normally a stainless steel or polypropylene mesh and the juices flow through the belt to be collected in trays and pumped to a juice storage tank. Clearly the weave of the belt affects both the yield and the amount of solid material present in the juice. This style of belt press then wraps the two belts around a series of cylinders; each wrap can be as much as 270°. The cylinders become progressively smaller in diameter and there can be as many as 16 rollers involved in each press. Depending upon the thickness of mash between the belts a shear action takes place which continually creates new channels for the juice to flow out from the mash. Each cylinder wrap is opposite to the other creating a speed differential between the two belts on each change in roller, with a subsequent increase in yield. There have been significant advances in this style of press over the last few years, notably the ability to add water to the first pressed mash and press it again on a following modified belt press. This has been achieved as an automatic continuous process (Figure 7.3).

(2) A simpler but less efficient way to extract juice is to have a system with belts as above but to use rollers each side of the belt to press the mash between the belts, and thus extract the juice. This was developed by the Dutch company, Ensink. Obviously with a fixed bed of mash the flow of juice is inhibited and the yields are therefore a little lower. The addition of filter aids such as rice husks or wood chippings has been tried to create channels for juice flow. This has helped, but then makes the pomace less suitable for further use.

(3) Belt presses have been developed where a single web of porous belt has been fed with a film of fruit mash. The belt is then folded and passed between rollers where the juice is pressed out and the belt then opens up to discharge the pomace before recharging. This style has not proved popular.

Screw presses. These are normally vertical tapering screws where the mash is introduced at the top and as it moves down with the rotating screw, it is compressed in the taper and the juice flows out through slatted conical walls. The problem of this style of press is the closing of juice flow channels as the pomace is compressed. Press aids to help juice flow are extensively used in this type of unit. These presses built by Reitz have been used extensively in North America; they have never gained popularity in Europe.

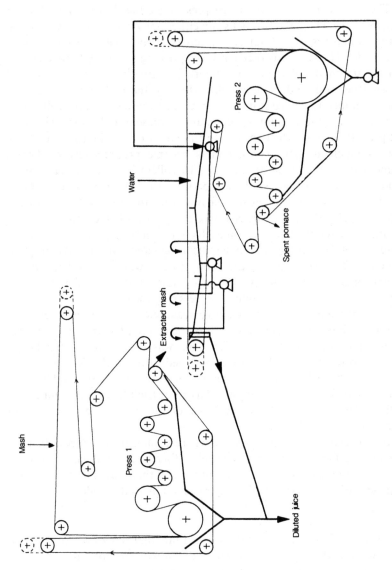

Figure 7.3 Continuous juice extraction using two belt presses with water extraction. Reproduced from Bellmer-Winkelpressen Kascade system.

Other presses. Many other methods have been used such as a continuous platten press based on a pack press with moving lathes, also methods using belts but it is generally accepted in Europe that the horizontal rotary presses and the belt presses give the best extraction rates. Methods of diffusion similar to those used by the sugar beet industry have also been tried but have not generally found acceptance. Diffusion presses for fruit have been built by Amos in Europe and by Bio-Quip in Australia.

7.2.4 Comparison of pressing systems

Yields. The methods described above give different juice extraction rates and juice release differs between and within fruit types. Different varieties of apples milled in their fresh state which have not been stored or enzyme treated may give widely varying yields. As a rough guide using dessert apples the following yields can be expected by weight:

	%
Pack presses	80
Horizontal rotary presses	84
With secondary water addition	92
Belt presses wraparound	70
With secondary water addition	92
Belt presses straight through	60

It must be noted that juice from belt presses generally has a much higher solids content. Other differences are noted in Table 7.1.

Table 7.1 Comparison of pressing systems

Pressing method	Advantages	Disadvantages
Pack presses	Cheap	Labour intensive
	Simple	Batch system
Horizontal rotary presses	High yield	Batch system
	Low solids	Expensive
	Good juice quality	
	Can automate	
	Low labour costs	
Belt presses	High throughput	High solids
	Low labour costs	Low yield
	Cheap	
	Continuous operation	
Screw presses	Continuous operation	Low yield
		Press aid needed

7.2.5 European grape pressing

The processing of grapes for the purpose of extracting juice for wine production is a totally different system and a whole series of machines have been developed over many centuries to achieve the best results. Grapes present different problems to most fruits in that they have a reasonably tough outer skin and a very soft and fleshy interior with easy flowing juice extraction properties. The skins contain many substances which are not desirable in the finished product and therefore a much softer pressing is required. The machinery most commonly used in Europe today falls into two categories.

Mechanical pressing system.   The grape mash is introduced gently into a horizontal rotary cylinder made from stainless steel with a perforated exterior. A screw runs through the centre on which pressure plates are mounted. By rotating the screw, the plates slowly compress the grape mash and the juice flows out through the perforations. The residue is mainly skins which are then discarded. A similar system can be used where hydraulic pressure pushes the pressure plates.

Pneumatic pressing system.   This is at present the most popular system and involves a stainless steel horizontal cylinder as above but with the perforations covering only 70% of the outer surface. Inside the remaining 30% of the area is a diaphragm which when receiving the grape mash is totally deflated allowing the mash to fill the volume of the cylinder. The pressing action takes place by inflating the diaphragm with compressed air allowing it to pressurise the mash thus forcing the juice through the perforations. The pressure at only 2 bar is therefore sustained over a much larger surface area and is much lower than that obtained from a piston.

In each case the cylinder can be rotated between pressing cycles to move the mash, create new flow channels and hence obtain the best juice extraction. For residue discharge the cylinder has hydraulically operated doors which are opened during the cylinder rotation to allow the residue to fall into a suitable container or conveyor system.

7.3   Pre-treatment with pectolytic enzymes

Some fruits will not easily release juice just by simple milling and pressing. Enzymes have been developed over the last few decades to assist in this process. Apples and pears can normally be pressed fresh without any assistance but after a period of storage the cell structure changes and juice is not released very easily. In the case of many other fruits such as raspberries, blackcurrants, strawberries, cherries and plums the juice is 'locked' into the cell structure and needs a helping hand to be released.

When added to fruit in the correct quantity and under the correct conditions, pectolytic enzymes can break down the cell structure and/or the dissolved pectins in the juice and so allow the juice to flow more easily. The main disadvantage of enzymes is their cost in relation to the cost of the fruit and the sale value of the extracted juice. If a fruit yields 60% juice with conventional pressing and 80% with addition of enzymes but the cost of the enzymes exceeds the value of 20% juice it is not commercially viable to use the enzyme unless a further technical advantage can be shown.

The fruit is received either frozen or at ambient temperature. If frozen it needs to be thawed, generally by passing through a heated thermal screw. Enzyme is added directly after milling and the mash heated to the optimum temperature for the enzyme. The cell structure breakdown is not instantaneous and may take from 1 to 2 h. It is therefore necessary to provide reaction tanks with insulation, agitation and possibly heating coils to ensure optimum enzyming conditions and maximum pressing efficiency. Alternatively a continuous process may be used in which the enzyme is injected at the milling stage and the mash heated slowly before feeding into an insulated holding hopper or directly into the press.

Enzyme action tends to increase the amount of free flow juice but can make the resulting mash slimy and difficult to handle. When pressing 'cheeses', layers of mash tend to slide sideways and instability can occur. Belt processes tend to push the mash layer outwards under pressure so that mash eventually spills over the belt edges. It is normally better to press this type of mash in an enclosed space.

Enzyme development is in its early stages and it is quite possible that within a few years enzymes will be capable of the total liquefaction of some fruits. In the future, economic considerations may well allow greater use of enzymes, particularly with the current trends towards 'healthy eating'.

7.4 Post-press clarification

7.4.1 Decantation

One of the simplest methods of clarification of a juice containing solids is to allow these solids time to drop from suspension and then decant or syphon away the partially clear juice for further treatment. Methods of creating a state in which solids drop, range from holding the product at a low temperature for long periods, to adding fining agents such as gelatine which combine with particles of opposite electrical charge thus creating larger uncharged particles of solids which settle more rapidly. A detailed description of fining is given in chapter 6.

7.4.2 Centrifugation

Normally the first stage of any clarification involves the separation of clouding components by centrifugal action. The juice containing the solids is fed into a conical shaped bowl which is spinning at high speed. This bowl is split using conical insert discs to produce several very thin layers of juice. Centrifugal force separates the light and dense components in each layer. The juice is collected and the unwanted solids are discharged to waste. These units can operate automatically and with in-place cleaning.

7.4.3 Earth filtration

If juice is passed through a cloth or paper filter, the fine suspended particles rapidly block the filter pores. Hence it is necessary to cover or 'coat' the filter with a suitable filter aid. The principal material used is Kieselguhr, a natural diatomaceous earth, but alternatives are available such as the synthetic Perlite. Filter aid is metered into the product prior to entering the filter to ensure that the inlet surface of the filtering layer retains porosity. As filtration proceeds, the layer grows in thickness, flow rate decreases and the pressure differential may rise.

Developments of filters have taken place in Germany by Seitz and Schenk and by Filtrox in Switzerland. In each case a range of equipment is available each applying similar principles. The main methods are described in more detail below (see also Brennan, 1969).

Plate and frame filter press. A series of plates of stainless steel with perforations leading to take off channels are fixed together alternately with hollow frames. Kieselguhr powder is mixed with water and pumped into the hollow frames and thus onto the surface of the plates. The water flows to waste through the perforations in the plates but the Kieselguhr is left behind forming a filter bed. Further Kieselguhr is suspended in the juice which is then pumped under pressure through the bed. Care has to be taken to ensure that the filter bed allows free flow of liquid; excessive solids can cause the bed to block or 'blind' with a subsequent breakdown in filtration. Judgement of this state is made by measuring the inlet and outlet pressure differential. If pressure builds up too quickly either the bed is becoming too full with filtered material or 'blinding' of the filter bed has occurred.

When the press frames are full of filter earth (or before, if the filtered juice flow rate becomes unacceptably slow) the press is opened and the spent earth removed. Sometimes the plates are covered with paper liners beforehand to facilitate quick release of spent earth. The plate cleaning operation has not been readily automated and the high labour content is a disadvantage of plate and frame pressing.

Horizontal plate filter. Using similar principles to the plate and frame, systems have been devised by several manufacturers in which the filter bed is built up on horizontal discs. These are normally about 1000 mm in diameter and there may be as many as 100 plates per filter. The disc construction varies depending on the manufacturer but normally the disc will have a perforated top surface through which the filtered product flows and thus out to a storage tank. The disc has a solid base and the outlet is normally down a central shaft, access to which occurs between the top and bottom plates.

In operation, the base coat of Kieselguhr is built up on the top surface of the discs in a similar way to the plate and frame. The juice to be filtered is then introduced with the addition of further Kieselguhr and the filter bed grows as the product is filtered. Differential pressure across the filter bed is again the guide for completion of a filter run after which spent filter earth is discharged. Plate cleaning can be carried out automatically on this type of filter as the principle is to rotate the plates at high speed to remove the spent earth by centrifugal action. Water sprays can then be introduced and the unit can be cleaned and sterilised in a very short period with minimal labour.

Candle filtration. Candle filters give results very similar to horizontal plate filters but use a whole series of stainless steel rods often wound with wire or using a series of washers around which the filter bed forms. This has the same advantage of simple automation as plate filters and gives a much larger filtration area for the size of vessel being used. Its disadvantage is its vulnerability to pressure shocks which may cause the initial bed to become disturbed and even fall off.

7.4.4   Rotary vacuum filters

Rotary vacuum filters have become popular for clarifying juice direct from the press and for extraction of surplus juice from tank bottoms (lees), after racking. They consist of a perforated steel drum rotating on a horizontal axis. The drum rotates in a half-filled tank of juice, and a vacuum is applied to the inside of the drum so that juice with added filter aid is sucked through the rotating filter bed and is thereby clarified. As the drum rotates, its outer surface is scraped by a knife which continuously shaves off the deposited solids. Fresh supplies of raw juice and filter aid are fed to the tank to maintain its level as the clarified juice is drawn off. Although this can be an efficient system, it requires skilled labour to prepare and maintain the filter layers. The juice is also exposed to oxidation and infection in the tank, and a continuous throughput of juice is required or the filter layer dries out irreparably.

7.4.5   Sheet filtration

Pre-formed disposable filter sheets offer an alternative to powder filtration. The labour requirements for sheet filtration are lower although the material costs are higher. By proper choice of sheet, near sterile conditions can be obtained, although the filters will soon 'blind' if much juice solid is present. For this reason, sheet filters tend to be used for 'polishing' to high clarity, following filtration by some other method.

In the past most filter sheets were made using asbestos as the filter medium as it has very efficient filtration properties; however, as asbestos is no longer an acceptable medium due to possible health risks in handling the dry material, combinations of cellulose and Kieselguhr have replaced it in many cases.

7.4.6   Cartridge filters

Cartridge filtration has become more popular in recent years as new methods of cartridge manufacture have been devised. The principle is that a cylindrical body houses a specially prepared 'paper' cartridge of selected porosity. When the liquid to be filtered is passed through this body all particles above a certain size are trapped. Once again when the pressure differential across the cartridge reaches a certain level it has to be discarded. This system is simple to operate and can filter down to very small particle sizes. It is normally used as a final sterile filter as it tends to have a very reduced life when attempting to filter a liquid containing large amounts of solids.

7.4.7   Membrane filtration

In recent years the development of filter membranes with very fine pore sizes but considerable toughness has made it possible to go beyond the normal limits of insoluble particle removal and to filter out colloidal materials or even larger molecules from solutions. In order of increasing fineness of pore size, these processes are referred to as microfiltration, ultrafiltration and reverse osmosis.

The main advantage of membrane filtration is that an acceptable quality of filtered juice can often be created by a single process whereas with the other methods described, it may be necessary to combine several systems together to achieve similar results. Considerable experimental work has been carried out in recent years by competing manufacturers such as Abcor, DDS, Romicon, PCI and by independent research establishments (Heatherbell *et al.*, 1977).

Ultrafiltration.   This is a low pressure (1–10 bar) membrane process for separating high molecular weight dissolved materials from liquids which is increasingly applied to fruit juice clarification. A semi-permeable membrane is used which can be made of ceramic, plastic or metallic substances incorporated into membrane modules. Water and all juice components pass through the membrane. Suspended solids, colloids and macromolecules are rejected by the membrane and are concentrated. Ultrafiltration membranes can retain material as low as 1000 molecular weight. The membranes may be flat, cylindrical or spaghetti-like in configuration.

In practice, an ultrafilter normally consists of a series of tubes containing the membranes down which is passed the raw juice. As it flows down the tubes filtration occurs and the remaining liquid is returned to the circulating tank where it is topped up by new raw juice. The process continues until the retentate becomes too concentrated and is dumped. The system is then flushed and sterilised in preparation for a further batch. All these actions can be easily automated.

In modern designs the juice flows across the membrane surface at high velocity and this cross flow characteristic differs from the perpendicular flow of ordinary filtration where 'cake' builds up on the filter surface requiring frequent filter replacement or cleaning. Cross flow prevents filter cake buildup resulting in high filtration rates that can be maintained continuously, eliminating the cost of frequent filter replacements and the need for filter earth.

In principle, all juice debris and macromolecular polymers such as pectin and protein remain behind, while clear juice passes through the filter. In practice it is necessary to degrade the pectin enzymically before ultra-filtration, to reduce viscosity and allow a satisfactory juice throughput. As a one-pass system, ultrafiltration appears to have many advantages. Although the membranes retain high molecular weight material, however, small haze-forming molecules can easily pass through which polymerise and cause hazes later (Nagel and Schobinger, 1985). Therefore ultrafiltration cannot always replace traditional fining procedures.

The main advantages of ultrafiltration for fruit juice are:

—high juice yield; in practice yields of 96–99% are realised compared with 80–94% yields with conventional filtration techniques;
—single system;
—cleaning and product change time is minimal;
—few moving parts.

7.5 Concentration/aroma recovery

The concentration of fruit juices is undertaken after pressing and clarification for several reasons:
—to reduce the liquid volume and subsequently reduce storage, packaging and transport costs;
—to increase concentration of soluble solids in the juice as an aid to preservation;
—to pre-concentrate a juice prior to further processing such as spray drying, drum drying or crystallisation.

The principal method of concentrating fruit juices in Europe is by evaporation. Other methods have been tried such as freeze concentration and reverse osmosis but these have not been widely used to date. The design and operation of fruit juice evaporators is a complex area of food engineering which lies outside the scope of this chapter (Mannheim and Passy, 1974), and has recently been reviewed (in German) by Schobinger (1987). The basic requirement for juice concentration is that the process should generate as little thermal degradation as possible, and that if necessary, volatile materials contributing to the aroma of the juice should be recoverable from the distillate. These criteria are best met by operating under vacuum at a high temperature for a short time, exemplified by popular types of modern plant such as the APV falling film evaporator or the Alfa Laval centrifugal evaporator.

Industrial evaporators normally consist of a steam fed heat exchanger to heat the raw juice to a suitable temperature for evaporation. The system includes a separator in which the vapour is separated from the concentrated liquid phase, a condenser to effect condensation of the vapour and its removal from the system, and a fractionating still for aroma recovery. There are many types and styles of evaporators available for fruit juice concentration all comprising these basic components (Spicer, 1974). However, most fruit juices are heat sensitive and designs giving short heat contact periods are the most suitable. These systems can be categorised as follows.

7.5.1   Rising film evaporators

Rising or climbing film evaporators normally consist of a heat exchanger with heated tubes between 3 and 10 m in length and diameters of 20–40 mm. The fruit juice pre-heated to near boiling is fed in at the bottom of the heat exchanger and rises within the tubes. The mixture of liquid and vapour from the tubes passes to an external separator in a continuous operation. Within the tubes there are three distinct regions. At the bottom under the static head of liquid no boiling takes place, only simple heating. In the centre region, the temperature rises sufficiently for boiling and vapour is produced. In the upper region the volume of vapour increases and the remaining liquid is wiped into a film on the tube surfaces resulting in good heat transfer conditions. One of the problems of this system is that the proportion of the liquid evaporated in a single pass is often insufficient to provide the required concentration so recycling is needed which extends residence time at elevated temperature; also scale and protein deposits can occur in the tube.

7.5.2   Falling film evaporators

To eliminate the restrictions caused by a static head of liquid and yet continue to take advantage of the high heat transfer rates obtained with a thin film of liquid, the falling film evaporator was devised. Higher efficiencies and thermal economy are achieved and residence times reduced. The preheated liquid is introduced at the top of the evaporator and passes down a tube or plate heat exchanger. It is important to ensure even distribution of liquid over the heated surfaces to avoid local hot spots and subsequent burning of the juice.

Falling film evaporators are particularly useful in heat sensitive fruit juice concentration due to their short heat-contact time, easy cleaning and rapid start-up and shut-down with minimum wastage due to very little liquid hold-up in plant.

7.5.3   Centrifugal evaporators

Centrifugal evaporators have very low heat contact periods and so are very suitable for fruit juices. There are several designs, both horizontal and vertical. One type comprises a vertical cylinder with a heated jacket containing a concentric rotor whose blades have a small clearance from the heated wall. In the other type, the jacketed tube is horizontally disposed and conical in shape. Within this tube rotates a concentric rotor, again with blades having a fine clearance from the heated wall and angled to ensure flow of the concentrating juice towards the smaller diameter end of the cone. For fruit juices, there are several disadvantages of both the systems described above, mainly economy of energy, cleaning difficulties and in the case of the horizontally mounted rotor, sufficient stiffness of rotor to ensure small wall clearances at varying temperatures.

In this context the expanding flow evaporator was specially developed to overcome many of these disadvantages and solve the problems of concentrating fruit juices in a fast continuous flow without harming the flavour or vitamin content. The unit consists of a stack of rotating, steam-filled hollow cones. The juice is sprayed onto the inside of the cones near the apex. Centrifugal force then spreads it across the inner cone surface and moves the resultant concentrate to the outer edges from where it is thrown to the static walls of the evaporator for collection. Centrifugal force similarly ejects condensate as soon as it forms on the outer surfaces of the cones so that the heating surface in contact with the product always has hot dry steam on the other side. Both juice concentrate and steam condensate are continuously removed. Secondary vapour is passed through a condenser. The resulting combination of thin film flow, high heat transfer and reduced boiling temperature due to partial vacuum inside the hood, can concentrate fruit juice in one pass, with contact time a matter of seconds. Such units can be cleaned in-place and can be fully automated in operation (Figure 7.4).

7.5.4   Heat recovery from evaporated water

The vapours removed from the boiling juice contain a significant quantity of heat and the thermal efficiency of an evaporator can be improved by its recovery. This can be done by arranging for only part of the concentration to be carried out in the first evaporator and providing later stages of evaporation operating at lower pressures to give the necessary temperature

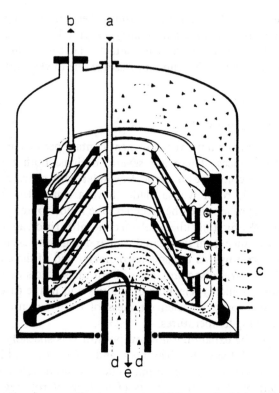

Figure 7.4 Expanding flow centrifugal evaporator. (a) Process liquid in; (b) concentrate out; (c) vapour to condenser; (d) steam in; (e) steam condensate out.

differential. This allows much of the latent heat of evaporation contained in the distilled vapours to be recovered in subsequent evaporative steps. Multiple effect evaporators using this principle have been highly developed, párticularly in the citrus industry (see chapter 8).

7.5.5   Aroma recovery

When evaporation takes place to concentrate the fruit juice, volatile aroma substances are removed with the water vapours. When the juice is recon- stituted the typical fresh flavour of fruit juice is best re-created if the volatile aromas are available for adding back to the juice. Therefore at the time of evaporation, the volatiles are recovered from the water vapours by fractional distillation. These aromas are then concentrated in liquid form and (usually) held separate from the concentrate to be added back to the juice at re- constitution.

If the concentration is done by a process which does not allow separate recovery of volatiles, such as freeze concentration, or if evaporated volatiles are added back to the concentrate before storage, deep frozen storage of the concentrate is normally required to obtain best flavour quality. At ambient temperatures chemically reactive components of the volatiles are lost as a consequence of the faster rate of chemical change due to the higher concentration, with adverse effects on flavour in some juices.

7.6 Pasteurisation

Pathogenic organisms will not grow and spoilage bacteria, moulds and fermentative yeasts are adequately controlled in acidic juices by pasteurisation.

7.6.1 Flash pasteurisation

Flash pasteurisation is carried out using a plate heat exchanger and a separate 'holding tube'. The heat exchanger consists of a pack of stainless steel plates separated from each other and arranged so that heating medium and juice can transfer their heat through the plates without contact.

Modern plate pasteurisers have four main parts:
  (i) A generative section in which hot, already pasteurised juice is fed back through one side to preheat the incoming cold juice and to be cooled back to a lower temperature in the process, thus saving energy.
 (ii) A heating section supplied with hot water or steam to boost the juice to pasteurising temperature.
(iii) A holding section that may be accommodated within the heat exchanger or possibly a length of separate piping insulated to prevent heat loss which could lead to inadequate pasteurisation.
(iv) A cooling section using chilled water as the medium. This often is not needed as the regenerative section can be made so efficient that exit temperatures approaching ambient can be achieved.

7.6.2 Batch pasteurisation

Juice is heated to a given temperature and held for a given period of time before cooling. The system is thermally inefficient and can lead to damage to the juice due to prolonged heating.

7.6.3 In-pack pasteurisation/hot filling

This system pasteurises the pack as well as the product and ensures product integrity, unlike flash pasteurisation systems which can still allow micro-organisms to enter at the bottling stage. The method of heating is normally by hot water sprays within an enclosed tunnel through which the bottles or cans pass at controlled speed. The spray temperatures are carefully controlled and the tunnel is often divided into three areas. Initially the sprays slowly increase the pack temperature over say one-third of the tunnel. The next third has sprays set at the pasteurisation temperature and finally a cooling section is provided with progressively cooler spray until the cooled pack emerges. Modern machines use sophisticated ways of conserving heat by heat exchangers and temperature gradients.

A variant on this process is hot filling in which the juice is pre-heated as for flash pasteurisation and filled into the pack at a temperature high enough to ensure pasteurisation of the pack and closure without further heat input. After the pasteurisation period the pack is cooled as quickly as possible.

7.7   Fruit juice plant layout

It is normal to lay out a factory to give the optimum flow from fruit reception to packaging with each individual process plant adjacent to its predecessor and successor. It is sometimes desirable to have fruit reception on the same side of the site as finished goods despatch for reasons of labour location, security and optimisation of transport. However, microbiological considerations must always be taken into account.

7.7.1   Materials of construction

Most fruit is corrosive due mainly to acid and in many cases all metal in contact with the juice needs to be of good quality stainless steel. Plastic materials can also be used to prevent corrosion but are often affected by both temperature and chemicals used for cleaning. Glass-lined tanks and epoxy painted mild steel are also used. However, the effective life of these cheaper alternatives will be short should any damage occur to the surface.

7.7.2   Fruit reception

Most large fruit juice producers process one or two main juices but also may do ten or more other juices for special customers. Fruit is delivered either in bulk or in boxes. Apples float and can be transferred from the reception pit by water which also cleans. Pears and other similar fruit which do not float need to be conveyed to the process plant often by belts. Soft fruit are usually moved to the process machinery by screw conveyor although if frozen they need to be defrosted first. This is normally carried out by a steam-heated thermal screw but the use of microwave defrosters is becoming more popular.

When siting fruit reception bays it is necessary to assess how the fruit will arrive before detailed planning. The method of transport could be rail, lorry, tractor and trailer and even private car with sacks of fruit and therefore adequate access and turning facilities need to be made, including rail links where appropriate. As fruit is such a seasonal product the sizing of the reception area must reflect the peak supply for any given fruit and a reception controller must be employed.

7.7.3 Handling and washing fruit

The best fruit juice is produced only by pressing good fruit. As many impurities as possible need to be removed prior to milling and pressing, particularly rotted or damaged fruit. It is difficult to remove damaged or rotted fruit automatically, therefore the most common method is to spread the fruit out on a conveyor where labour can be used. Clearly the initial quality of the fruit purchased determines the problems that may occur and the proportion that has to be removed and therefore minimum quality levels must be established for fruit purchased. As fruit continues to ripen after harvesting, the journey time from the point of harvest to the processor must be taken into account to reduce wastage. In some cases freezing of the product directly after harvesting can arrest the ripening process but this will not be a success with apples and pears as the fruit structure is altered and juice yield reduced.

Modern harvesting methods, particularly for apples and pears, rely upon the ripened fruit being shaken from the tree onto the ground where they are sucked or rolled up into a wagon. This method although very efficient, is not very selective and relies upon mass harvesting at a given time with immature fruit mixed with overripe fruit. Even worse is the risk that grass, leaves, wood and stones will be swept into the wagon with the fruit. Washing and separation of these undesirable elements therefore becomes important and large amounts of water have to be used to achieve the required quality.

Soft fruit is seldom dirty enough to need cleaning although rotten fruit must be removed before pressing.

7.7.4 Seasonal problems

As previously stated, fresh fruit does present scheduling problems for processors as given varieties tend to mature all at the same time. Although freezing and cold storage can spread the load there is no doubt that freshly harvested fruit gives the best yield and quality of juice. One of the most difficult crops to handle is grapes, where the pressing season seldom encompasses more than 3 weeks during which time the whole of the crop has to be pressed. In addition yield of crop varies considerably from year to year due to weather conditions.

7.7.5 Effluent treatment

It is general to have to install an effluent treatment plant within the pressing site to enable all contaminated water to be processed for return to the environment. An example of effluent processing is given in chapter 8.

7.7.6   Juice storage

Having planned for fruit reception, washing, pressing and concentration and associated processing, provision must be made for juice storage in single strength or concentrated form. This is dependent upon volumes, further processing required, customers' delivery requirement and range of juices being processed. Often several million litres storage capacity will be needed, possibly with associated refrigeration.

7.8   Summary

This chapter has attempted to give a basic review of the equipment used in European soft fruit processing. Much progress has been made on this subject in the last 20 years both in terms of mechanical/electrical equipment and in thermal efficiency. It should be recognised that progress has not stopped; in fact it has probably increased over the last 5 years, with the commercial introduction of belt presses for juice extraction, and ultrafiltration for juice clarification. It is felt certain that enzyme performance will become more advanced and cost effective in the near future.

With current trends towards healthy eating and drinking and less use of additives, greater pressure is being put on the producers of food products to develop methods of manufacture and packaging to ensure as natural products as possible. Fruit processors and their equipment suppliers are responding to this challenge.

Acknowledgements

Thanks are given to the following for help in the compilation of this chapter: Dr A. G. H. Lea who kindly gave some of the information used in this chapter; Bucher Guyer of Niederwenigen, Switzerland who helped with information and illustrations regarding fruit pressing; Alfa Laval Engineering Ltd, who provided much information on fruit juice processing.

References

Brennan, J. G. et al. (1969) Food Engineering Operations. Elsevier, London.
Gierschner, K., Hang, M. and Wirner, H. (1978) Dtsch. Lebensm.-Rundsch. 74, 338–343.
Heatherbell, D. A., Short, J. L. and Struebi, P. (1977) Confructa 22 (5/6), 157–169.
Mannheim, C. H. and Passy, N. (1974) In: Advances in Preconcentration and Dehydration of Foods, A. Spicer (ed.), Applied Science Publishers, London, pp. 151–194.
Nagel, C. W. and Schobinger, U. (1985) Confructa 29, 16–22.
Schobinger, U. (1987) Frucht und Gemusesafte, 2nd edn., Ulmer Verlag, Stuttgart.
Schobinger, U. and Dürr, P. (1977) Flüss. Obst. 44, 275–283.
Schobinger, U., Dousse, R., Dürr, P. and Tanner, H. (1978) Flüss. Obst. 45, 196–216.
Spicer, A. (1974) In: Advances in Preconcentration and Dehydration of Foods, A. Spicer (ed.), Applied Science Publishers, London, pp. 151–194.

8 Processing of citrus juices

H. M. REBECK

8.1 Introduction

The citrus growing areas in the United States are located in the States of Florida, California, Texas and Arizona. The largest crop is harvested in Florida where over 90% of the oranges and approximately 55% of grapefruit are processed into juice products. Brazil's crop is larger than Florida's, where even larger percentages of oranges are processed for juice. Other citrus growing areas in the Western Hemisphere include Mexico, Central America, Puerto Rico, Jamaica, Dominican Republic and countries on the northern part of South America. The machinery used for the processing of citrus juices in these countries and in other regions, such as Spain, Italy, Israel and around the Mediterranean is quite similar.

Most of this chapter deals with the processing of oranges in the State of Florida, USA, with which the author is most familiar. The handling of grapefruit, tangerines, lemons, limes, etc. is quite identical in most ways to that of oranges, though some of these citrus varieties require additional process equipment for certain by-products.

The basic unit in Florida for describing the size of the orange crop is the fruit box. A box of oranges, by definition, weighs 90 lb and a box of grapefruit weighs 85 lb. The harvesting season crosses over the New Year and has a duration of 7–10 months, depending on varieties; 123 100 000 boxes of oranges (61 550 US tons) were harvested during the 1986–1987 season of which 92% went to processing, the balance going to the fresh fruit market. In the same season, 49 800 000 boxes (2 116 500 tons) of grapefruit were harvested with 56% of the crop processed for juice. Of the approximately 113 million boxes of oranges processed in the 1986–1987 season, the juice from 96 million boxes was concentrated to make frozen concentrated orange juice (FCOJ).

By-products resulting from the processing of citrus fruit include dried peel for livestock feed, molasses concentrated from liquid pressed from the peel, commercial d-limonene, which is distilled peel oil, and 'cold pressed' oil, the processes of which are discussed later in this chapter. The production of livestock feed from the processing of all varieties of citrus in Florida for the 1986–1987 season was 600 626 tons of dried peel and 27 811 tons of molasses. d-Limonene production was 13 483 000 lb.

8.2 Fruit harvesting and transport

The harvesting of citrus fruits in Florida begins when the fruit reaches maturity standards set by the United States Department of Agriculture (USDA) and the Florida Department of Citrus. For juices, these regulations have to do with Brix-acid ratio (see Glossary), color, oil content, etc. and in general are set to ensure quality products.

The picking of fruit for the orange juice market begins in September with most of the juice going to the single strength market. There are four main varieties of oranges growing in Florida for the juice (and fresh fruit) market. The earliest oranges consist of Hamlin and Parson Brown varieties. These early fruits are harvested mostly from October to December. Mid-season fruit (called Pineapple oranges) mature during the first 3 months of the year. Late season fruit (Valencia oranges) are harvested from March to June.

The production of FCOJ usually begins early in December when the soluble solid (sugars) content is around 12% (12° Brix). With evaporators operating at full capacity, concentrate production drops by 5% if infeed Brix is 11.5° instead of 12°.

Hand picked and mechanically harvested fruit are brought from the groves to the roadside and loaded into trucks (tractor-trailer type), which hold 500–550 boxes of fruit. The trailers are then trucked to the processing plant.

8.3 Unloading and storage of fruit

These operations are shown diagrammatically in Figure 8.1. The trailers are approximately 8 ft wide × 40 ft long. Trailer sides extend approximately 5 ft above the bed of the trailer. After weighing, the trucks are hauled to the

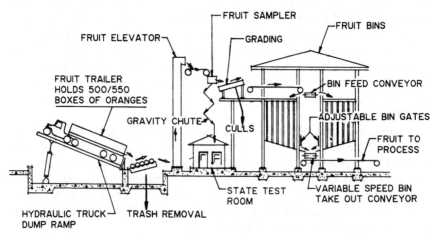

Figure 8.1 Schematic diagram of fruit unloading and storage.

unloading ramps where they are inclined and unloaded through a gate at the rear of the trailer. There are various types of unloading ramps used, the most historic of which is a fixed 'back-over' or fixed 'back-down' ramp. As the crop size grew and processing plants became larger, a need for quicker unloading became necessary and the trend now is to convert to hydraulic ramps for lifting and tilting the trailers. Of these, there are two types, the back-into and the drive-through arrangement.

The older fixed back-into ramps (Figure 8.2) are concrete slabs with a 10° slope from horizontal. Because of the fixed slope of the incline and associated clumsy traffic patterns, a good average unloading rate for one ramp is about three loads per hour. This would allow an infeed fruit flow of 1500 boxes/h or 65–70 tons raw oranges per hour. Other disadvantages include the varying height of different trailer beds above the road which sometimes requires manual fixing of fill-in ramps or wheel elevating shims before the back gate of the trailer can be opened.

The 'drive through' type hydraulic ramp allows for the most favorable traffic patterns, adjustable tilting to accelerate the rolling of fruit near the end of the unloading cycle, and an adjustable hydraulic back stop and fill in ramp which can be completely lowered to facilitate driving through. Unloading ramps of this design (Figure 8.3) can handle up to six trailers per hour (3000 boxes of oranges per hour).

Figure 8.2 Fruit unloading: back-down ramp.

One of the problems facing the unloading staff is the accumulation and disposal of leaves, stems, dirt and even small branches that arrive with the load. The amount of trash has increased over the years as plants have grown and as the percentage of mechanically harvested fruit increased. In the path of flow from the trailers to the storage bins, the fruit may travel over a gravity bar grate, roller spreaders, belt conveyors, chutes, elevator, etc., and in each of these transitions, leaves and stems must be contended with. Bar grates and roller spreaders will drop some of the small, loose pieces through their open spaces to a collection pit which needs periodic sweeping and shoveling. Conveyor transfer gates and elevator chutes sometimes become clogged. The clean-up and manual disposal of this trash requires considerable man hours throughout the production day. Many attempts have been made to reduce the labor required for this clean-up by mechanically conveying the trash from under bar grates and roller spreaders. Screw conveyors are not completely suitable for elevating the trash out of the pit as the longer stems will wind their way around the shaft, accumulate other pieces and stop the flow.

The latest solution to the trash problem is the installation of machinery similar to that used in the corn industry for husking corn (Figure 8.4). These units consist of feeder and distribution belts which put the fruit in single file through a belt and roller arrangement that not only drops out the loose trash

Figure 8.3 Drive through type by hydraulic unloading ramp. Courtesy of Gulf Machinery Co.

but will pull off stems that are attached to the fruit. The use of this type of equipment requires 10–12 ft head room to complete all the gravity transfers of fruit and trash through it to other conveyances.

Layout of this type of equipment depends on availability of plant space. Florida surface water tables are not too far below ground level so deep pits are not always feasible and careful attention needs to be paid to the type of conveyors, elevators and transfer points. Dirt and small pieces of trash will pass through transfer points, especially under the fruit wipe-off on belt conveyors and on the bottom of bucket type elevators, so there still needs to be access under these conveyors for periodic clean-up.

On its path to the storage bin the fruit passes through a sampler, a device which takes an 'on-line' sample of the incoming fruit. The sampler is usually located after the fruit is elevated, and the fruit drops through chutes to the State Test Room. An approximate 40 lb sample of raw fruit is run through a state test room extractor and tested for yield, soluble solids content, Brix and acidity.

Next in the path of fruit flow is a grading station, where unsuitable fruit is 'culled out' and attached stems that have passed through the trash-removal equipment are removed. The grading stations consist of roller spreaders with grading personnel picking out the culls and dropping them through a chute to

Figure 8.4 Fruit receiving trash eliminator. Courtesy of Alcoma Packing Co.

a conveyor (usually a screw conveyor) that transfers them to the cattle feed production part of the citrus plant, referred to as the 'feed mill'.

An unloading station that can handle six loads per hour requires a grading station capable of removing 'culls' (unsatisfactory fruit) from fruit that passes at a rate of fifty 90 lb boxes of oranges per minute. One box of oranges lying single height requires approximately 13 ft^2 of area on a conveyor belt and grading table. A grading table (Figure 8.7) consists of a roller type conveyance that spreads the fruit across the table and conveys them forward at about 60 ft/min. The widest tables are 52 inch across the rollers with grading personnel on each side. Grading personnel are spaced about 48 inch apart with drop chutes between them. A single 52 inch wide grading table × 10 ft long with six people will ideally handle 20 boxes of oranges per minute. However, since the fruit will be graded again on its path to extraction, the unloading grading tables are pushed harder, allowing for two tables per unloading ramp, each handling approximately 25 boxes per minute. After grading, the fruit travels through belt conveyors and elevators to the fruit holding bins (Figure 8.5).

The holding bins are constructed with steel columns, beams and braces with either wood slats for siding or flat mesh expanded metal. They are usually constructed in parallel rows with each bin having a capacity to hold

Figure 8.5 Fruit storage bins. Courtesy of Gulf Machinery Co.

one truck load. Each bin has approximate dimensions of $10 \times 10 \times 25$ ft high and the succeeding bins have common walls. The two rows of bins are fed with a belt conveyor equipped with manually set wipe-offs and gates for feeding the individual sections. The fruit enters the bins through inclined ramps that more or less cause the fruit to roll down in a spiral motion so that dropping and bruising is minimized. The inclined ramps are located so that the total weight of the load is distributed throughout the height of the bin to prevent squashing of the fruit at the bottom.

8.4 Fruit transfer from storage bins to extractors

The introduction of fruit into the processing plant begins with the bin

Figure 8.6 Fruit washer.

operator opening gates at the bottom of the bins where fruit rolls onto a conveyor belt in the center of the bin rows. The rate of flow out of the bins is manually controlled by the bin gate opening and by an adjustable vertical gate at the end of the bin take-out conveyor. The processing buildings are located at some distance from the fruit bins in order to be away from the flies and other insects that tend to inhabit the bin area. Belt conveyors and bucket elevators bring the fruit to a height suitable for gravity flow through washing, grading, sizing and extractor feed.

The elevated fruit usually drops into a fruit surge bin located just outside the extractor room with a volume capable of holding 5–10 min of fruit. The surge bin is equipped with level switches which send signals to the fruit bin operator and extractor operator, for purposes of regulating flow. The surge bin take-out conveyor has a variable speed drive that is controlled by the extractor operator so that fruit flow rate coincides with juice demand.

From the surge bin, the fruit is discharged onto a roller spreader and from there it enters a brush washer (Figure 8.6). The washer consists of a series of rotating cylindrical brushes turning in the direction of the fruit flow. Detergent is added at the fruit inlet end and water is sprayed over the spinning fruit. Most Florida citrus processors use evaporator condensate (the water removed from the juice in the concentrate process) for washing fruit.

Figure 8.7 Fruit grading.

Just downstream from the washer are located further grading tables (Figure 8.7) where unsuitable fruit is graded out by personnel on each side; the culls are dropped through chutes and conveyed to the feed mill.

After grading, the fruit travels through another roller type device similar in design to the grading tables but used to spread the fruit into a pattern feeding the total width of the fruit sizer (Figure 8.8). The sizer consists of several narrow belts running in parallel and tilted so that the fruit travels in single file along the low edge of the belt and against a rotating roller that can be adjusted up and down. Juice extractors require larger and smaller cups for different size ranges and the actual construction of the sizer depends on type and number of juice extractors. Each roller bay of the sizer is adjusted to suit the size variations of the incoming fruit. Extractors have constant speeds and only meet maximum capacity when their feed chutes are kept full. The extractor room people must make periodic checks and adjustments of sizer roller spacing, gates and lane dividers to ensure that the extractor room is not the 'bottle-neck' of the process.

The sized fruit drops through chutes to a distribution belt that feeds the extractors. This belt conveyor travels the full length of a row of extractors plus a few more feet to accommodate the drive and drop-out for overflow fruit. The top belt is tilted toward the extractors and separated into lanes for

Figure 8.8 Fruit sizer.

conveying the larger fruit past the smaller cupped extractors as the sizer above is not as long as the total row of extractors. The distribution belt is equipped with gates and wipe-offs that can be set to maximize the extraction efficiency.

The return side of the distribution belt is usually a few feet lower than the top or feed side of the belt. Along the feed flow between extractors that change cup size, is a wipe-off and an overflow chute that drops fruit to the return side of the distribution belt. The overflow fruit travels back towards the front end of the system, is wiped off, conveyed and elevated back to the fruit surge bin. Fruit sizing must be set to ensure that the fruit surge bin does not accumulate too many of one size which would limit the capacity of the extraction operation.

8.5   Juice extraction and finishing

8.5.1   Extractors

There are two types of extractors in common use in the Citrus Processing industry, both originating in the United States. One type is manufactured by FMC and the other by Brown Citrus Machinery. Extractors are leased on a royalty basis and the equipment is owned and serviced by the extractor manufacturer. In Florida there is about an even split between the two types and the advantages one has over the other are best extolled by the respective suppliers. Both types produce a juice yield and quality sufficient to meet high technical standards and profitability.

Extractors are lined up along the length of the distribution belt on a platform 8–10 ft above the room floor. The maximum number of machines per row is ten (maximum of eight preferred) on the FMC extractors and eight on the Brown. Additional in-line extractors would require wider and more unwieldy conveyor belts.

The FMC extractor (Figure 8.9) is equipped with a feeder that runs fruit into the extractor cups in five rows on the five head machine. Eight head machines are used for small fruit such as lemons, limes and tangerines and three head machines are used for larger oranges and grapefruit. A five head FMC extractor, operating at 100 rev./min with 90% of the cups full, will handle 450 fruit/min. During the Florida Valencia orange picking season (March to May) most of the fruit will be of a size range that averages 250 fruit per 90 lb box with a yield of 6 gallons of 12° Brix juice when finished. This is 10.8 US gallons/min per extractor. However, not all the cups will always be 100% filled (or even 90% filled), especially the larger cup machines on the end of the line.

Fruit is deposited by the feeder into the bottom cup of the FMC extractor. The upper half of the cup descends and presses down on the fruit, and as

Figure 8.9 FMC juice extractors.

contact is made, the sharp end of a round stainless steel tube, located inside the bottom cup, is inserted in the bottom of the orange, cutting a plug. As the fingers of the top and bottom cup halves mesh, the fruit is pressed inward forcing the juice into the tube which is perforated with small holes in the cylindrical section and has a restriction in the bottom to prevent loss of juice. The resulting internal pressure forces the juice and some pulp through the perforations in the tube wall and strains out the seeds and larger pieces of pulp. Through precise timing and as the upper half of the cup is in the pressing position, the strainer tube containing the restriction, plug, pulp and seeds, rises to further press the contents of the tube and to eject the pressed plug.

As the two halves of the FMC cups come together, the oil cells in the skin of the fruit are ruptured, forcing the oil out of the skin where water sprays can be mounted which wash the oil and small pieces of skin ('crumb') down the outside of the cups and via a screw conveyor to a cold pressed oil recovery system. The extracted juice drops into a juice manifold connected to the line of extractors which is sloped in the direction of flow toward the final finishing. Ejected peel drops through chutes to a screw conveyor located under the extractors which conveys the peel to the feed mill.

The extractors manufactured by Brown Citrus Machinery (Figure 8.10) use a reaming action to extract juice from citrus fruit, except for the model 1100

Figure 8.10 Brown citrus machinery juice extractors.

machine which is described later. Of the reaming type, there are three models, the most used of which are models 700 and 400. Of these two types, the model 700 machines have the highest speed and will handle up to 700 fruit/min. In a line of these extractors, most will be model 700 when processing oranges, with one or two model 400s at the end of the line to handle large fruit. The model 400 can be equipped with larger cups and reamers and is also used on grapefruit lines. The Brown model 500 extractor has cups and reamers of a smaller size for handling limes and lemons.

The model 700 extractor has a rotating feeder wheel with a horizontal shaft and paddles spaced apart on the perimeter of the wheel. Fruit rolls from the feed chute into the spaces between the paddles which are timed to rotate and feed fruit cup halves that are connected to traveling chains. The cups accept the fruit from the feeder wheel and pass over a knife that halves the fruit. The fruit then passes across a slider plate which holds the halves in the cups as they spread apart to feed the reamer wheel.

The reamer wheel also has a horizontal shaft and is equipped with reamers at its perimeter that are spinning through gear action in the rotating wheel. The cups are so spaced on the chains to match the position of the reamers on both sides of the reamer wheel. As cups and reamers rotate around, the juice is extracted along with pulp, rag and seed and it drops down into a juice trough that feeds the juice finishers. The peel is ejected from the cups as the chain returns to the feeder wheel and is dropped to screw conveyors to transport the mass to the feed mill.

The Brown model 1100 extractor has a feeder that places the fruit into three single lanes as it enters the extractor. As the fruit drops into the extractor it is caught by a series of rotating discs with wide angle, nearly flat conical shapes on horizontal shafts. The fruit is wedged between pairs of discs which forces the fruit across a knife and cuts the fruit in half with the skin side of the halves against the discs. As the discs rotate the open sides of the halves are forced across a stationary screen. Juice, pulp, rag and some seeds pass through the perforations down to the juice trough that feeds the finishers. Peel is ejected by the rotation of the discs after about 320° of circular path.

8.5.2 Finishing

From both FMC and Brown machines, the extracted juice needs 'finishing' to separate cloudy but otherwise 'clean' juice from pulp, rag, seed and pips. Extracted juice enters the finishers from headers or troughs that are sloped toward the direction of flow. The finishers separate the pulpy matter from the juice by the action of a rotating auger inside a cylindrical screen. The spinning auger forces the pulp out the end of the finisher through a valve which is either spring loaded or loaded by use of an air cylinder. The valve seat or clearance area is conical and is about the same diameter as the screen cylinder (approximately 14 inch in diameter). Screen hole sizes range from approximately 0.020 inch to 0.030 inch in diameter, depending on the condition and 'softness' of the fruit.

The 'tightness' of the finish results from the force applied by the auger against the pressure of the valve. Correct juice yields are controlled by adjustable pressures applied in both the extractors and finishers. Higher pressures give greater juice yields, however excessive pressure can result in 'off-flavors' by forcing too much peel and pulp matter into the finished juice.

Two finishers are often placed in series at the end of the extraction line, the upstream finisher referred to as the 'primary finisher' and the downstream called the 'secondary finisher'. The primary finisher is not set as tight as the secondary unit and so will have a higher flow capacity. Sometimes large plants require two secondary finishers in parallel as they will have higher finishing pressures thus less throughput capacity. The juice from the primary finisher will be lower in pulp matter and its juice stream can be directed to a process that may specify low content of insoluble solids (Figure 8.11).

The amount of pulp allowable in the final processed juice is set by plant quality standards and the maximum is set by USDA and Florida regulations. Pulp content is tested using a clinical centrifuge which spins a sample placed in a graduated glass tube. 'Sinking' pulp is forced to the bottom of the tube and 'floating' pulp appears at the top. The target is to achieve 12% or lower

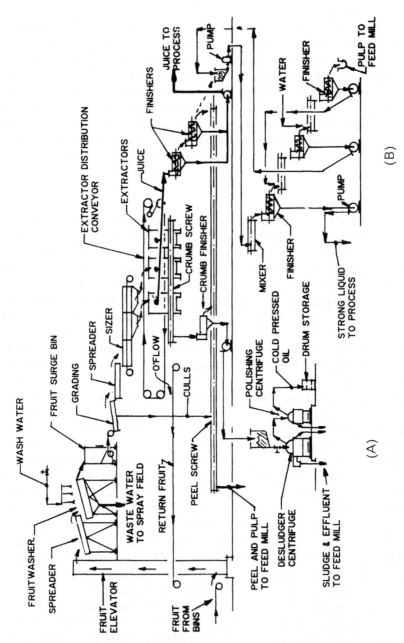

Figure 8.11 Schematic diagram of extraction and finishing.

total pulp, but the measurement is by volume, not weight, as the cellulose matter on a dry basis is only a very small fraction of the total weight of the sample.

The pulp discharge off the finisher is transferred to the feed mill drying operation, or to a pulp wash operation which yields further soluble solids by the action of counter current water leaching. Washed or unwashed pulp is also sometimes directed to a pulp recovery system which through further equipment removes seeds and reject material. The 'clean' portion of the pulp is pumped through heat exchangers, pasteurized, cooled and chilled by refrigeration. This product is packed in 40 lb containers and stored in freeze rooms for future use.

8.6 Juice processing for pasteurized single strength

The canning of pasteurized juice begins in Florida in August and September and coincides with the harvesting of fruit for the fresh fruit market. Overflow from the packing houses is sent to processing plants, and some fruit is delivered directly from the groves. Blending of fruit varieties and control of ratios is usually accomplished by the fruit storage bin operator. Juice is pumped from finishing to process receiving tanks that are equipped with agitators to keep the pulp suspended. These tanks can also be used for adding sugar or other ingredients when canning sweetened or fortified juices. From the tanks the product is pasteurized by hot filling at 195–200°F.

Peel oil enters the juice in the extraction process. Higher extraction pressures yield more juice but also force more peel oil into the juice stream. Too much oil gives 'off-flavors' to the juice and in the early days the amount of oil could be controlled by extractor pressure only. Since then de-oilers have been developed as an integral part of the pasteurizing process.

In the de-oiler the juice temperature is elevated above the can filling temperature. The juice is then flashed into a vacuum chamber controlled by a flash condenser and vacuum system to the can filling temperature. As a rule of thumb, a flash of 10°F, say from 205 to 195°F will remove approximately 1% of the mass as evaporated water vapors. This flash evaporation strips out some of the oil and the greater the difference between the heater and flash chamber, more evaporation occurs and more oil is removed. Since the oil has a higher boiling point than the juice, the removal of oil is an example of steam distillation or 'steam stripping'. Some lower boiling constituents in the juice which contribute to the juice essence are also removed in the de-oiling process. This is not considered objectionable in processing 'hot pack' canned products. The juice entering the can must be hot enough to pasteurize it (about 195°F). However, there is a limiting high temperature for taste reasons which, therefore, limits the amount of oil which can be removed by flashing and has an influence on extractor pressure settings.

Some de-oilers use heat exchangers to bring the juice up to filling temperature and heat the juice further by direct injection of live steam. The condensing of steam adds water to the juice but in the vacuum chamber an equal amount of water is flashed-off. It is the author's opinion that there is a danger of adding 'essence of boiler' or 'essence of steam pipe' to the juice with the use of this type of de-oiler. An advanced design uses indirect heating only. The flash chamber is connected to a stripping condenser through which the water and oil vapors pass. The water phase is condensed and returned to the juice and the oil exits through a vent line connected from the condenser to an oil recovery system.

The heated juice is pumped from the pasteurizer–de-oiler to the can filler bowl which has a demand float that regulates the juice flow through the system. Can fillers rotate and accept empty cans through mechanical timing devices driven by the filler which space the cans to synchronize with the filler valves. Under the rotating filler bowl, the cans are lifted by cam action and forced upward toward the bottom of the bowl. The cans contact the valves located in the bowl, push the valve stems up, opening the valves and the hot juice flows by gravity into the cans.

Filled cans exit the filler and enter the can closing machine which is driven by the filler and thus synchronized. A small stream of live steam is injected into the head space at the top of the can to expel air that would otherwise be trapped between the lid and the juice. Lids are applied and seamed and the cans exit the closer into a 'can twist' which inverts them so that the hot juice pasteurizes the lid.

The cans are then fed into a water spray can cooler. Cans enter the cooler at about 195°F and exit the cooler at about 100°F. They are stored in warehouses at ambient temperature.

Single strength orange juice is also filled into glass containers at 190–195°F, after which they are cooled first by warm water to avoid thermal shock, followed by chilled water. The coolers in this type of operation carry the bottles on a wire mesh belt (as compared to the spinning of the cans) with recirculating water sprays in sections along the length of the cooler, the warmest water at the inlet end. The outlet section uses chilled water (about 35°F) and cools the product to around 50°F after 1 h residence time in the cooler. This product is stored and distributed at cool temperatures and given a shorter shelf life than the canned product due to the inevitable color and taste changes with age.

8.7   Juice processing for concentrate

The obvious advantages of making concentrate from fruit juices are: (i) reduced volumes for storage and shipping, and (ii) if enough water is removed from the juice, the concentrate can be kept at ambient temperatures without spoiling. The most widely used technique is to apply heat to the product in order to evaporate water by boiling, though other processes, such as removing the water by freezing or ultrafiltration, have been tried on citrus juices without any wide acceptance.

Citrus juices are quite sensitive to heat and if exposed to elevated temperatures or even kept at ambient temperature for too long, flavor and color changes will occur. With high temperature–short time (HTST) pasteurizers orange juice may be heated quickly to 185–200°F, held for 30 s and quickly cooled without any color change. This process stops enzymatic reactions which would otherwise cause cloud separation and flavor and color changes.

There is, however, a non-enzymatic reaction that takes place in juice and concentrate that in time causes 'browning' due to sugars reacting with proteins. This reaction is retarded at low temperatures. For this reason all orange juice concentrate produced for reconstitution into juices is stored at temperatures of 20°F or lower. This allows for holding inventories of concentrate for more than 1 year if necessary.

The earliest orange concentrate in Florida and California was produced on steam-driven evaporators, sometimes only open kettles. The concentrate produced was not suitable for juice, but was used for drink bases and confections. By 1935 several processors were producing concentrate on vacuum evaporators designed by various equipment manufacturers originally for use on other products. These were medium to high temperature evaporators which by today's standards produced an inferior product. The concentrate was then heated and hot-filled into cans. The canned product was cooled by water spray and stored at ambient temperatures. The market for this product was in Europe and England and for the armed forces during World War II.

It was during the period between 1935 and the middle 1940s that considerable research and development was done on producing concentrate by using high vacuum–low temperature evaporators. Research and development showed that by the use of a low temperature (below 160°F) concentration process followed by frozen storage a much fresher tasting concentrate could be produced. The trade name for this product became 'Frozen Concentrate Orange Juice' or 'FCOJ'.

By 1950 equipment manufacturers had successfully developed commercial 'low temperature–high vacuum' evaporators, some of which used refrigeration 'heat pump' designs and others which used large recompression jets. Concentrate was pumped out at 58° Brix at temperatures around 55–60°F. Unpasteurized, single strength 'cut back' juice was blended in to make 42° Brix for canning, which partly compensated for the loss of volatiles in the concentrate process.

The concentrate produced by this low temperature technique was of very good quality and was packed mostly in 6 oz cans which were sent through a blast freezer and stored at −10°F.

It became apparent after some time that heat treatment was necessary to stabilize the canned product, as frozen storage temperatures could not always be guaranteed after leaving the plant refrigeration facilities. Microbial spoilage and/or cloud separation due to pectolytic enzyme action would sometimes occur. Around the mid-1950s, high temperature–short time pasteurizers were added to the concentrate process which would bring the temperature of the juice to around 195°F with enough 'holding time' to kill the microorganisms and de-activate the pectinesterase.

8.7.1   Characteristics of 1950s evaporators

To understand the significance of the advance taken by later designs, it is necessary to examine the first commercially successful evaporators. These early 'low temperature' evaporators were of the recirculating type, shell and tube 'falling film' design. Stainless steel tubes were installed vertically in a configuration referred to as a 'tube nest' or 'evaporator body'. Juice entered the top of tubes and flowed downward as a film against the inside wall of the tubes. Evaporated vapors traveled downward through the center of the tubes and exited with the juice through a vapor–liquid separator. The juice was pumped back to the top of the tube nest and into the tubes through some sort of distribution device. On the outside of the tubes, referred to as the 'shell side' of the tube nest, steam condensed under vacuum giving up its heat to the boiling juice on the 'tube side' of the tube nest.

Tube nests were arranged in 'stages' and 'effects' (these terms are described later in this chapter) and each stage held a considerable amount of product. When the correct concentration was reached in the final stage, concentrate was pumped out and replaced from earlier stages under liquid level control. Brix control of the product from these evaporators was good because of the large stabilizing effect of the bulk of product in process.

Maximum evaporator operating temperatures varied with manufacturer's design, up to 120°F for steam driven types with recompression jets. Refrigeration heat pump evaporators' highest process temperature was around 75°F (not including the pasteurizer).

These evaporators required considerable heat transfer surface because of low temperature difference across the tube walls and because of high resistance to heat transfer caused by cold thick product inside the tubes. Vapor–liquid separators were large as were vapor transfer ducts by virtue of the large volume of the evaporated water vapors. These large spaces were almost impossible to clean-in-place (CIP) and after 30–40 h of operation the separator manways and top distribution chambers needed to be opened and manually cleaned. Cleaning required a considerable amount of down-time and man hours. Recharging these evaporators with juice, restarting and reaching steady operation at 58° Brix pump out, also added to loss of production time.

8.7.2   Modern evaporators for citrus fruit

The low temperature evaporators had been designed on the basis that low evaporation temperatures would compensate for the long residence time in the process; and avoid heat damage. The successful use of high temperature–short time pasteurizers demonstrated that no significant heat damage was done to the product provided the time of elevated temperatures was kept short.

This experience along with the aforementioned deficiencies of the 'low temperature' evaporator design gave rise to the development of the thermally accelerated short time evaporator (TASTE), which at the present time is used almost exclusively in the world-wide citrus industry. These evaporators are

Figure 8.12 TASTE evaporators (three units). Courtesy of Gulf Machinery Co.

of the multi-effect, multi-stage, single pass design and the first one was installed in Florida in 1958 (Figure 8.12).

For the reader who is not familiar with evaporator terminology it might do well here to explain the difference between the evaporator 'effect' and evaporator 'stage'. Effect defines the heat flow through an evaporator, the first effect always the warmest which receives the energy for driving the evaporator. In refrigeration low temperature 'heat pump' evaporators, which were usually double effect units, the first effect received heat as the condenser side of a refrigeration system. The water vapor from evaporation inside the tubes of the first effect traveled through vapor lines to the shell side of the second effect, condensed and gave up its heat to the boiling liquid inside the tubes of the second effect. The water evaporated in the second effect entered the shell side of a vapor condenser, condensed and transferred its heat to a boiling refrigerant inside the tubes. The refrigerant vapors were boosted by the compressors and the refrigeration condensing heat was absorbed by the first effect, completing the cycle.

The TASTE evaporator is driven by boiler steam directed to its first effect shell side. For each pound of steam condensed, an equal amount of water vapors are evaporated and sent to the next (second) effect, from the second to the third and so on, with the evaporated vapors from the last effect condensed by cool water, usually recirculated from a cooling tower.

The stages of an evaporator define the product flow, the first stage receiving the single strength feed juice. Any effect of an evaporator can be designed to receive feed liquid. A 'forward flow' evaporator feeds the first effect and product flows parallel with the heat flow through the effects. 'Reverse flow' feeds product to the last effect and pumps out of the first effect. 'Mixed flow' loosely defines an evaporator which is staged as neither of the above.

The number of effects used determines the energy efficiency of the evaporator. A single effect steam driven evaporator will evaporate approximately one pound of water for each pound of boiler steam condensed. In a two effect unit, the first effect is the boiler for the second, thus two pounds of evaporation per pound of steam is achieved, and so on. This pound for pound ratio does not exactly apply to citrus evaporators as it does not include the heat required for preheating (heating the juice to the first effect temperature). Other forms of evaporator design have been offered to the citrus industry, such as the use of vapor recompression, in order to reduce the steam and/or energy requirements. However, the evaporation to steam ratio for a seven effect TASTE evaporator is 6.1:1. That is, 1 lb of steam will cause 6.1 lb of evaporation (including the preheating load). This energy efficiency is more favorable than that achieved with recompression, especially when considering the greater capital investment for the recompression plant. Multi-effect steam driven evaporators are much more forgiving to upsets caused by infeed flow and Brix variables and are thus easier to control. They may also be operated at lower than design capacities simply by throttling the steam valve.

The earliest TASTE evaporators were typically three or four effect models having five or six juice stages with the final stages in the last effect. In single pass evaporators (no recirculation), the liquid flow rate through the stages diminishes as water is removed. When concentrating orange from 12° Brix to 65° Brix, the liquid leaving the last stage has a flow rate of about 15% (by volume) of the liquid feeding the first juice stage. The heat transfer rates in the high Brix stages are much lower than in the low Brix stages. Additional heat transfer surface is therefore required to match the heat flow from the forward effect yet there is much less liquid to wet the inside tube surface. For this reason, the last effect of the TASTE evaporator is broken up into a number of stages, in series on the juice side with all stages receiving the same vapor on the shell side. Four effect evaporators typically have three stages in the last effect.

The escalation of fuel prices in the 1970s gave rise to the development of the seven effect TASTE evaporator shown in Figure 8.13. This diagram demonstrates a seven effect, seven stage, mixed flow, TASTE evaporator. Juice is fed to the evaporator through inter-effect preheaters and enters the first stage which is located in the fifth effect. At this point, and at full capacity, the feed juice has been heated to about 160°F quickly through preheaters and it immediately flashes to the tube side temperature of the fifth effect (155°F). One-seventh (approximately) of the total water removal takes place in this stage along with removal of dissolved and entrained air, and most volatiles, referred to as essences. This 'mixed flow' design permits the removal of these constituents from the juice at lower temperatures than would be experienced in a forward flow design. The reasoning is that if the juice were heated to the first effect temperature (about 200°F) while containing air and other volatiles, oxidation would be accelerated by the high temperatures and flavor changes would result.

Juice flows by gravity to the second stage which is the sixth effect and is then pumped through further inter-effect preheaters, through the final steam-driven preheater-pasteurizer and into the first effect (which is the third stage). The final preheater raises the juice to its highest temperature, around 205°F. This preheater is sometimes referred to as the 'final stabilizer' as, besides pasteurizing, it also stops pectolytic enzyme action, thus stabilizing the cloud. After the third stage, no further preheating (nor reheating) is necessary and the juice flows forward through the next three effects and finally to the last effect where the concentrate enters a flash cooler at around 110°F and is cooled to 55°F.

The steam flow rate to the TASTE evaporator determines the evaporation rate and adjustments of the steam and juice feed valve are used to control pump-out concentration. Steam and juice flows can be held steady by simple control devices and if, in these single pass evaporators, the infeed Brix remains constant, the pump-out Brix will not change. Changes of infeed Brix do occur and this is the uncontrollable variable. With constant steam and feed flows a

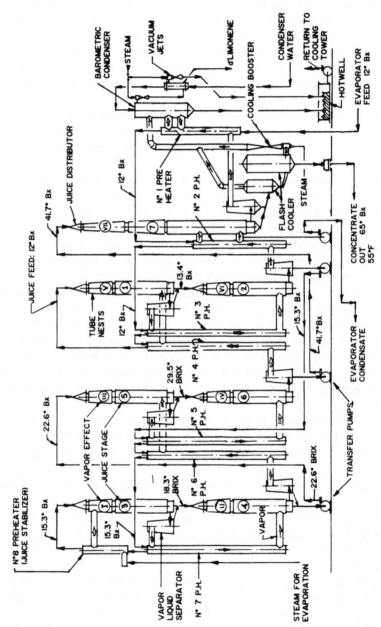

Figure 8.13 Schematic diagram of seven effect, seven stage TASTE evaporator. Courtesy of Gulf Machinery Co.

change in feed Brix from 11° to 12° would cause the pump-out to rise from 65° to 70.9° Brix. However, because of citrus harvesting practices with respect to fruit ripeness, evaporator infeed Brix variations are very small, usually much less than 1° Brix and control of pump-out Brix is easily achieved by manual adjustment of flow valves after periodic checking of Brix or measurement of temperature changes occurring in the evaporator with Brix changes.

TASTE evaporators are 'cleaned-in-place' and need not be opened between major overhauls. Periodic cleaning coincides with general plant clean-up, about twice per day. Evaporator cleaning takes 45 min to 1 h, so 22 h of production is available for each day.

Concentrate is pumped from the evaporator flash cooler to batch blending tanks with stirring type agitators where the batch of concentrate is standardized and pumped through chillers to large stainless storage tanks located inside a refrigerated space. 65° Brix concentrate is chilled and stored at about 15–18°F. Concentrates from different citrus varieties and ratios are separated into several tanks and inventoried. This allows future blending to ensure a uniform product for canning or shipping bulk. Automated 'tank farms' allow pumping from several storage tanks to a blending station, also on a batch basis, where cut-back juice and/or essences can be added depending on the market requirement.

Concentrate is canned at 44.8° Brix by piston fillers, capped, conveyed through a blast freezer and warehoused before shipping at 0°F. Bulk concentrate is usually loaded into tanker trucks at 15–20°F. Significant quantities are delivered to bottlers who do their own blending and reconstituting for the reconstituted single strength dairy shelf market throughout the United States.

8.8 Essence recovery

Citrus essences are components of the juices that are boiled off in the evaporation process. When condensed, they form liquid aqueous and oil phases. Over 140 compounds have so far been identified in aqueous and oil orange essences. Both phases contain compounds with boiling points lower than water. The two phases are separated by decanting and/or centrifuging, graded and stored separately in tanks for future use. Essence (aqueous and oil) is added back to the concentrate in the blending operation just ahead of canning and freezing. The use of high quality water and oil essences has more or less replaced the use of cut-back juice.

The development of essence recovery systems for citrus juices began in the late 1950s and various designs were offered by the evaporator manufacturers to fit their respective processes. These early units were not too successful in producing consistently good products, partly because of lack of experience in using the essence and partly due to off-flavor produced in the evaporators. In the early 1960s, an essence system was developed that would fit onto, and be an integral part of the TASTE evaporator process. By 1964, citrus processors were producing high quality essences for use and for sale.

When the feed juice enters the first stage of an evaporator, the volatile components evaporate with the water. The evaporated vapors enter the shell side of the next effect (not necessarily the next juice stage) and are condensed. As the water condenses, volatile essence vapors which are 'low boilers' pass through the shell side of this effect to the essence system. The essence system consists of a series of fractionaters, condensers, scrubbers, and chillers that produce water and oil phase essence at rates dependent on the concentration required. Essence drains from the system by gravity through a barometric leg to a seal-decanter where oil is separated and both phases are stored in batches for grading and subsequent use.

Aqueous phase essence strength is controlled by the evaporator-essence operator and is measured with an alcohol hydrometer. Strength is adjusted at 12–15% apparent alcohol. The US government considers the essence equipment an alcohol still, though the process does not produce ethanol. However, alcohol taxes are not levied if the aqueous phase is produced at 15% or lower as read on an approved measuring device. Some processors check strength and recovery by measuring the quantity recovered of a certain compound, such as acetaldehyde or ethyl acetate, using sophisticated laboratory equipment.

The quality of the essence is best graded by experienced noses and taste panels. The processing of Valencia oranges, which make up a large portion of the crop, produces the bulk of the orange essence manufactured. Early variety fruit produces little essence and sometimes of low quality. Although concentrate flavor can be enhanced by adding high quality essence, poor essence can downgrade flavor and is not used.

The amount of essence material blended into the concentrate is information that is proprietary to each respective manufacturer. More essence is produced than is used in frozen concentrated orange juice (FCOJ). Other markets for essence are flavoring houses and beverage manufacturers.

8.9 Chilled juice from concentrate

The market for natural drinks increases every year. With the improvement of processing techniques of concentrate and essences, bottled and carton filled single strength juices have gained considerable favor with the public. Milk bottlers in all cities throughout the world have most of the equipment necessary for blending, pasteurizing and bottling juices.

A considerable amount of the citrus concentrate produced in the world's citrus areas is shipped in bulk form as 65° Brix orange concentrate, the final destination being a reconstitution plant. Inland shipments within the United States are made with insulated bulk tanker trucks or in polyethylene lined

55 US gallon drums. Shipments from Brazil are carried in refrigerated sea tankers to tank farm terminals at various sea ports. Smaller concentrate producers export in 55 US gallon drums.

Concentrate is pumped from cold storage tank farms to blending stations at the concentrate plant and blended to meet customer specifications, which may or may not require the addition of essences at this point. In Florida, it is generally pumped from blend tanks into 'dairy tanker trucks' for shipment, though some users' facilities are too small and receive shipments in drums. The dairy tankers hold approximately 4500 US gallons. Tankers are filled at about 18°F. Travel time to the reprocessors may take as long as two days in the United States and the temperature rise in the unrefrigerated tankers may be 3–5°F.

The 'reprocessing' facility receives the concentrate into large cold wall tanks similar to those used at the juice plant; these are cooled by circulating chilled propylene glycol solution through their external jackets. Concentrate is then pumped to blending tanks where treated water is added. Essence and oil is blended in, if this has not already been done at the concentrating plant. Pulp, which is supplied by the juice plant in 40 lb frozen packages, is sometimes added to give the juice a 'fresh feel'. The reconstituted juice is pasteurized, then chilled and filled into bottles or cartons and shipped to the market.

8.10 Pulp wash

As mentioned previously, flavor consideration dictates the 'tightness' of extraction and finishing processes. Higher pressures give higher yields but at the same time force more macerated fruit cells through the screen holes into the juice. In Florida, juice yields are set by the state test room data to ensure fairness throughout the industry and to maintain high quality FCOJ. High pulp levels require centrifuging before concentration, which results in yield loss. High finisher pressures also force pectinous materials through screens along with the macerated pulp, which results in high viscosity concentrates that are hard to handle and difficult to reconstitute. For these reasons, not all the juice is extracted from the pulp leaving the juice finisher.

In some processing plants the pulp is treated to remove this juice by back wash leaching, 'pulp washing'. Mixers and finishers are set up in series to wash juice solids from the pulp flow. Four stage systems are in general use, one stage consisting of a mixer and finisher, and finisher pressures are set lower than prime production to avoid further size reduction of the pulp.

The liquid produced by pulp washing usually contains 5–6% soluble solids. Before concentrating, it must be subjected to enzyme treatment to break down pectinous matter that would otherwise cause the concentrated product to become too viscous. The liquid is then centrifuged and sent to the evaporator, where it is concentrated to around 60° Brix. The yield of soluble solids from the original fruit can be increased up to about 10% with the pulp wash equipment. Pulp wash concentrate is regarded as inferior quality and is not normally used in citrus juice for drinking as such. In the United States, it must be labeled 'pulp wash' and cannot be added back to FCOJ. However, with careful processing high quality, good flavor concentrate can be produced which is used in various types of citrus drinks.

8.11   Frozen pulp processing

Adding whole pulp sacs to blended citrus juices adds a 'fresh feel' to the drink and this type of product has gained wide acceptance. This has given rise to the development of pulp recovery systems to furnish the market. The amount of pulp sacs required for the market is small compared to the volume of juice extracted so only a small portion of the extracted juice is directed away from the primary finisher to a 'pulp recovery system'.

The pulp sacs are the small pieces of pulp which in the FMC extractor pass through the extractor strainer tube. This juice is directed to a separate finisher, where excess juice is strained with a very loose finish, with the pressure so low that the pulpy fraction separated from the juice is still pumpable. The juice from Brown extractors contains more pulp, rag and seed and a portion of this juice is directed to a special finisher with larger than normal screen holes. Large seeds and rag are screened out and the flow draining through the screen contains the pulp sacs. This juice–pulp fraction is further finished similarly to the FMC process to produce a pulp fraction which is still pumpable.

The pulp fraction from either process, referred to as 'pulpy juice', is pumped through a cyclone type centrifugal separator to remove small immature seeds and defects and is then sent to a regenerative type pasteurizer and through a chiller to bring the final temperature to around 45°F. The pasteurized, chilled pulp juice is then sent through another finisher for removal of excess juice. The pulp fraction is dropped by gravity and packed into 40 lb containers and frozen. The juice fraction is pumped back to juice processing.

Pulp sacs are usually added back to the juice during the reconstitution to single strength, though may sometimes be added to the bulk concentrate at the concentrate facility. The frozen 40 lb portions are passed through a 'chopper' or 'shaver', which breaks the block into small pieces for quicker melting. Some reconstitution facilities mount the chopper directly above the blend tanks and drop the pieces into the tank to melt and be mixed into the juice (or concentrate) by the agitators. Other facilities melt the frozen shavings before pumping to the blend tanks.

8.12   Manufacture of citrus cold-pressed oil

The skins of citrus fruit contain oil which has flavor fractions that can enhance the taste of the reconstituted juice made from concentrate. Aqueous essence, essence oil and peel oil are added during concentrate blending or single strength blending operations to replace volatiles lost in the concentration process. Essence oil is mostly peel oil which enters the juice during extraction and is distilled from the juice by the evaporator and collected in the essence system.

Unlike essence oil cold-pressed oil receives no heat treatment, being extracted by mechanical action only. In the FMC extractor, oil cells in the skin are ruptured, forcing some oil outward from the fruit where it is washed down by water sprays. The reaming action of the Brown juice extractors does not break up the peel so for recovering oil, Brown offers two options: (i) The Brown oil extractor (BOE) machines or (ii) the Brown 'peel shaver'. The BOE is set in the raw fruit line ahead of the juice extractors. Its action is similar to the historical scarifier, where fruit is passed over spinning abrasive rollers with water sprays to wash down the oil and pieces of skin. The rollers on the BOE machine, however, puncture the oil cells without removing skin which could impart off-flavors to the oil. The Brown 'peel shaver' is a device fitted to the peel discharge of each Brown juice extractor. A knife quarters the peel halves, and the peel drops through a series of rollers and blades which flattens the pieces, separates the albedo from the flavido and presses the oil from the flavido (outer, or colored part of the rind containing the oil cells). Oil is washed down from the final pressing rolls with water sprays.

The oil–water mixture from either FMC or Brown equipment is screened and centrifuged and the oil sent to storage. Water is recycled through the sprays but since other materials besides oil are expressed from the peel into the water, continuous dilution of the recycled water is necessary. The waste water outflow from the system contains soluble and insoluble solids and this effluent may read as high as 2° or 3° Brix on a refractometer. The effluent flow is usually about 3–5 lb water per 90 lb box of oranges run and it is not readily bio-degradable. It is therefore sent to the 'feed mill' where it is concentrated along with 'press liquor' in the feed mill evaporator.

8.13 Manufacture of livestock feed from citrus peel

The term 'feed mill' is used for this operation as the product is used as livestock food and is not to be confused with the word 'feed' as it is used in 'evaporator feed' or 'dryer feed'. The feed mill carries out a drying and evaporating operation where water is removed from peel, pulp, rag, seed and waste effluents (Figure 8.14). Though its product is not for human consumption, the feed mill is a necessary part of a citrus juice operation and also produces a saleable product.

Peel and pulp from juice extraction and finishing is transferred through screw conveyors to a 'peel bin' sized to hold about 4 h material at maximum

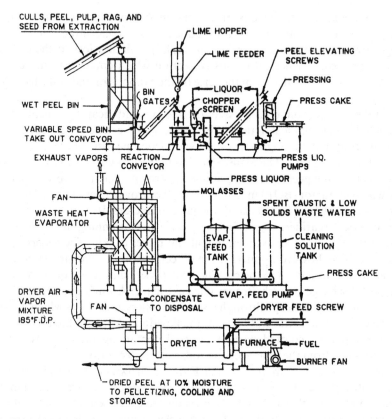

Figure 8.14 Schematic diagram of citrus feed mill. Courtesy of Gulf Machinery Co.

plant rate. This mass should not be held too long before drying, especially during the warmer days of spring in Florida, because fermentation and other reactions take place which causes difficulties in the process. Density of the mass in the bin can be figured equal to water ($\pm 62 \, lb/ft^3$) and about 40–45 lb per box of oranges run. The bottom of the bin is sloped at 50° from horizontal and the mass is gravity fed to the 'bin take-out' screw through adjustable gates. The take-out screw is equipped with a variable speed drive which regulates the flow into the feed mill.

The moisture content of the mass ranges from 76% to 82% depending on the condition and variety of the fruit and also upon how much water has been carried over from the oil recovery and pulp wash operations.

The mass from the peel bin is conveyed to a hammermill (shredder or chopper) where it is cut into pieces up to 1 inch. Lime (calcium hydroxide or calcium oxide) is added at the chopper to raise the pH to between 6.5 and 7. The peel is dropped into a mixer-reaction conveyor where the lime is thoroughly mixed. Reaction time for the mixture should be about 12 min at ambient temperatures. The higher alkalinity occurring after the lime addition brings about a rapid degrading and demethylation of the pectins in the peel and allows for easier separation of the liquid.

From the reaction conveyor the peel is pressed to reduce moisture to 72% or lower. The soluble solids content of the press liquid will range from 8% to 12% (oranges). The press cake is directed to the dryer and the press liquid is fed to an evaporator running on waste heat from the dryer.

8.13.1 Peel dryer

The peel dryer (Figure 8.15) consists of an oil- or gas-fired furnace connected to a rotary kiln. This is a horizontal rotating cylindrical drum. Hot air dries and conveys the mass through the drum entering from the furnace at about 1200°F and exiting with the dried peel at around 275°F. The peel travels through a series of baffles and tumblers in the rotating drum and exits the dryer at about 10% moisture content.

The peel dryer used in Florida citrus processing plants is a design which produces superheated exhaust vapors at 240–275°F and with a dew point of 185°F. Each pound of dry air that exits the dryer contains 0.83 lb of water vapor with usable heat value. This air–vapor mixture is directed to the feed

Figure 8.15 Peel dryer. Courtesy of Gulf Machinery Co.

mill 'waste heat evaporator' where it passes through the shell side of the evaporator first effect.

8.13.2 Waste heat evaporator

The earliest peel dryers used in the citrus industry were fed unpressed peel and the total water removal load was accomplished by the dryer. As plants grew in capacity, presses were added along with steam driven 'press liquor' evaporators. Removing liquid from the peel by pressing reduces dryer loads by 30–50%, depending upon the 'wetness' of the mass entering the feed mill. Removing water from the press liquid by multi-effect evaporation is more fuel efficient than removing water in the dryer. Thus by adding presses and

Figure 8.16 Waste heat evaporation. Courtesy of Gulf Machinery Co.

evaporators, capacities of feed mills could be increased without additional drying equipment and production energy requirements could be reduced with respect to overall feed mill load.

In the mid-1960s a considerable amount of development was done to design an evaporator which would use as its driving force the available heat energy from the dryer stack which normally discharged into the atmosphere, hence the name 'waste heat evaporator' (Figure 8.16). The development also included a new dryer design which, by recycling some of the dryer exhaust back to the furnace gave a high dew point in the air–vapor mixture (AVM) leaving the dryer.

As the AVM passes through the shell side of the first effect of the waste heat evaporator, it is cooled by transfer of heat through evaporator tubes. In the larger waste heat evaporators, the AVM passes through the shell side of a series of tube nests, all of which are first effect bodies. In these units, the AVM exhausts from the waste heat evaporator at 140°F. The largest dryers in Florida and Brazil remove up to 60 000 lb/h of water from the peel. With 60 000 lb/h water vapor entering the waste heat evaporator at an AVM dew point of 185°F and exhausting at a dew point of 140°F, 49 000 lb/h of vapor is condensed, transferring its heat to the boiling liquid inside the tubes.

The dew point drops as the AVM passes through the tube nests and each succeeding first effect body operates at a lower temperature. The first, or hottest in the series, has a high enough operating temperature to be connected to two more effects. The boiling liquid inside its tubes is around 165–170°F and the evaporated vapor is directed to the shell side of a second effect, and its boiling vapors to a third effect. The boiling vapors from the third effect are directed to a barometric type condenser. By arranging effects and liquid stages in a manner which efficiently uses the heat available, a waste heat evaporator can have a water removal capacity twice the rate of the peel dryer. This gives the citrus processor the ability to concentrate other plant effluents, such as oil mill wastes, spent cleaning solutions and high solids wash water in the feed mill which would otherwise require another form of waste treatment.

Liquid from the pressing operation is screened and pumped to a tank from which the evaporator is fed. The press liquor along with other effluents is concentrated to about 50% solids, this figure being limited by viscosity, and is returned to the pressing operation where the concentrate is combined with the press cake and fed to the dryer.

The dried hot peel at about 12–15% moisture is conveyed to a pellet mill where it is extruded into pellets and then sent to a cooler. The cooler passes air through the mass and discharges the pellets at about 100–110°F, 10% moisture. The cooled pellets are sent to storage where they are usually deposited in piles on a concrete floor inside a ventilated building. The fresh pellets need some curing time and further ventilation to prevent overheating from chemical reactions that can cause spontaneous combustion. The piles of bulk fresh pellets are kept small and usually moved around by a front end loader after sitting about 24 h, by which time the chemical reactions have finished.

The pellets are sold to the livestock feed market and usually transported in bulk. Analysis shows that the dried peel, pulp, rag and seed is high in carbohydrates (typically about 63%) and low in crude protein, fiber and fat (typically 13%, 6.2% and 3.5%, respectively). It is used as a 'filler' by combining with other grains of higher nutrient value.

Further reading

Hendrickson, R. and Kesterson, J. W. *By-Products of Florida Citrus*, Bulletin No. 698. Agricultural Experiment Stations, Institute of Food and Agricultural Sciences, University of Florida, Gainesville, FL, USA.

Hugot, E. *Handbook of Cane Sugar Engineering*, Elsevier, London.

Kern, D. Q. *Process Heat Transfer*, McGraw-Hill, New York.

Kesterson, J. W., Hendrickson, R. and Braddock, R. J. *Florida Citrus Oils*, Technical Bulletin No. 749, Agricultural Experiment Stations, Institute of Food and Agricultural Sciences, University of Florida, Gainesville, FL, USA.

Nagy, Shaw and Veldhuis. *Citrus Science and Technology*, Vols. 1, 2, AVI Publishing Company, Westport, CT.

Papers of Annual Citrus Processors Meeting, University of Florida, Agricultural Research and Education Center, PO Box 1088, Lake Alfred, FL 33850, USA.

Proceedings of the Annual Short Course for the Food Industry, Cooperative Extension Service, Food Science and Human Nutrition Department, Institute of Food and Agricultural Sciences, University of Florida, Gainesville, FL, USA.

Quality Control Manual for Citrus Processing Plants, Intercit Inc, 1575 Tenth St. South, Safety Harbor, FL 34695, USA.

Transactions of the Citrus Engineering Conference 1955–1989, Florida section, American Society of Mechanical Engineers.

9 Juice enhancement by ion exchange and adsorbent technologies

S. I. NORMAN

9.1 Overview

This chapter describes the various processes employed by industry to enhance the quality of a juice or juice by-product. These enhancement processes can be broadly described as being either ion exchange and/or adsorption technologies. The various juices include, but are not limited to, apple, pear, white grape, orange juice and pineapple mill juice, a juice by-product. Juice enhancement refers to the improvement in a juice's characteristics as dictated by consumer demand for a particular juice or juice related drink. An enhancement typically refers to reduction of a juice's acidity, color, odor, flavor, improvement in shelf stability or a combination of these characteristics. In order for the reader to better understand the various enhancement processes, a brief historical summary of the technology along with a description of the various media is offered.

9.2 History

Ever since man satisfied his 'sweet-tooth' with honey and plants containing sugar, we have been on a quest to grow, harvest and purify sugar and sugar related products. The first use of adsorption is lost in antiquity, although records date back to the time of Aristotle (Kunin and Myers, 1950; Kunin, 1958; Helfferich, 1962). Perhaps it was associated with the observation that water tasted differently when it was treated with charred wood (Kunin and Myers, 1950; Kunin, 1958; Helfferich, 1962; Keller et al., 1987). Sugar cane, sugar beets and fruit wines all came under man's scrutiny well before the Christian era. Materials used to decolorize solutions were known in the fifteenth century. Bonechar was commercially introduced to remove color from sugar solutions in the late eighteenth century (Keller et al., 1987). The earliest research in the field of ion exchange was reported by two English chemists, Thompson and Way in 1850. They observed that a soil sample exchanged ammonia for calcium (Thompson, 1850; Way, 1850, 1856; Kunin and Myers, 1950; Schubert and Nachod, 1956; Mindler, 1957; Kressman, 1957; Kunin, 1958; Helfferich, 1962; Wheaton and Seamster, 1966; Applebaum, 1968; Keller et al., 1987). However, they thought that the reaction was irreversible. Eichorn in 1858 discovered that the reaction was indeed reversible (Applebaum, 1968). The first suggestion for an application of ion exchange occurred in 1896 by Harm. Harm's German patent outlines a method of treating sugar solutions with a calcium form exchanger in order to remove melasigenic sodium and potassium ions (Harm, 1896; Cantor and Spitz, 1956; Schubert and Nachod, 1956; Mindler, 1957).

In 1935, Zeo-Carb, a sulfonated coal, was introduced as the first organic cation exchanger (Mindler, 1957; Applebaum, 1968). Shortly after this, an anion exchanger was developed that could remove acids. Thus, by using a cation exchanger in the hydrogen form followed by an acid adsorbing anion resin, one could demineralize a solution (Mindler, 1957; Applebaum, 1968). In the same year, Adams and Holmes set the stage for modern ion exchange technology by discovering that certain synthetic resins had ion exchange capabilities (Kunin and Myers, 1950; Cantor and Spitz, 1956; Mindler, 1957; Kunin, 1958). Austerweil, in 1938 postulated that with the acid adsorbing anion resin one could deacidify various fruit juices (Austerweil, 1957). Later, others laid claim to recovering the adsorbed acids from the anion resin (Austerweil, 1957). Work continued on developing higher capacity, more stable resins. This resulted in several commercial fruit juice applications in the 1940s. Pineapple mill juice, the juice from the fruit hulls, was de-mineralized and de-colorized using cation and anion resins to yield a syrup suitable for canning (Cantor and Spitz, 1956; Austerweil, 1957; Diamond Shamrock Chemical Company, 1972). This technology is still in practice today. Apple juice during the 1940s was de-ionized using cation and anion resins. The resultant juice was used as a tobacco humectant (Cantor and Spitz, 1956; Mindler, 1957). The basic practices today differ very little from 40 years ago except that the resins have been improved by the resin manufacturers. This continuous improvement has been the driving force behind broader industry implementation of the technology as a means of offering fruit juices and fruit drinks to the consumer.

9.3 Resin chemistry

Ion exchange may be defined as the reversible interchange of ions between a solid and a liquid phase in which there is no permanent change in the structure of the solid (Wheaton and Seamster, 1966; The Dow Chemical Company, 1985). Ion exchange resins are a special class of polyelectrolytes. They consist of a cross-linked polymer matrix which is functionalized to provide ion exchange capability. An important feature of ion exchange resins is the ability to reversibly exchange ions and ionizable compounds without a permanent change in resin structure (Kunin and Myers, 1950; Kressman, 1957; Kunin, 1958; Helfferich, 1962; Wheaton and Seamster, 1966; The Dow Chemical Company, 1985). These resins are also insoluble in the process stream. They are important in the purification of water and food as well as finding utility in chemical synthesis, mining, catalysis, pharmaceuticals and medical research. The information presented in the next few pages has been taken from Helfferich (1962), Kunin (1958) and from Wheaton and Seamster (1966).

9.3.1   Polymer matrices

Four classes of commercially significant ion exchange resins are synthesized from the polymerization of styrene (vinyl benzene) and divinylbenzene (Figure 9.1). A strong acid cation resin is used to exchange cations. The two strong base anion resins are used to exchange anions. The fourth class, weak base anion resin is used to remove free acidity.

Due to the chemical nature of these resins, very strong interactions can

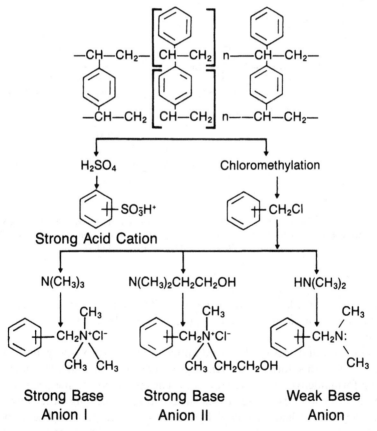

Figure 9.1 Styrene/divinylbenzene ion exchange resin structures. Redrawn and adapted from The Dow Chemical Company (1985).

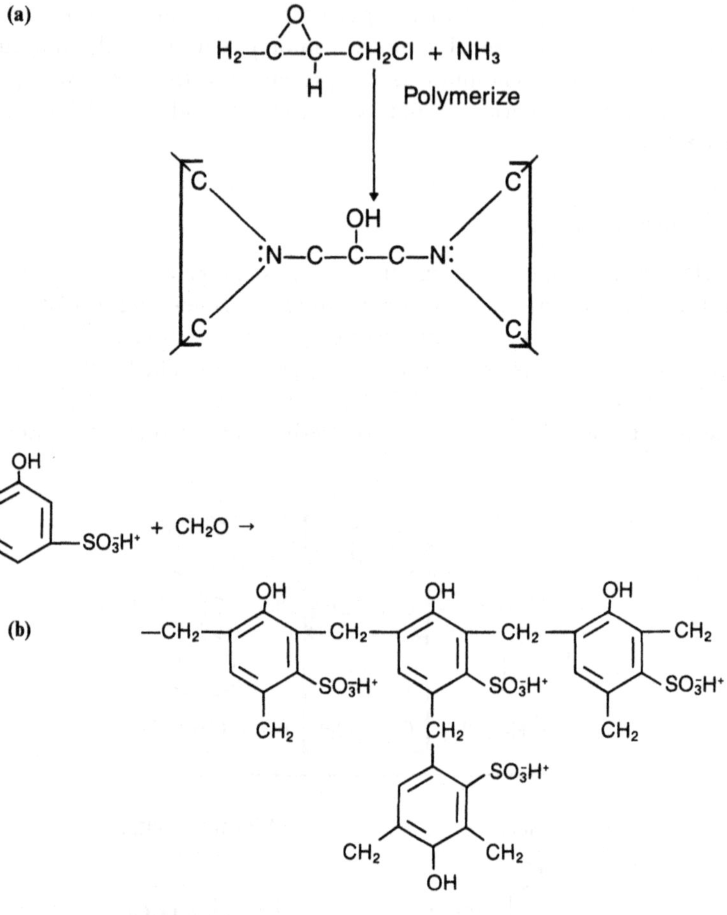

(a)

$$H_2-C-C-CH_2Cl + NH_3$$

Polymerize

Figure 9.2 Other resin structures. (a) Weak base epoxyamine resin structure. Redrawn and adapted from The Dow Chemical Company (1985). (b) Strong acid cation phenol-formaldehyde resin structure. Redrawn and adapted from Kunin (1958).

develop between like aromatic structures. This interaction is useful when these resins are used to remove polyphenolic color bodies from various food streams.

Other resin matrices used in food applications are epoxy-polyamine weak base anion resins formed by reacting epichlorhydrin and ammonia and phenol-formaldehyde resins (Figure 9.2). The epoxy-polyamine resins have a high capacity for adsorbing strong acids without affecting other desirable food properties. Phenol-formaldehyde resins can be produced as either a cation or as an anion resin. The weak base form is widely used in the food industry to remove strong acids. The weak base phenol-formaldehyde resin

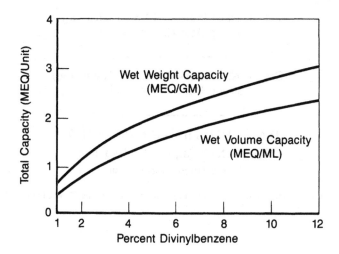

Figure 9.3 Total ion exchange capacity, in milliequivalents, of a styrenic strong acid cation resin in the H$^+$ form as a function of cross-linking with divinylbenzene. Redrawn and adapted from The Dow Chemical Company (1985).

also adsorbs phenolic color bodies due to the interaction between like aromatic structures. However, these resins have been largely replaced by styrene-divinylbenzene resins which have longer life and higher thermal stability.

Variations in the properties of the resin synthesized from styrene-divinyl-benzene co-polymers are possible by varying the amount of cross-linking agent, divinylbenzene. Figure 9.3 shows how the exchange capacity of a strong acid cation resin can vary depending upon the amount of divinyl-benzene used to make the co-polymer matrix.

Macroporous resins can be produced by the use of inert diluents in-corporated into the monomer mixture prior to co-polymerization. The use of inert diluents results in a resin matrix having higher porosity or void volume. By varying the amount of divinylbenzene and inert diluent, one can change the resin's physical strength and porosity. Since resins made with inert diluents contain some void volume, there is actually less functionalized polymer per unit volume. Therefore, these 'macroporous' resins have less ion exchange capacity than the gel-type resins, where no diluents are used during co-polymerization.

Macroporous resins are most useful when their high porosity is desirable from the standpoint of processing materials that contain large molecules which need to permeate into the resin. Since most food streams contain quite large amounts (4–20% by wt) of both large and small water soluble organic molecules (sugars, vitamins, proteins, phenolic color bodies, etc.),

$$R\text{-}H^+ + NaCl\ (aq) \leftrightarrow R\text{-}Na^+ + HCl\ (aq)$$
Strong Acid Cation Resin

$$R\text{-}OH^- + NaCl\ (aq) \leftrightarrow R\text{-}Cl^- + NaOH\ (aq)$$
Strong Base Anion Resin

$$R_3N\text{:} + HCl\ (aq) \leftrightarrow R_3N\text{:}HCl + H_2O$$
$$R_3N\text{:} + NaCl\ (aq) \longrightarrow No\ Reaction$$
Weak Base Anion Resin

Figure 9.4 Ion exchange reactions. Redrawn and adapted from The Dow Chemical Company (1985).

macroporous resins are generally preferred by the food industry for their processing needs.

Ion exchange reactions are equilibrium reactions which are stoichiometric, reversible and are possible with any ionizable compound. The subsequent reaction that occurs between an ion exchange resin and the solution depends upon the contact time, the selectivity of the resin for the ions or molecules involved and the kinetics of that particular reaction.

The stoichiometric nature of ion exchange reactions allows the prediction of resin and equipment requirements based upon lab scale testing. The reversibility of ion exchange reactions allows the re-use of the resin. In column operation, the resins can be selectively and repeatedly converted from one ionic form to another. The three important ion exchange reactions for juice enhancement are shown in Figure 9.4, where R represents the ion exchange resin.

9.3.2 Equilibrium of ion exchange

Due to the reversibility of the ion exchange reaction, it is necessary to understand the chemical equilibrium of such reactions. The chemical equilibrium of any ion exchange reaction dictates the quantities of regenerant chemical needed to reverse the reaction, thereby allowing the resin's re-use. Figure 9.5 illustrates a typical strong acid cation resin reaction. The figure also depicts the equation for the equilibrium constant, K. If K is a large number, the reverse reaction is less favorable than the forward reaction. Therefore, a large K value means more regenerant chemical is needed to drive the reverse reaction in order to regenerate the resin and make it suitable for re-use.

An ion exchange resin's selectivity, i.e. relative affinity for various ions is affected by many factors. The degree of cross-linking, the ion's valence and

$$R\text{-}H^+ + Na^+Cl^- \rightleftharpoons R\text{-}Na^+ + H^+Cl^-$$

$$K = \frac{[R\text{-}Na^+]\,[H^+Cl^-]}{[R\text{-}H^+]\,[Na^+Cl^-]}$$

Figure 9.5 Strong acid cation resin chemistry. Redrawn and adapted from The Dow Chemical Company (1985).

size, the solution's polarity and specific interactions between the resin and the ionizable species are the key factors determining a resin's affinity for a particular ion. To a lesser degree, the solution's concentration, the counter ion's participation in complex formation and temperature can also influence a resin's selectivity. As indicated before, the degree of cross-linking affects the density of the resin's active exchange sites. This proximity has an effect on the selectivity which a given functional group has for a particular ion. As the amount of cross-linking or fixed ion concentration is lowered, the affinity for an ion approaches unity.

A resin's affinity for one ion over another is related to the ion's chemical and physical properties within the solution. The exchange phenomena are related to the electrostatic interplay between the resin and the ions in solution. Therefore, it is quite important for the resin and the ionizable species in question to be in a solution that yields a high degree of ionization. Since the ionic potential and activity coefficient of any particular ion are directly affected by the polarity of the solvating solution, it becomes apparent that the greater a solution's polarity, the greater the ionization of the electrolytes.

Specific interactions between the ion exchange resin and the ionizable species also affect the affinity of the resin for a specific ion. These interactions involve ion-pair formation, electrostatic attraction, London and dipole–dipole interactions. The ion exchange resin prefers the counter ion which forms a stronger ion-pair or bonds with the functional group. Electrostatic interactions are described in subsequent paragraphs. London interactions are the weak attractive forces that similar organic structures display for one another (Helfferich, 1962). An ion exchange resin will prefer to adsorb organic structures which are similar to its matrix. The dipole–dipole interactions between the polar solvent molecules and the polar groups of the solute also play a major role in organic adsorption. Both the London and dipole–dipole interactions favor local adsorption of the hydrocarbon groups of the solute onto the matrix of the resin and thus favor the sorption of a non-electrolyte. The London and dipole–dipole interactions are important in the removal of polyphenolic color bodies found in many fruit juices.

In dilute concentrations and ordinary temperatures, the extent of the resin's exchange potential increases with increasing valence of the exchanging or counter ion ($Na^+ < Ca^{2+} < Al^{3+}$).

In dilute concentrations and ordinary temperatures, the exchange resin's selectivity also increases as the atomic number for ions of similar valence increases ($Li^+ < Na^+ < K^+$; $Mg^{2+} < Ca^{2+} < Sr^{2+} < Ba^{2+}$).

In concentrated solutions, the differences in selectivity of ions with different valences diminish. The regeneration of a strong acid cation resin to the H^+ form for use in the de-mineralization of water and various food streams is dependent upon this shift.

Ion exchange equilibria can be strongly influenced by the interactions between the counter ion and its respective co-ion. Since the co-ion is normally excluded from the resin phase, its relative size and interactive strength with the counter ion can result in either an increase or a decrease in the ion's affinity for the functional group. Typically, the ion exchange resin prefers a counter ion which associates less strongly with the co-ion.

As in most chemical equilibria, the temperature at which the ion exchange or adsorption occurs greatly influences the kinetics of a reaction. However, since most resin–ion affinity involves association or aggregation processes, increased temperature usually lowers the resin–ion affinity while increasing the rate of the reaction. This effect is even more pronounced when working with food streams where the inherent higher viscosities of the food, which result in lower ion mobility, can be the rate determining factor.

The speed at which ion exchange reactions take place is affected by many other conditions: the solution concentration and viscosity, the counter ion, the resin's cross-linkage, the resin's exchange group, the ion's size and valence, the resin's surface area. At higher concentrations, the rate of reaction tends to increase, provided the solution's viscosity does not become the governing factor with respect to diffusion.

The counter ion, as discussed previously, can affect the kinetics, since the diffusion rates of the exchangeable ion are dependent upon the size and possible complexation with the co-ion. The ion's size can also limit the diffusion into the resin's structure. The kinetics of the reaction are also dependent upon the ion's valence in as much as the valence affects the selectivity of a particular ion versus another.

The resin's cross-linkage influences the kinetics because of its influence on the pore size and charge density within the resin's matrix. The amount of surface area presented by the exchange resin also affects the kinetics since increased surface area reduces the diffusional path.

There are two important fundamentals that need to be covered with respect to juice enhancement using ion exchange resins. The first relates to the removal of amino acids and proteins from various juices. The second concerns the use of weak base anion resins for acid reduction in juice streams.

Proteins, amino acids and other nitrogenous materials commonly associated with juices need to be reduced or removed from solution due to their tendency to break down and re-combine with the reducing sugars found in the juice to form browning compounds. The chemistry has been investigated for more than 70 years by Maillard and numerous others. A detailed description of the chemistry is beyond the scope of this text. For further information, see Kawamura (1983).

Many nitrogenous materials are amphoteric, capable of either accepting or donating a proton, depending upon the pH of the solution. Due to this phenomenon, a strong acid cation resin in the H^+ form can protonate many nitrogenous materials. It is in this instance that the protonated species can be removed from solution by a strong acid cation resin. However, as the pH of the solution rises during treatment, the nitrogenous materials will become more anionic and can return to solution. The loss of nitrogenous materials during the process can be prevented by designing the proper resin matrix so that the nitrogenous materials are first ionically attracted and later adsorbed in the pore structure of the resin. In this way, a juice can be de-colorized and made shelf-stable for later blending purposes.

A weak base anion resin is preferred in the food industry for acid reduction. This is due in large part to the fact that a weak base anion resin does not exchange ions but adsorbs acids. Therefore, a weak base anion resin does not change the alkalinity content of the juice. The chemical equilibria for weak base anion resins are controlled by the ionization or dissociation constants of the particular acid, the pH of the solution and the kinetics of the resin.

A weak base anion resin's affinity for acids is directly related to the acid's ionization or dissociation constant. The stronger the acid, the higher the affinity for the weak base site. This is dependent upon the degree of ionization of the acid in the fluid. As the pH of the solution approaches and exceeds 7, the dissociation of the acid becomes weaker. Therefore, the weak base site's affinity for that acid is subsequently reduced.

9.4 Adsorbents

Adsorption refers to the use of a solid material to which solutes of interest bind (Belter *et al.*, 1988). The adsorbent materials typically possess high surface areas and large porosities such that large molecular weight color bodies and other impurities can permeate into the material, become entrapped, and are thereby removed from solution. Typical adsorbents include activated carbons (bonechar, various carbons), natural and synthetic zeolites, clays, ion exchange resins and their co-polymers.

9.4.1 Activated carbon

This is an amorphous form of carbon specially treated to produce a very large surface area (Doying, 1965; Keller *et al.*, 1987). The activated carbon can be derived from either animal, vegetable or mineral sources. The most common sources of decolorizing activated carbons are bones, wood, peat, lignite, lignin and coconut shells (Doying, 1965). Typical activation processes include treating the carbon source with various inorganic chemicals prior to charring. Bonechar has been widely used in sugar decolorization due to its limited ion exchange capacity plus its color removal properties (Cantor and Spitz, 1956; Mindler, 1957; Doying, 1965; Applebaum, 1968; Keller *et al.*, 1987). Activated granular carbon (vegetable) has recently been displacing bonechar based on its higher color removal efficiency, and its ability to use the same rotary kilns for activation (Zemanek, 1984). The fruit juice industry typically uses either powdered activated carbon or polymeric adsorbents. The quantities of juice processed through a juice plant are typically insufficient to justify the expense of a granular carbon system. The powdered activated carbon is normally added to the juice in large, agitated tanks. After an appropriate amount of contact time, usually less than an hour, the juice is filtered to remove the carbon. The powdered carbon is discarded after one use.

9.4.2 Zeolites

Zeolites or molecular sieves are inorganic aluminosilicates. The naturally occurring zeolites are clays. Both natural and synthetic zeolites have a crystalline structure composed of AlO_4 and SiO_4 tetrahedra which are linked together by the sharing of an oxygen ion (Kunin and Myers, 1950; Cantor and Spitz, 1956; Kunin, 1958; Helfferich, 1962; Applebaum, 1968; Keller *et al.*, 1987). The empirical formula is

$$M_{2/n}O \cdot Al_2O_3 \cdot ySiO_2 \cdot wH_2O$$

Where M equals the cation (Na^+, K^+, Ca^{2+}, Mg^{2+}), n equals the cation's valence, y equals 2 or greater and w equals the amount of water contained in the void spaces (Kunin and Myers, 1950; Kressman, 1957; Kunin, 1958; Helfferich, 1962; Applebaum, 1968; Keller *et al.*, 1987). Since the cations and the water fill the void spaces within the structure, cation exchange and/or adsorption can take place provided the ion or molecule can permeate into the crystalline framework (Kunin and Myers, 1950; Kunin, 1958; Helfferich, 1962; Applebaum, 1968; Keller *et al.*, 1987). Natural clays find wide application in the deodorization of edible oils and as filtering aids in the food industry.

9.4.3 Polymeric adsorbents

These are similar in nature to the ion exchange resins discussed earlier. For juice enhancement, the polymeric adsorbents that are employed today consist of either co-polymers of styrene and divinylbenzene or of phenol and formaldehyde (Figure 9.6).

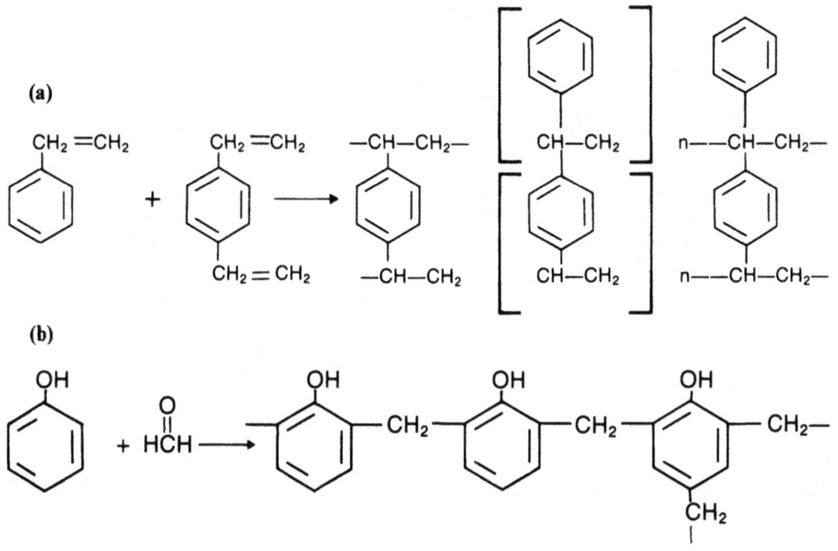

Figure 9.6 Polymeric adsorbent structures. (a) Styrene-divinylbenzene adsorbent structure. Redrawn and adapted from The Dow Chemical Company (1985). (b) Phenol-formaldehyde adsorbent structure. Redrawn and adapted from Kunin (1958).

The main difference between these adsorbents and their ion exchange counterparts is that most of these materials are not functionalized and consequently do not possess any ion exchange capacity. In order to make effective adsorbents, the polymerization is carried out in such a way as to optimize the adsorbent's pore volume and surface area. This enables these synthetic materials to emulate activated carbon. Unlike activated carbon which offers only physical entrapment as a means to remove organics, these polymeric adsorbents also have the ability to chemically bind the entrapped organic species through London interactions, as discussed earlier. This added advantage makes these polymeric adsorbents quite useful in the treatment of various juices such as apple, pear, grape juices and pineapple mill juice, where the predominant color bodies are condensed phenolics.

An important advantage of these adsorbents over activated carbon is that they are regenerated with common chemicals instead of a rotary hearth furnace or discarded after one use. One drawback to these materials relates to their strong interaction with phenolic color bodies. A strong London interaction between the adsorbent and a phenolic color body can cause irreversible fouling of the adsorbent, resulting in premature adsorbent failure. Therefore, it is important that one chooses an adsorbent that has a backbone and pore structure suitable for the type of juice being processed. An example of a weakly anionic phenolic color body is shown in Figure 9.7.

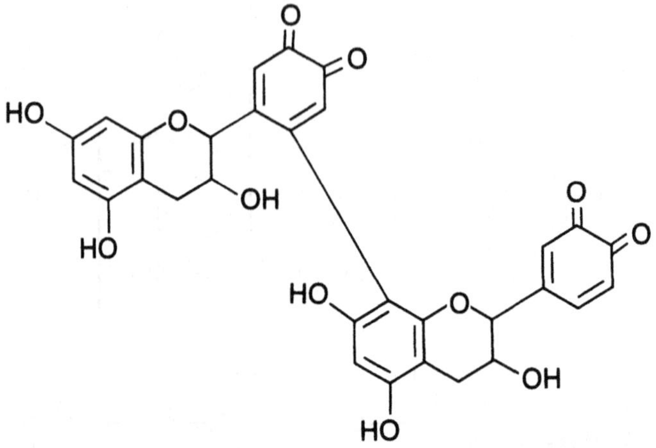

Figure 9.7 Typically weakly anionic phenolic color body. Redrawn and adapted from Ribereau-Gayon (1974).

9.5 Applications

The basic enhancement processes involve either adsorbing impurities or exchanging less desirable ions for more desirable ions. Both processes use a solid matrix to accomplish their respective task. In the case of ion exchange resins, these materials can not only exchange ions but they can also function as adsorbents. Typically, these enhancements occur in a batchwise manner. The method of contact between the adsorbent and the juice is either in a stirred tank or in a fixed bed column. Usually a stirred tank process is reversed for powdered activated carbon. Fixed bed column processes center around ion exchange and polymeric adsorbents.

A stirred tank adsorption is carried out by mixing the juice with the appropriate amount of powdered carbon for approximately 1 h. The juice/carbon slurry is then pumped through a filtration device, such as a plate and frame pressure filter, to separate the spent carbon from the juice. The spent carbon along with the filter-aid, typically diatomaceous earth, is subsequently discarded.

A fixed bed column (Figure 9.8) is predominantly used with the granular or spherical polymeric adsorbents and ion exchange resins. This design usually passes the juice and the chemical regenerants down through the packed bed of material. In this way, the maximum capacity of the material is utilized prior to regeneration. An advantage of this system over the stirred tank process is that the fixed bed system gives multiple stages of equilibrium and therefore much higher levels of purification are obtainable. Complete automation and the potential recovery of the adsorbed species are other advantages of a fixed bed system.

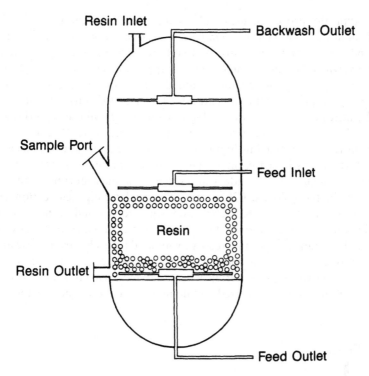

Figure 9.8 Schematic of fixed bed column. Redrawn and adapted from Streat and Cloete (1987).

Today, most industrial applications for enhancing fruit juices are for the recovery of the fruit sugars. Most fruit juices contain approximately 10–15% sugars with varying proportions of sucrose, glucose and fructose depending upon the type of fruit.

These processes allow juices which would otherwise be unacceptable for reasons of taste or color, to be used in other foods for their natural sugar content. The common enhancement processes include de-colorization, de-flavorization and de-mineralization. De-acidification and discrete organic particles removal are also practised in selected industries. The predominant juices used for blending purposes are derived from bland fruits such as white grape, pears and apples. Pineapple mill juice has been purified and used as a canning syrup since 1947 (Diamond Shamrock Chemical Company, 1972).

9.5.1   Grape juice

Grape juice is a 'bland', slightly colored fruit juice typically containing in excess of 15% sweet carbohydrates. Therefore processed grape juice is some-times used as a substitute for other sugars (cane and beet). White grape juice is typically processed using adsorbents and ion exchange resins. The adsorbent materials are used to remove any residual color and flavor from the juice, and the ion exchange resins are used to remove the ionic constituents.

In Italy, when the sugar content of the grape is not sufficient to yield a high quality wine, processed grape juice is added to the must prior to fermentation (G. Somaruga, private communication). This gives the must the additional carbohydrates necessary to yield a high quality wine without disturbing the delicate flavor, color and aromatic balance of the wine. The processing scheme in Italy is outlined in Figure 9.9. The strong acid cation resins are in the H^+ form during juice treatment. The weak base anion resin is the free base form for treatment. The secondary anion resin is a macroporous strong base resin used in the chloride (Cl^-) form. The lead strong acid cation resin exchanges H^+ ions for the K^+, Ca^{2+} and Mg^{2+} ions found in the juice. The weak base anion resin removes the free acidity and some color bodies from the juice. The grape juice is now relatively mineral free. The strong base anion resin removes any residual color bodies. The last strong acid cation resin picks up any nitrogenous materials which may have leaked from or through the other resins. The de-colorized and de-mineralized grape juice is then sent

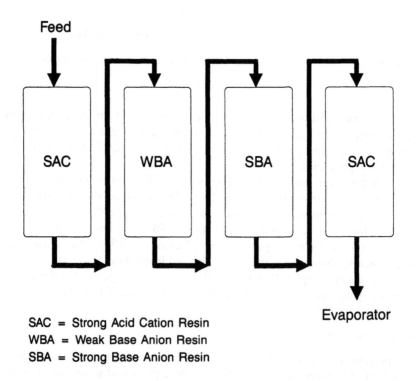

SAC = Strong Acid Cation Resin
WBA = Weak Base Anion Resin
SBA = Strong Base Anion Resin

Figure 9.9 Flow schematic for grape juice enhancement.

to an evaporator for concentration. The resultant juice is now suitable for blending into a low sugar grape must for fermentation. Similar practices are carried out in the United States, where the final product is not only used for wine but also for blending stock in fruit juice drinks and carbonated juice beverages.

9.5.2   Apple and pear juice

Apple and pear juice are also used as blending juice stocks for many of today's carbonated juice drinks. Upon extraction, apple and pear juice contain approximately 12% sugars (sucrose, glucose and fructose). Apple and pear juices also contain a class of color bodies, polyphenolics, which must be removed prior to their use. Upon extraction, these color bodies are formed either enzymatically or by rapid oxidation during exposure to air. It is this same brown color that occurs when one cuts an apple. Within minutes, the apple surface takes on a rather unpleasant brown hue.

These two juices can be enhanced by activated powdered carbon for color and flavor reduction. As alluded to earlier, an activated powdered carbon system is composed of several stirred tanks and a pressure filter device. The juice is first de-pectinized using an enzyme. This treatment coupled with filtration removes most of the insoluble solids. The enzymatically clarified juice is then blended with the appropriate amount of activated powdered carbon. After approximately 1 h of contact time with the carbon, the juice/carbon mixture is passed through a pressure filter which contains a pre-coat of diatomaceous earth to remove the carbon from the juice. This process yields a bland, nitrogen reduced, de-colorized juice suitable for blending with many of today's fruit juice-containing soft drinks.

Another means of accomplishing much the same thing is to use polymeric adsorbents and ion exchange resins. With the use of these materials, one can achieve a much purer source of carbohydrates than the carbon system mentioned above. The processor will also use less labor and have less solid waste with a polymer-based system. However, the polymer system will generate a liquid waste stream during the regeneration procedure. A typical flow schematic is illustrated in Figure 9.10. The lead macroporous adsorbent removes color bodies and color precursors. The first weak base anion resin removes the free acidity found in the juice along with other color bodies which leak through the adsorbent. The strong acid cation resin is used in the H^+ form to exchange with the K^+, Ca^{2+} and Mg^{2+} ions found in the juice. The resin also removes the nitrogenous compounds, which can be major contributors to off-flavors and browning products. The final weak base resin removes any mineral acidity associated with the cation effluent. This resin also acts as a final color polisher. The final juice is then sent to an evaporator for concentration. The final product is now suitable to be used either as a canning syrup or for blending.

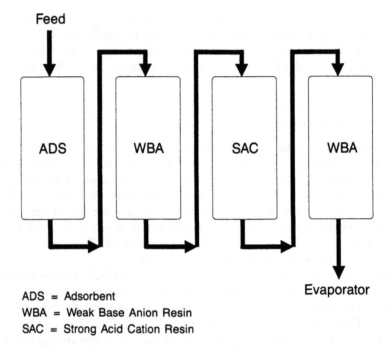

ADS = Adsorbent
WBA = Weak Base Anion Resin
SAC = Strong Acid Cation Resin

Figure 9.10 Flow schematic for apple and pear juice enhancement. Redrawn and adapted from Industrial Filter and Pump Manufacturing Company (1987).

9.5.3 Pineapple mill juice

Pineapple mill juice is the juice extracted from the waste trimmings during the canning operation. This juice contains roughly 10% sugars. The juice also contains relatively high amounts of citric acid and nitrogenous materials. As mentioned previously, the purification of pineapple wastes was developed in the mid-1940s (Cantor and Spitz, 1956; Diamond Shamrock Chemical Company, 1972). This process was also one of the first commercially successful food purifications for ion exchange resin technology (Cantor and Spitz, 1956; Diamond Shamrock Chemical Company, 1972). The basic process was developed by Dole and is illustrated in Figure 9.11.

The primary objective of the process is to improve the quality of the sugar source so that it can be used as a canning syrup. The two strong acid cation beds remove the cationic species from the juice, namely the ash (K^+, Ca^{2+}, Mg^{2+}) and the nitrogenous materials. The weak base resins remove the subsequent acids from the cation effluent plus de-colorize the stream. The resultant mill juice is essentially devoid of mineral content and color. After evaporation, the syrup is suitable for canning purposes.

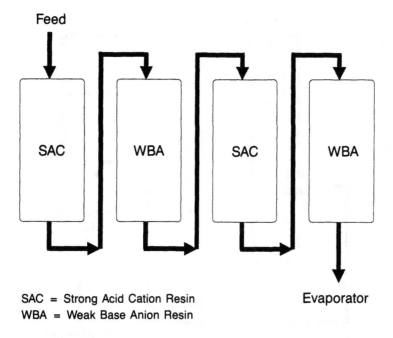

SAC = Strong Acid Cation Resin
WBA = Weak Base Anion Resin

Figure 9.11 Flow schematic for pineapple mill juice purification. Redrawn and adapted from Diamond Shamrock Chemical Company (1972).

9.5.4   Reduced acid frozen concentrated orange juice

Reduced acid frozen concentrated orange juice is an example of a specialized adsorption to meet customer needs. It has been recognized in the citrus industry that approximately 20% of the US population does not drink citrus products due to the high acidity associated with them. Early efforts to remove the citric acid from citrus juices dates back to the 1960s, where electrodialysis was tried on an experimental basis (Keller *et al.*, 1987). Other attempts were also tried with little success. In the late 1970s, The Coca-Cola Company Foods Division in Plymouth, Florida developed and later commercialized an acid reduction process utilizing a weak base anion resin (Assar, 1979; Varsel, 1980). This process was approved by the US FDA and a standard of identity was given to this particular orange juice product. This product has been commercially available for the last 7 years in the United States. The basic flow schematic is illustrated in Figure 9.12. The process can handle either freshly extracted, stabilized juice or concentrated juice which has been reconstituted back to 15° Brix (Assar, 1979; Varsel, 1980). The juice is then pumped downflow through the weak base anion resin where the resin preferentially removes citric acid over the ascorbic acid, folic acid and vitamin E.

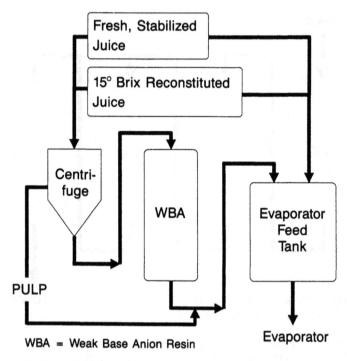

Figure 9.12 Flow schematic for reduced acid orange juice (Assar, 1979; Varsel, 1980).

As outlined in a previous section regarding weak base anion chemistry, it is important to process enough juice so that the pH of the effluent out of the weak base anion resin falls below 4.6. This ensures that the ascorbic acid and the folic acid have been displaced off the resin by the stronger organic acid, citric acid (Assar, 1979; Varsel, 1980). After the resin column, the juice is pumped into a holding tank where the pulp from the centrifuge can be added back. Also at this stage, fresh, untreated juice or juice concentrate can be added to the de-acidified juice in order to maintain a uniform taste and flavor to the product. The resultant juice blend is then sent to an evaporator for concentration (Assar, 1979; Varsel, 1980).

9.5.5 De-bittered orange juice

De-bittered orange juice is another example of a highly specific adsorption utilizing a polymeric adsorbent. The Navel orange is primarily grown for the fresh fruit market. However, during the harvest approximately 20% of the fruit are culled due to skin blemishes and the like (Kunin, 1958). The culled fruit is later sent to a processor to be turned into juice and juice by-products. All oranges contain a chemical species called limonoic acid a-ring lactone in

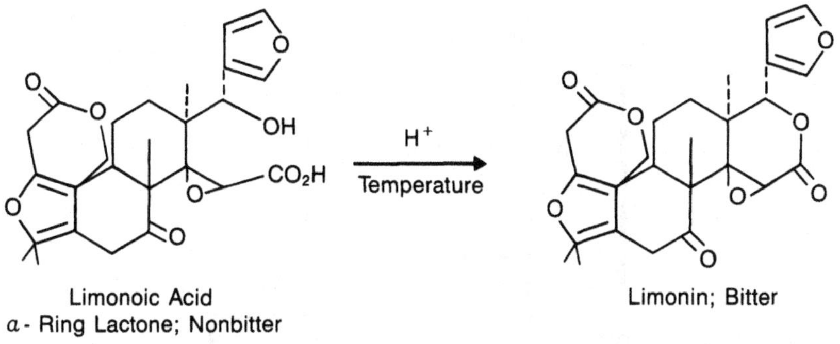

Limonoic Acid
α - Ring Lactone; Nonbitter

Limonin; Bitter

Figure 9.13 Limonin formation. Redrawn and adapted from Maier *et al.* (1980).

their phloem. The distribution of this lactone in the fruit is related to the quantity of seeds any particular variety has, since the seeds contain the largest concentration of the lactone (P. Lobue, private communication). Since Navel oranges contain no seeds, there is a large preponderance of this lactone distributed throughout the various membrane structures in the Navel fruit (Kilburn and Drager, 1965). Upon expression, this lactone esterifies to limonin (Figure 9.13), which is highly bitter (P. Lobue, private communication).

The general population can taste limonin down to levels averaging 6 ppm. Navel orange juice averages 20–30 ppm (P. Lobue, private communication). Due to this intense bitterness. Navel orange juice has historically been used as a blending juice where sufficient quantities of sugar can be added to the final product in order to mask the taste.

Various methods to de-bitter Navel orange juice have been attempted since the early 1970s. These methods include enzyme treatments, growth regulators, chemical masking and adsorbents (Schubert and Nachod, 1956; Mindler, 1957; Chandler *et al.*, 1968; Chandler and Johnson, 1977, 1979; Mitchell and Pearce, 1985; Industrial Filter and Pump Manufacturing Company, 1987). The use of a styrene divinylbenzene polymeric adsorbent has been commercially practised since the mid-1980s in Australia. The first commercial plant in the United States was brought on line in 1988 using a process and adsorbent developed by The Dow Chemical Company. This process can treat either fresh or reconstituted juice and is illustrated in Figure 9.14. After the treated juice has been concentrated, it is suitable for use in standard orange juice products.

9.6 Summary

As one can see from the above examples, fruit juice enhancements are quite

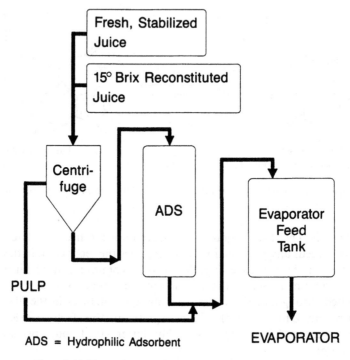

Figure 9.14 Flow schematic for de-bittered Navel orange juice.

diverse and serve the producer and the consumer in a variety of different ways. From the treatment of waste streams (pineapple mill juice), to the selective adsorption of bitter compounds, the use of adsorbents and ion exchange technology makes vital contributions to the juice and beverage industry. These techniques make fruit juice products and their respective beverage products more palatable and available to the consumer at reasonable prices. Even though this technology is rather new to the food processing industry, the benefits derived from its use have given the producer and the consumer a wider selection of fruit juice products.

References

Adams, B. A. and Holmes, E. L. (1935) *J. Soc. Chem. Ind.* 54, 1.
Applebaum, S. B. (1968) *Demineralization by Ion Exchange*, Academic Press, New York, Chap. 1.
Assar, K. (1979) *Proceedings of the 19th Annual Short Course for the Food Industry* 13, 114.
Austerweil, G. V. (1957) In: *Ion Exchangers in Organic and Biochemistry*, Wiley Interscience, New York, Chap. 33.
Belter, P. A., Cussler, E. L. and Hu, W. S. (1988) *Bioseparations, Downstream Processing for Biotechnology*, Wiley Interscience, New York, Chap. 6.
Cantor, S. M. and Spitz, A. W. (1956) In: *Ion Exchange Technology*, Academic Press, New York, Chap. 18.

Chandler, B. V. and Johnson, R. L. (1977) *J. Sci. Food Agric.* 28, 875.

Chandler, B. V. and Johnson, R. L. (1979) *J. Sci. Food Agric.* 30, 825.

Chandler, B. V., Kefford, J. F. and Ziemelis, G. (1968) *J. Sci. Food Agric.* 19, 83.

Diamond Shamrock Chemical Company (1972) *Duolite Ion Exchange Resins in the Treatment of Sugar Solutions*, Functional Polymers Division, Cleveland, OH, p. 34.

The Dow Chemical Company (1985) *Dowex Ion Exchange Resins for Processing Foods*, Separation and Process Systems Department, Midland, MI.

Doying, E. G. (1965) In: *Kirk-Othmer Encyclopedia of Chemical Technology*, Vol. 4, 2nd edn., John Wiley, New York, p. 149.

Harm, F. (1896) *German Patent* 95, 447.

Helfferich, F. (1962) *Ion Exchange*, McGraw-Hill, New York, Chap. 1–6, 9.

Industrial Filter and Pump Manufacturing Company (1987) *Newsletter*, Vol. 5(2), Cicero, IL, p. 4.

Johnson, R. L. and Chandler, B. V. (1988) *Food Technol.* 42(5), 130.

Kawamura, S. (1983) *Seventy Years of Maillard Reaction*, American Chemical Society, Washington, DC, Chapter 1.

Keller II, G. E., Anderson, R. A. and Yon, C. M. (1987) In: *Handbook of Separation Process Technology*, John Wiley, New York, Chap. 12.

Kilburn, R. W. and Drager, H. P. (1965) *U.S. Patent* 3, 165, 415.

Kressman, T. R. E. (1957) In: *Ion Exchangers in Organic and Biochemistry*, Wiley Interscience, New York, Chap. 1.

Kunin, R. (1958) *Ion Exchange Resins*, 2nd edn., John Wiley, New York, Chap. 2–5, 9, 14, 17.

Kunin, R. and Myers, R. J. (1950) *Ion Exchange Resins*, John Wiley, New York, Chap. 1, 2, 10.

Maier, V. P., Hasegawa, S., Bennett, R. D. and Echols, L. C. (1980) In: *Citrus Nutrition and Quality*, Vol. 143, American Chemical Society, Washington, DC, Chap. 4.

Mindler, A. B. (1957) In: *Ion Exchangers in Organic and Biochemistry*, Wiley Interscience, New York, Chap. 32.

Mitchell, D. H. and Pearce, R. M. (1985) *U.S. Patent* 4, 514, 427.

Puri, A. (1984) *U.S. Patent* 4, 439, 458.

Ribéreau-Gayon, P. (1974) In: *Chemistry of Wine Making*, Vol. 137, American Chemical Society, Washington, DC, Chap. 3.

Schubert, J. and Nachod, F. C. (1956) In: *Ion Exchange Technology*, Academic Press, New York, Chap. 1.

Shaw, P. E. and Wilson, C. W. (1983) *J. Food Sci.* 48, 646.

Streat, M. and Cloete, F. L. D. (1987) In: *Handbook of Separation Process Technology*, John Wiley, New York, Chap. 13.

Thompson, H. S. (1850) *J. R. Agric. Soc. Engl.* 11, 68.

Varsel, C. (1980) In: *Citrus Nutrition and Quality*, Vol. 143, American Chemical Society, Washington, DC, Chap. 11.

Way, J. T. (1850) *J. R. Agric. Soc. Engl.* 11, 313.

Way, J. T. (1852) *J. R. Agric. Soc. Engl.* 13, 123.

Wheaton, R. M. and Seamster, A. H. (1966) In: *Kirk-Othmer Encyclopedia of Chemical Technology*, Vol. 11, 2nd edn., John Wiley, New York, p. 871.

Zemanek, L. A. (1984) In: *Proc. 43rd Annual Meeting of Sugar Industry Technologists*, Sugar Industry Technologists, Martinez, CA, p. 101.

10 Processing systems for fruit juice and related products
L. B. FREDSTED

10.1 Introduction

Processing of liquids has been carried out for many years on various types of equipment and especially that designed for low acid products such as milk. The processing of high acid products like fruit juices can be carried out using similar equipment adapted to the special requirements of juice products.

This chapter will concentrate on processing of ready-to-drink juices. The processes are described in general terms but can in some countries be limited in use because of legislative or patent constraints. The fruit juice market has undergone major changes during the last 10–15 years. Consumers now require better quality, more exotic varieties and convenient packaging systems. Consumer focus on health, freshness, no additives, less over-processing and less overpackaging together with the growing strength of larger retailers and supermarkets, has set new standards for the juice industry. The industry must supply higher quality, larger variety and better packaging at even more competitive prices. As a consequence all suppliers of equipment (processing/packaging) also face the challenge to supply new systems that (i) are optimally designed for the specific product, (ii) fulfil the consumers' requirements for clean and safe process and finally (iii) are cost efficient for the juice industry.

10.2 Design parameters

Before finalising design of processing equipment, relevant process parameters must be obtained and evaluated. Non-compliance with these will in many cases lead to incorrect designs and waste of both time and money.

Typical considerations include:

—What type of juice product is to be processed? Juice, nectars and drinks often require different design parameters (Figure 10.1).
—What is the acidity of the product? Fruit juices have various acidities and

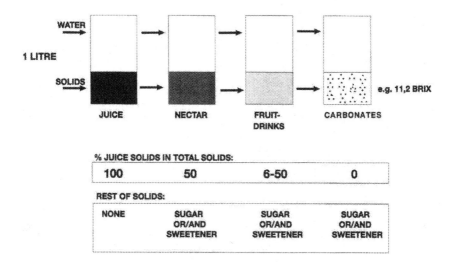

Figure 10.1 Different requirements for processing juice, nectars and fruit drinks.

Brix/acid ratios (e.g. Pineapple about 25 and Grape about 9). Time/ temperature combinations in relation to possible harmful micro-organisms should be observed (Figures 10.2 and 10.3).

—Pulp content. (Weight of fine fruit particulates < 1 mm). Normal values

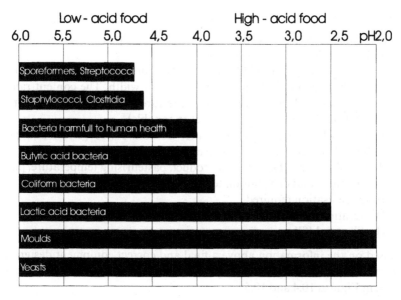

Figure 10.2 Limiting effect of acidity on growth.

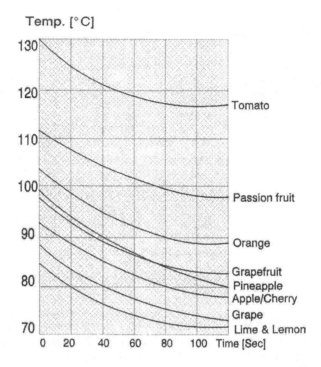

Figure 10.3 Pasteurisation of fruit juices, requirements of temperature against time.

are found between 2 and 10%. More than 6% pulp increases the viscosity and makes the product shear sensitive.

—Content of floating cells/fibres. The fibres are larger particles of fruit, normally 1–20 mm in length and 3–5 mm in width. The content ranges from 1–2 g/l to 45 g/l. Larger values are rarely found in ready-to-drink juice (about 100 g/l makes a product spoonable).

—Viscosity. Depends on temperature. Soluble solids (°Brix) and pulp content. Only products with pulp are shear sensitive.

—Required shelf life. Depends on product, pasteurising temperature, packaging system and distribution temperature.

—Homogenisation required.

—De-aeration required.

—Hotfill or coldfill.

Based on the information above a final and correct equipment design can be carried out. Normal process diagrams for fresh and reconstituted juice are shown in Figures 10.4 and 10.5.

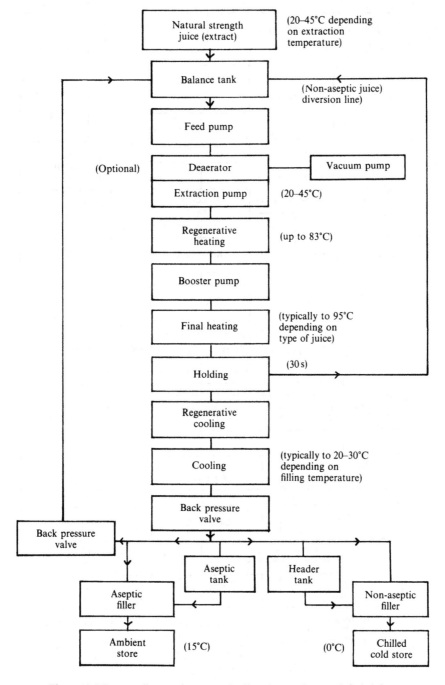

Figure 10.4 Process diagram for pasteurisation of natural strength fruit juices.

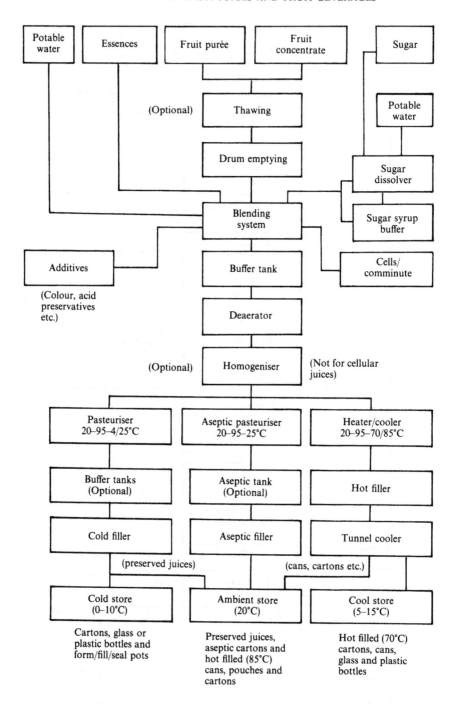

Figure 10.5 Process diagram for reconstitution of fruit juices and manufacture of nectars.

10.3   Plant configuration

10.3.1   Plate heat exchanger (PHE)

The PHE (Figure 10.6) offers the possibilities of varying flow patterns and

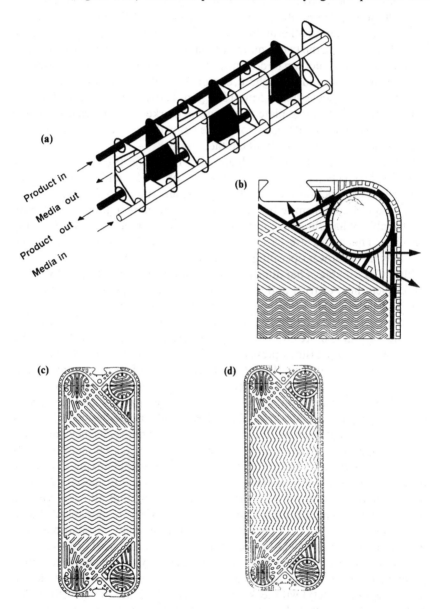

Figure 10.6 Plate heat exchanger. (a) Typical plate configuration, (b) leak detection/protection around cornerholes, (c) thermal long, (d) thermal short.

using multi-channel passes. Asymmetric counter-current and co-current flows are also possible. These features allow for an optimal balance between thermal transfer and pressure loss in any given situation.

Further the plates are available with numerous flow profiles, e.g. thermally short (soft) vertical profile and thermally long (hard) horizontal profile. The thermally long plate produces greater turbulence and thereby higher transmission values than the thermally short plate; the latter offers lower pressure drop. By combining the two, a more optimal heat transfer/pressure drop ratio suitable for the product can be designed. Plates are supplied in a wide range depending on product and throughput, comprising different plate surfaces (measured in m^2), thicknesses (0.5–1.5 mm) and gaps between plates (0.3–6 mm).

There are mainly three plate types on the market; high efficient plate (A) with multi-contact points, easy-flow plate (B) with fewer contact points and self-supporting plates (C) without contact points.

Plate A is cost-effective even at high regenerative effects ($> 90\%$). It can be used for products having a pulp content of $< 5\%$ and without fibres. Typically it will operate at relatively high pressure (up to 16–25 bar).

Plate B is a compromise between A and C, meaning a good heat transfer and capable of handling the most common juices with normal pulp and fibre contents. The plates are easier to clean compared to plate A, however, generally have a lower operating pressure (up to 12–16 bar).

Plate C is capable of handling juices with high fibre contents, usually with a less efficient heat transfer and with a lower maximum operating pressure (up to 6–12 bar). High viscosities and pressure drops should be avoided.

Plate-type juice pasteurisers normally use product-to-product regeneration making the plants more cost effective. The PHE however is regarded as more sensitive to high pressures/pressure shocks and requires more cleaning and maintenance. Cleaning-in-place (CIP)-backflushing of PHE systems is quite common, but also critical, and requires easy inspection. Inspection is improved by today's more commonly used non-glued gaskets. New detergents based on enzymes are also available. These products are designed to break down fibres so that cleaning time is reduced and cleaning efficiency improved.

When operating within today's industry there is increased focus on safety and hygiene. One of the main safety issues is the possible mixing of raw and pasteurised product. Most PHE systems are equipped with special gaskets at the corner holes, ensuring protection against mixing and allowing for leak protection. New systems should however also consider the possible demands set by present and future legislation. Furthermore, many installations must produce high acid as well as low acid products (i.e. milk-based products).

Pressure differential and flow control can easily be installed into juice pasteurising systems. The use of a water intermediate and also special double plates (i.e. Duo-Safety) with leak detection can eliminate the need for differential pressure control. For any safe system easy inspection and process control/documentation is essential.

10.3.2   Tubular heat exchanger (THE)

Tubular heat exchangers (Figure 10.7) have undergone and are undergoing extensive developments to improve heat transfer, lower costs and give more compact and flexible construction.

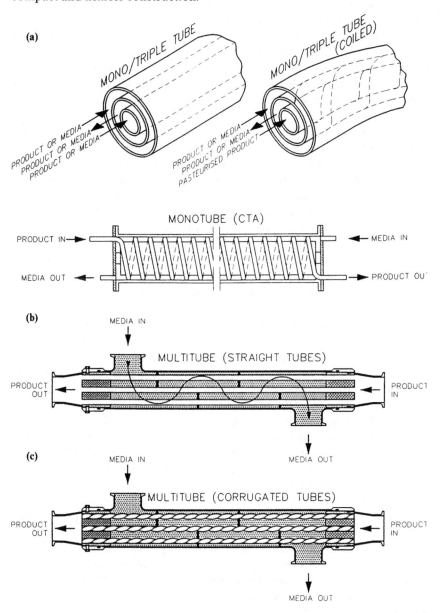

Figure 10.7 Tubular heat exchangers. (a) Mono/triple tube, (b) multi-tube, (c) multi-tube with corrugated tubes.

There are, in principle, three types of tubes on the market: (a) mono/triple tube, (b) multi-tube and (c) multi-tube with corrugated tubes.

Mono-tubes (a) are normally used for products with larger particles and/or very high pulp/fibre contents. Mono-tubes are manufactured as straight tube-in-tube or as coiled tubes. The latter is more cost effective (fewer connections and joints) and has a higher heat transfer (straight/axial flow compared with straight flow in the tubes). Straight mono-tubes however allow for inspection and higher flexibility (interchangeable outlets). Product-to-product regeneration is possible with the mono-tubes. Juice with fibres should be limited to straight tubes or coiled tubes with an intermediate water circuit, thus avoiding installation with distance wires, wall–wall contacts and lack of inspection points.

Multi-tubes (b and c) have won most recognition; a high degree of flexibility at lower costs is one of the reasons. Multi-tubes are normally designed for an intermediate heat transfer with a water loop, resulting in a larger regeneration surface than by ordinary product-to-product regeneration. It is therefore very important that the product requires a THE (because of the specific pulp and fibre content and/or the range of products to be produced). If this is not the case, then other systems are preferred (e.g. PHE).

The wide range of tubes available (size, lengths and number of tubes) allows for designs similar to the PHE. Parallel flows, number of passes, flow rates, turbulence and pressure drops can be calculated for specific applications, and the right plant configuration can be obtained. This flexibility is a distinct advantage for the tubular heat exchanger. As the tubes are generally very pressure resistant (up to 60–100 bar), increased heat transfer based on higher flow rates can be obtained even at higher pressure drops. The individual products must be able to stand these higher pressures.

Within the last 5–10 years corrugated tubes have been introduced. Corrugation mainly increases the heat transfer for products in the laminar flow area. Non-viscous juices do not require corrugation as the small increase in heat transfer does not support the extra equipment costs. For more viscous products (i.e. juice concentrates, dressings etc.) corrugation combined with a reasonable pressure resistance (up to 30–60 bar) is useful. As for PHE systems, tubular systems also require particular consideration and cleaning.

Most THE installations use a water intermediate circuit and are therefore already built to fulfil requirements safeguarding against cross-contamination. CIP backflush may also be necessary, depending on the specific product and plant design.

The use of coiled mono-tubes with product-to-product regeneration should be avoided for juice with pulp/fibres due to cleaning problems, lack of inspection opportunities and differential pressure. Straight mono-tubes can be used for product-to-product regeneration, but the differential pressure should be kept in mind.

10.3.3   Direct steam heating system (DSH)

The direct steam heating type of system (Figure 10.8) is not common within the juice industry, however the system is widely in use for UHT treatment of milk products, baby food, liquid eggs and similar. Two systems are available: injection and infusion, the difference being steam-to-product (injection) and product-to-steam (infusion).

Infusion is the most widespread system for new products and juices and is given a brief description here. Infusion technology is based on a relatively small quantity of product being led to an infusion chamber full of steam. The quantity of steam is relatively high in relation to product. With a controlled and even distribution of product a very fast and gentle heating will occur during the free fall through the chamber.

The steam temperature is almost the same as that of the final product so that overheating of the product is avoided.

The system uses vacuum steam or steam under pressure depending on the pasteurising temperature (operating range is between 75 and 160°C). After reaching pasteurising temperature within 0.5 s or less, the product can be held for a set holding time prior to cooling in a PHE, THE or a scraped surface heat exchanger (SSHE). The product can also be hot filled immediately after the infusion heater. As all steam condenses into the product, a certain dilution will take place, and a final adjustment of °Brix/water content with sterile cold water may be necessary. The infusion system is normally used for heat treatment of reconstituted products from various concentrates.

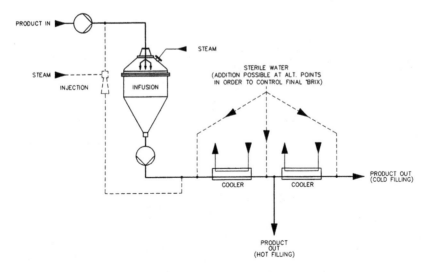

Figure 10.8 The direct steam heating system.

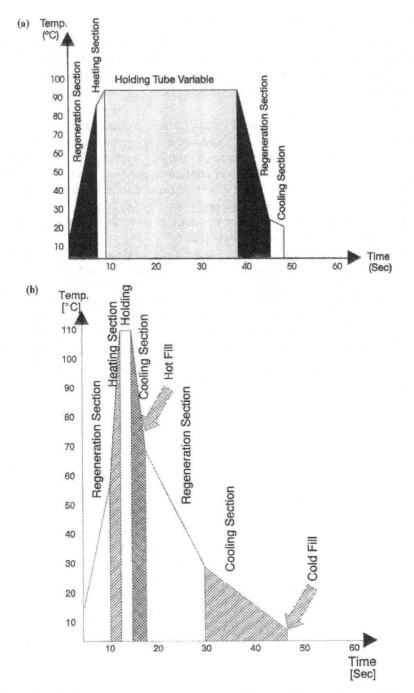

Figure 10.9 Typical temperature profiles for juice processing. (a) Indirect heating, (b) direct heating.

The direct heating system allows for higher pasteurising temperatures (higher reduction of microorganisms) combined with lower chemical changes compared to any indirect system. Furthermore high viscosities can be processed successfully.

Temperature profiles for an indirect heating system as well as for a direct system are shown in Figure 10.9.

10.3.4   Scraped surface heat exchanger (SSHE)

SSHEs (Figure 10.10) are not common for juice processing. The main interest for SSHEs are juice concentrates and various paste products (e.g. tomato). The specific product and capacity should decide the correct SSHE plant configuration.

It is important that the SSHE is carefully designed with regard to vertical/horizontal mounting, shell diameter and available services (e.g. water, steam or ammonia). The SSHE can operate with products with very high viscosities:

—vertical: up to   100 000 cps
    small capacities   up to   7 bar
    (depending on viscosity) up to 2000–8000 l/h
—horizontal/vertical: up to   1 000 000 cps
    high capacities   up to   12–30 bar
    (depending on viscosity) up to   10 000 l/h

Note: that direct expansion cooling with ammonia/Freon is possible.

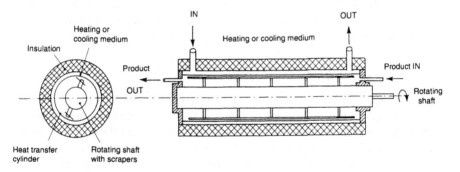

Figure 10.10 Scraped surface heat exchanger.

10.3.5   Sterile tank (aseptic tank)

Today, aseptic tanks are very often used in the juice industry. The need for a higher utilization of the processing plant/filling machine(s) is evident, and can be covered by use of an aseptic tank. The tanks are usually designed for both

2–3 bar overpressure and full vacuum. This allows for a high temperature sterilisation (about 120–130°C) of the equipment if necessary.

Quality aspects of recirculation of product are also addressed with use of a sterile tank. No recirculation means less overprocessing of product. Tanks are normally installed with a valve arrangement that allows for production from the tank whilst cleaning the pasteuriser.

When products with larger quantities of fibres, cells or particles, are produced the tanks must be fitted with agitators. For most contents, a simple aseptic agitator can be used (top or bottom mounted). For more difficult products either a special wide-wing propeller should be installed or, in some cases, a horizontal, aseptic tank with a full length horizontal-type agitator will be necessary.

For some juice products the use of sterile air in the tank can cause taste and/or colour changes. In such cases nitrogen can be used.

10.4   Other parameters for juice processing

10.4.1   Raw material

Control of raw material quality is essential. The production of freshly squeezed/extracted juice by cold chain distribution (2–6 weeks shelf life) is dependent on the quality and selection of fruits, fruit washing and cleaning, peeling and further processes. The better the raw product, the higher the quality that can be produced. Lower temperatures than the normal long life juice temperatures can therefore be used (e.g. 75–85°C instead of 90–95°C for orange). If low temperatures are used, enzymes are not completely in-activated; this is another factor in determining shelf life.

10.4.2   De-aeration

De-aeration of juice and juice concentrates is normally carried out at 20–50°C, depending on product (e.g. citrus at low temperatures and tomato at higher temperatures). De-aeration is desirable to maintain the vitamin C level and to reduce colour changes during storage. However, after about 2–3 weeks' storage the vitamin C level becomes more dependent on storage temperature and enzymatic degradation.

Several de-aeration methods are used. A normal de-aerator will operate with vacuum at temperatures below boiling, thus reducing the possible vapour/flavour extraction. A balance between the air removal and the acceptable flavour losses is normally achieved (low de-aeration temperatures and careful product handling).

During evaporation and concentration of juices, essence and flavour recovery systems are in place. These processes require primary/secondary condensers and specially designed scrubbers, and will not be dealt with in this chapter.

For de-aeration of more 'difficult' juices (higher viscosities and/or containing volatile aromas), temperatures and pressures closer to or above the boiling point are necessary. Cooling systems (vapour condensers) can in these cases be installed in the top of the de-aerators, or externally in the vacuum line. An external installation will require a second pump to return condensate.

De-aeration is less important during hot filling as the saturation level of air/oxygen is very low at high temperatures.

This natural feature and the possibility of adding pulp and fibres before adding hot liquid juice, thus limiting processing equipment, makes this process interesting. An evaluation of alternative processing methods, product quality, investment and running costs should be considered carefully.

10.4.3 Homogenisation

Homogenisation is not widely used within the juice industry, however some products like tomato and fruit purée can be enhanced by the technique. Otherwise the use is limited to products where a more homogeneous, viscous and/or economical product can be achieved. When processing juices like mango, guava etc., care must be taken to avoid large pressure drops and homogenisation. Too high a pressure drop can produce gel-like products instead of the desired pourable, viscous products.

10.4.4 Water quality

As for other reconstituted products the quality of water is critical with respect to chlorides, metals, salts and, not least, the oxygen content. The level of oxygen should be low ($< 2–3$ ppm), however for sensitive and/or longer shelf life products, less than 1 ppm is advisable. In carbonated soft drinks and beer production, de-aerated water with oxygen levels between 0.1 and 0.5 ppm is used. De-aerators for water use either traditional vacuum treatment with recirculation of the water (oxygen level about 0.3–0.5 ppm), or vacuum treatment together with CO_2 stripping (about 0.1–0.2 ppm). General flow diagrams for juice processing (see Figure 10.11).

10.5 New developments and technologies

10.5.1 High pressure pasteurisation

High pressure pasteurisation is a new technology using pressures from 1000 to 10 000 bar. After the product is filled into flexible containers and placed in a pressure chamber, the chamber is filled with liquid. Filling the chamber with

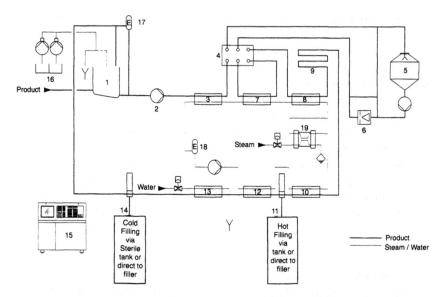

Figure 10.11 General flow diagram for juice processing. (1) Balance tank, (2) feed pump, (3) pre-heater at 30°C, (4) selection of hot or cold de-aeration, (5) de-aerator, (6) homogeniser, (7) pre-heater, (8) final heater at 95°C, (9) 30s holder, (10) regenerator cooler, (11) hot filling, (12) regenerator cooler, (13) final cooler, (14) cold filling, (15) control panel, (16) CIP system, (17) pressure vessel (only when sterilising temperature > 100°C), (18) water circulation system, (19) water heater.

liquid allows for high pressures without crushing the product or container. The high pressure induces gelification and inactivation of many vegetative microorganisms by disruption or alteration of the cell membranes. The high pressure is normally maintained for 30 min at 25–50°C depending on product and microorganisms.

Inactivation especially of enzymes and yeast and mould is very important for the production of, for example, jams and juice concentrates. Vitamins, colours and natural flavours are normally maintained during the process. Due to incomplete inactivation of all bacteria spores and enzymes, cold storage of the final product is required. The high pressure process is clearly suitable for high acid products, but the processing costs are very high at present and therefore of no further interest for ready-to-drink juices.

10.5.2   Electro shock

This new method is still under extensive development at leading universities and major companies supplying equipment to the food processing industry (e.g. APV). It is a non-thermal process utilising pulsating high voltage (20 kV) through the product. This technique should disrupt the cell membrane of the microorganism. As no major effect on spores is observed, the technology is again a possible process well designed for high acid products in the future.

Acknowledgements

Thanks are due to Bodil Simonsen, Ella Isaksen, Dorthe Berthelsen, Kirsten Degnbol, Helle G. Rasmussen and Paul Said Fredsted for their kind and efficient assistance in the writing and preparation of this chapter.

Further reading

APV Publications relevant to this chapter from: APV Baker, P.O. Box 4, Gatwick Road, Crawley, West Sussex, RH10 2QB, UK.

Burton, H. (1988) Ultra-high Temperature Processing of Milk and Milk Products, Elsevier, Amsterdam, Chap. 4, pp. 77–129.

Butz, P. and Ludwig, H. (1986) Pressure inactivation of microorganisms at moderate temperatures. *Physica* 139, 140B, 875–877.

Butz, P. and Ludwig, H. (1991) Hochdruckinaktivierung von Hefen und Schimmelpilzen. *Pharm. Ind.* 53(6), 584–586.

Butz, P., Ries, J., Traugott, U., Weber, H. and Ludwig, H. (1990) Hochdruckinaktivierung von Bakterien und Bakteriensporen. *Pharm. Ind.* 52(4), 487–491.

Hayashi, R. (ed.) (1989) *Use of High Pressure in Food*, San-EI, Kyoto, Japan.

Kessler, H. G. (1981) *Food Eng. Dairy Technol.*, 149–173.

Mannheim, C. H. and Passy, N. (1979) The effect of deaeration methods on quality attributes of bottled orange juice and grapefruit juice. *Bd.* 24(5/6), 175–187.

Massaioli, D. and Haddad, P. R. (1981) Stability of the vitamin C content of commercial orange juice. *Food Technol. in Australia*, Vol. 33.

Robertson, G. L. and Samaniego, C. M. L. (1986) Effect of initial dissolved oxygen levels on the degradation of ascorbic acid and the browning of lemon juice during storage. *Journal of Food Science*, Vol. 51, 1, 184–187.

Trammell, D. J., Dalsis, D. E. and Malone, C. T. (1986) Effect of oxygen on taste, ascorbic acid loss and browning for HTST-pasteurized, single-strength orange juice. *Journal of Food Science*, Vol. 51, 4, 1021–1023.

11 Packaging systems for fruit juices and non-carbonated beverages

H. B. CASTBERG, J. I. OSMUNDSEN and P. SOLBERG

11.1 Introduction

The dramatic growth of fruit juice and non-carbonated fruit beverage markets world-wide has been made possible by the development of new packs and packing systems and improvements in traditional packaging. Packaging systems for long life juices have been reviewed (Wiesenberger, 1987). This chapter is concerned with other packaging systems. The intention is not to discuss all packaging used for juices but to select some of the most common, including Pure-Pak type gable top laminated board packs, containers for frozen concentrated juices, glass bottles and developments in plastics packaging.

Key parameters to be considered when selecting a packaging system are:

  (i) processing;
  (ii) distribution, shelf life requirements, legislation;
  (iii) product composition and quality as produced and at full shelf life;
  (iv) product protection required during storage, distribution and retail sale;
  (v) positioning in retail market;
  (vi) pack size, size combinations, printing options, display, etc.;
  (vii) packing system concept, automation options, ability to integrate with existing and or future systems;
  (viii) consumer appeal, image of product and package.

The difference between cold and hot filling systems extends beyond the filling temperature to include the process system, the product, the filler specification, the packaging material, its closure or seal and, last but not least, the product shelf life required.

11.2 The cold fill fresh system

This packaging system gives a product for distribution and sale through a refrigerated distribution system at 0–5°C with a shelf life of 4–6 weeks. Fresh

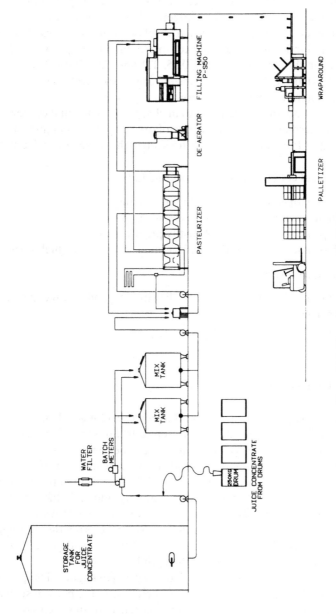

Figure 11.1 Flow diagram for cold fill juice.

cold fill systems permit retention of flavours of the juice as packed whether this is prepared from freshly squeezed fruit or recombined from concentrate with the addition of pulp and aroma. The cold distribution minimises loss of aroma and vitamins. The need for refrigerated warehousing and distribution facilities and sale through refrigerated display cabinets makes this system most suitable for branded, 'up market' premium priced products. Typical packs used for this application are 0.25, 1.0 and 2.0 l Pure-Pak gable top cartons and plastic containers and cups in various sizes from 0.18 to 5.0 l. The system flow diagram and layout is shown in Figure 11.1.

Homogenisation, though referred to in chapter 12 of the first edition of this book, is not necessary to reduce the size of fibrous particles in juice products to be packed in Pure-Pak shaped cartons since these are not sealed through the product.

Advantages of cold fill system for juices include:

(i) well developed processing and filling technology;
(ii) economical investment and processing costs;
(iii) high product quality able to justify premium price;
(iv) high profile in retail outlets (refrigerated section has a high turnover).

Disadvantages include:

(i) requires facilities for refrigerated distribution and storage (particularly expensive with only one product);
(ii) requires frequent distribution.

11.3   The hot fill system

Hot filling has been known as a method to achieve a longer shelf life for juices since long before the introduction of aseptic technology. The objective for both hot fill and aseptic processing must always be to minimise the microbial count present in the raw materials and to avoid reinfection of the finished product. Some aerobic bacillus spores are able to develop in products with a low pH value and thereby affect product quality and shelf life, as are some non-spore-forming bacteria, e.g. *Lactobacillus plantarum*. However the main spoilage organisms are yeasts and moulds. Nevertheless with good hygiene, the hot fill system is one of proven suitability for acid juices and non-carbonated beverages, but still requires careful control.

Hot filling is a simpler operation than aseptic packaging. The product is pasteurised at 92–95°C, filled above 82°C, held hot in the pack in such a way that all parts of the pack are adequately pasteurised and then cooled. The filled containers whether glass, cans, laminated board cartons or plastic, are cooled by passing the containers through a tunnel in which jets of a fine mist of cold water envelope the containers. The cooling tunnels work on a counter

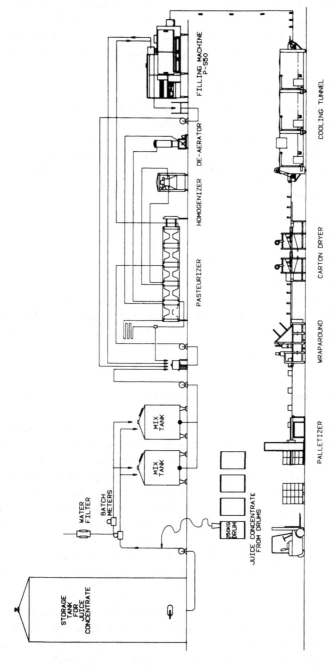

Figure 11.2 Flow diagram for hot fill juice.

flow principle for reasons of heat efficiency and to minimise thermal shock. Some designs incorporate water vapour exhaust fans to increase cooling efficiency by speeding water evaporation rate. When discharged from the cooling tunnel the containers are passed through an air blower for drying and are then ready for date coding, labelling, nesting or group packing (Figure 11.2).

Advantages of hot filling include:

(i) less complicated process and filling line than aseptic;
(ii) no chemical sterilisation of containers;
(iii) high degree of line/container/pack size flexibility;
(iv) high degree of flexibility in product filling, e.g. viscosity, pulp content, ability to add ingredients before and after in-line pasteuriser.

Disadvantages include:

(i) larger requirement for floor space;
(ii) requires product with adequate thermal stability.

Other factors to be considered are:

(i) energy consumption is high on old line designs but new line designs with optimised energy regeneration can recover up to 75–80% energy;
(ii) shelf life and product quality; in many tests, aseptically filled and hot filled products show no significant difference in organoleptic, chemical or bacteriological qualities.

11.4   Filling equipment for gable top cartons

The filling equipment used originates from the United States where it was originally developed for milk products. Re-design for juices and non-carbonated beverages required adaptation of the carton and the structure of the board from which it is fabricated. Modern filling equipment is available in a range of capacities from 1200 to 15 000 cartons/h and can be obtained with programmable logic computer (PLC) control, automatic carton size change over, an automatic feeding system for carton flats and line efficiency monitoring systems (Figure 11.3). This equipment is available for 2 l (0.5 gal), 1 l (0.25 gal) and for portion pack sizes. Changes to the original specification of the filling machine for the dairy industry required to fill juices were as follows:

(i) *Product related changes:* the product tank was constructed to withstand more corrosive juices and to incorporate agitation, filler valves modified to handle particles and general changes to accommodate the higher viscosity of juices without reducing filling speeds.
(ii) *Machine related changes* were required to the bottom pre-breakers,

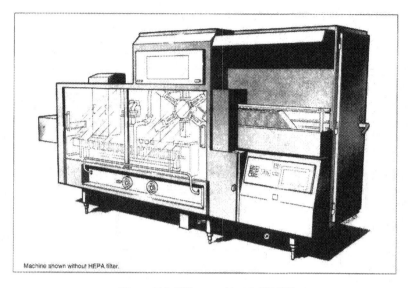

Machine shown without HEPA filter.

Figure 11.3 Filling machine P-S50 HF.

sealing and cooling pads, top pre-breakers and top sealer.
(iii) *Technical changes* were needed for hot filling of juices to provide: heat
resistant gaskets, O-rings and seals; a product tank with insulation,
temperature sensor and display; to insulate the drain and product
return system from the product tank; and to provide additional
cooling for carton conveyor parts.
(iv) *Carton-related changes* are those relating to bottom folding, side seam
skiving and oxygen barriers.

The general machine function can be described as follows:

(i) a magazine of carton flats holds approx. 10 min run;
(ii) the cartons are opened and loaded onto rotating mandrel with six
arms;
(iii) as the mandrel moves from position to position the following
sequence is followed:
  (a) loading onto mandrel;
  (b) bottom heating and folding;
  (c) bottom sealing;
  (d) cooling of bottom seal;
  (e) transfer from mandrel to conveyor chain;
(iv) the bottom sealed cartons are then transferred by an indexing carton
chain to the top pre-breaker;
(v) the carton is filled;
(vi) the top is folded and sealed;

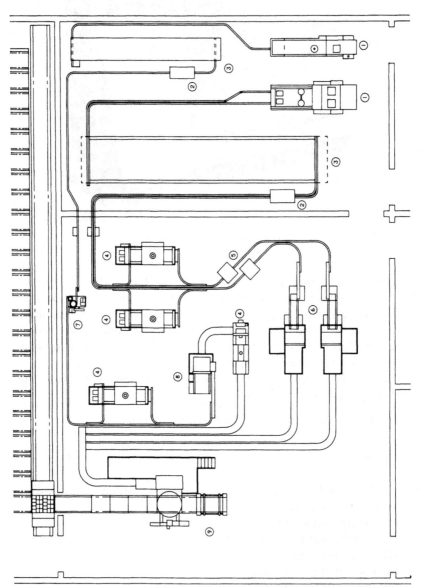

Figure 11.4 Hot filling line for juice products 1: filling machine; 2: drying unit; 3: cooling tunnel; 4: shrink wrapper; 5: handle applicator; 6: wrap around machine; 7: straw applicator; 8: tray packer; 9: palletiser.

(vii) the pack is discharged from the filling machine onto the carton conveyor.

A typical cold fill juice line consists of the filling machine, a conveyor line and a caser or roll container loader (see Glossary) or in some instances a wrap-around outer packer. Cases are stacked before being transported to the cooling room. Roll containers are handled directly to cold store and wrap around boxes are normally palletised. All non-aseptic cold filled juices are 'fresh products' and require chilled distribution. A typical hot fill line consists of the filling machine, a cooling tunnel, a carton dryer and a wrap-around unit (Figure 11.4).

Hot fill systems require more space due to the cooling tunnel. Hot filled products can be stored under ambient temperature, as for aseptic filled products. For individual pack cartons, a typical line will include a straw applicator and a display-tray packer.

11.5 Aseptic Pure-Pak

The Aseptic Pure-Pak system (Figure 11.5) was launched world-wide in

Figure 11.5 Aseptic Pure-Pak U-S80A System.

November 1992. The Aseptic Pure-Pak system differs both in carton shape and in machine technology design compared with the traditional Brik system. The system concept is based on an 'all-in-one' idea, which includes clean-in-place (CIP), pre-sterilisation, and product and utility valve filters, which simplifies installation work.

The Aseptic Pure-Pak system is based on a three line machine working with flat Pure-Pak blanks. The system has an automatic carton size change over for three of the five carton sizes (1/1, 3/4, 1/2, 1/3, 1/4 l) of standard cross-section. Bottom sealed cartons are sterilised in a pressurised sterile tunnel by spraying a 2% hydrogen peroxide solution with UV treatment and hot air carton ventilation prior to product fill and top sealing. For oxygen sensitive products there is a standard option for nitrogen flushing the headspace.

The system operation is controlled by the operator via the operator control unit. Production performance, CIP and pre-sterilising are all monitored by the machine monitoring system, including calculation of efficiency rate and waste, if any. Further flexibility includes the system's capability to operate within the product temperature, 0–40°C range, and different board laminates include the latest development in non-foil barriers like X-board and EVOH material.

11.6   Packing materials for gable top cartons

The basic material for gable top cartons is a paper board coated on both sides with a thermoplastic material, normally polyethylene (PE). The board can be fully bleached, semi-bleached or unbleached. The raw board is mainly made of sulphate pulp and is normally a mixture of long and short fibres. The long fibres are derived from pine and spruce, the short fibres from birch, oak and eucalyptus. The mixture gives a finished board with good mechanical properties, good printability and good performance during carton assembly and filling. In the pulp mill, the wood chips are cooked with alkali under high temperature and pressure. During this process the lignin holding the fibres in the wood is dissolved. Up to 200 different substances are also extracted during the cooking process; some of these, together with chemical compounds created during cooking, adhere to the fibres. Many of these compounds contribute to the strong odour and taste typical of unbleached board. After washing and refining the pulp is bleached with chlorine. This dissolves and de-colours the remaining brownish lignin and other chemical compounds adhering to the fibres. In modern bleaching plants chlorine has been largely replaced by other oxidising agents which do not give rise to chlorinated organic components whose safety has been questioned. A typical modern bleaching process for pine fibres consists of five steps as follows:

(i) chlorine + chlorine dioxide;

(ii) alkali (sodium hydroxide);
(iii) chlorine dioxide;
(iv) alkali (sodium hydroxide);
(v) chlorine dioxide.

For birch fibres the bleaching process would be similar but no chlorine is needed at all. The bleaching process thus makes a clean paper board since unwanted chemical components which would give an off-taste are removed and there are significant reductions in the bacterial count. The fully bleached board is an excellent base for printing. Before feeding the pulp to the board machine small additions of proprietary materials are made to improve wet strength to prevent penetration of liquid at any raw edges of board in the finished pack.

Board for packs to contain liquids ranges from 200 to 450 g/m² (0.3–0.7 mm thickness). At one time made as one homogenous layer, the trend now is to make carton board in two or three layers to improve stiffness to weight ratio. To further increase stiffness, fibres in the middle layer can be dimensionally oriented (e.g. Enso Multilayer Board) or can contain a specially treated pulp, chemical thermomechanical pulp (CTMP) (e.g. Billerud Triplex Board). One or two layers can be made of unbleached pulp. Modern three layer boards are also made with an improved surface smoothness giving good printing results.

The next step in the production process is to coat the board on both sides with PE. For long life products such as hot or aseptic filled juices, water and wine, a thin aluminium layer is incorporated into the laminate to provide a barrier to volatile flavourings, oxygen and light. A typical structure is shown in Figure 11.6.

Packing material for gable top cartons should have the following properties:

(i) high standard of hygiene;
(ii) good mechanical strength, including internal bonding of fibres;
(iii) liquid-tightness;
(iv) barrier to light;
(v) low migration of components into packed product;
(vi) barrier to gases and flavours (for long life products);
(vii) good sealability;
(viii) good performance in carton assembly and filling.

In the converting process to prepare the carton flats the PE-coated board is printed, provided with score lines and cut into blanks. Printing processes can be flexographic using rubber or photopolymer typeface 'clichées' with alcohol- or water-based inks. Offset lithographic or rotogravure processes can be used where high quality printing is desired. A typical converting operation will have up to four production lines and will automatically stack carton flats on the pallets ready for the next process which is sealing of the

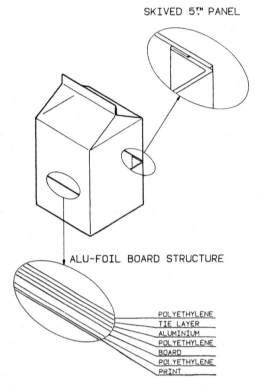

SKIVED 5™ PANEL

ALU–FOIL BOARD STRUCTURE

POLYETHYLENE
TIE LAYER
ALUMINIUM
POLYETHYLENE
BOARD
POLYETHYLENE
PRINT

Figure 11.6 Pure-pak shaped carton.

side seams in a flame sealer to make the blanks required by the filling machine.

For this the flat cartons are fed from a magazine in a continuous stream through the machine supported by belts and guiding rails. All vertical score lines are pre-broken and the longitudinal sealing zone is heated to make the polyethylene sticky. In the final folding section pressure is applied to the heated area to form the longitudinal seal. Blanks are counted and automatically packed, the pack being marked with its content and the customer's name. The boxes are then automatically palletised and normally protected with stretch film.

Blanks for long life products which incorporate aluminium foil are sealed in a special way. The thickness of part of the longitudinal sealing flap (the 5th panel) is reduced to 50% with a grinding wheel and half this strip is folded over. The final folding of the 'skived' 5th panel is then made with PE meeting PE in such a way that no raw edge remains inside the package as shown in the inset of Figure 11.6. In the filling machine the flat side-sealed blanks are opened and fed onto a mandrel. The bottom is formed and sealed, product is filled and the top is formed and sealed.

Further development of packaging materials continues with the aim of providing:

(i)   even more environmental friendly board types;
(ii) more cost effective boards with better stiffness to weight ratio;
(iii) improved barrier materials for long life products and special product applications.

For best performance throughout the distribution chain the package must be fully integrated into the distribution system. This requires consideration of costs, technical factors and ergonomics. After processing and filling, filled cartons are grouped in an outer package designed to provide handling, display and protection in the proposed distribution system. For transport, several distribution units are grouped together in a transport unit which normally fits a standard pallet (Figure 11.7).

Several needs and preferences, often contradictory, must be considered when evaluating and designing optimal distribution systems. Depending on the relative location of production facilities and the market, different distribution systems will be required, from long distance transport by ship or rail (including transport to and from terminals) to traditional distribution direct from production site to retailer. All these distribution systems must fit into existing modular systems, pallets and containers. All must meet the different physical regulations and limitations of different transport systems and all must permit simple and efficient handling. This, in combination with the need for cost efficient transport, has led to the development of different standards for pallet and container systems and the packaging to fit them. Unfortunately the resulting standards are not all compatible with each other.

The needs of a transport system also have to be reconciled with the requirements of retailers for easy handling, display and accessibility of pack outers. Outers should be designed for easy display. Price tagging and waste handling should require minimum effort. If the outer is to be placed directly on the shelf it must be compatible with the standard shelf designs (Figure 11.8).

Therefore the design of the distribution unit must meet the following requirements:

(i)   compatibility with the different pallet standards;
(ii) compatibility with standards at the retailer;
(iii) within the weight limits permitted (for handling);
(iv) appropriate for the frequency of deliveries to the retail outlet.

This last requirement can lead to the use of large units for packs with high rate of sale (see Figure 11.9). Such outer packs can be exposed anywhere in the shop and handled with minimum effort. They are used mainly for large supermarkets and/or in connection with special promotional activities.

A system that meets most of these demands is based on a corrugated wrap-

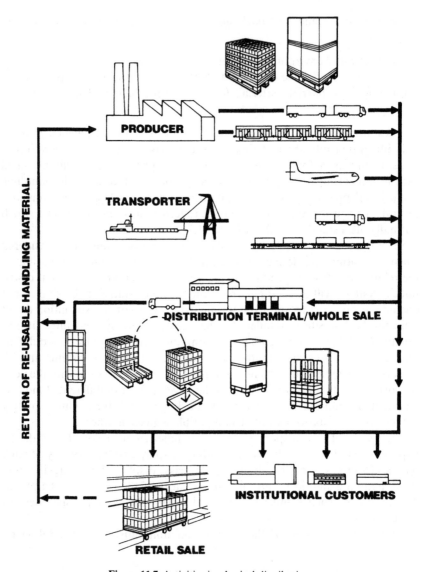

Figure 11.7 Activities in physical distribution.

around case or a tray with full height side walls protected from dust and moisture by shrinkwrapping. The units are stacked in a pattern which gives high utilisation of the pallet and are shrink or stretch wrapped on the pallet. High pallet utilisation is essential not only for economy but to reduce transport damage (Figure 11.9).

When frequently purchased products are supplied to a market close to the production unit, a closed distribution system is normally used, with return-

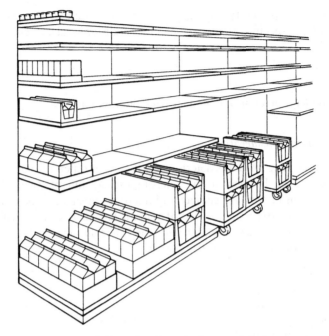

Figure 11.8 Examples of standards at retail dealers.

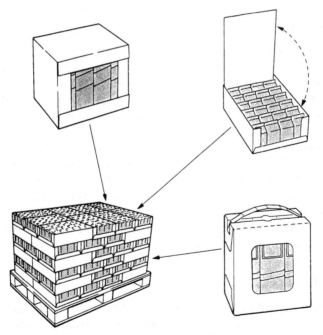

Figure 11.9 Example of non-returnable distribution system.

able packs. Compatibility with general transport standards is not so important and the specific system can be tailor-made for local cost efficiency and performance.

11.7   Product protection and product/pack interaction

11.7.1   General considerations

Like all foods, juice products deteriorate with age at different rates according to their chemical, biomechanical, microbiological and physical characteristics. The main factors are spoilage by microorganisms and flavour/colour defects caused by oxidation. In recent years it has also been recognised that flavour components may interact with the inner surface of the packaging. Measures important for good shelf life are:

   (i) good hygiene;
  (ii) low air entrainment;
 (iii) good quality raw materials;
 (iv) good barrier in pack to gases and flavourings;
  (v) control of temperature in distribution;
 (vi) minimum absorption at food contact surface.

11.7.2   Cold filled juices

Normally juices pasteurised at 90–95°C for 20–30 s and cold filled non-aseptically into cartons with a barrier of aluminium foil will maintain a satisfactory quality for 6°C for 4–6 weeks. During that time colour, vitamin C content and sensory properties remain acceptable with marginal changes. Using PE-coated cartons with no aluminium foil, shelf lives longer than 14 days are not recommended due to the high rate of oxygen permeation and penetration of flavourings from the juice into the pack.

Care should be taken to minimise air entrainment during juice processing. Juice oxygen content should preferably be below 5 ppm immediately after filling. For juice stored at 6°C for 4–6 weeks the influence of head air space is minor. As an extra precaution nitrogen gas may be used in the head space to replace air but this is not normally necessary if proper measures are taken in the reconstitution and processing steps.

11.7.3   Hot filled juices

The microbiological status of hot fill juice products allows storage at ambient temperatures for several months. It is therefore mandatory that packs have a high barrier against gas and aroma permeation. This barrier can be aluminium foil or other polymer materials tailor-made for the purpose. The oxygen permeation rate should be in the range 0–10 ml oxygen/m^2 per 24 h. Note that the critical factor is total permeation of oxygen into the filled cartons, not just the barrier characteristics of the laminate. Thus it is essential to obtain good seals and avoid cracks in barrier materials.

When hot filled juices are stored at ambient temperature all chemical processes, including those not requiring oxygen, will occur faster than in chilled juices. Thus even with packs made from high barrier materials, discoloration and vitamin C degradation will take place quickly. If controlled temperature storage is not available, a 'best before date' of 2–3 months from production is recommended.

11.7.4 Aseptic filled juices

Commercially sterile juice products (high acid) are obtained through heat treatment by pasteurising plate or tubular system at 85–95°C for 4–20 s. The process equipment and tube system for transfer of the product to the aseptic filling system must comply with requirements and safety for aseptic filling standards.

Aseptic products can be filled at ambient temperature to obtain 6–8 months' shelf life. The single largest benefit with aseptic packaging is ambient storage and distribution, which is easier to operate than refrigerated distribution. The demand on the total packing material barrier to oxygen/light is however, so much larger, and must secure product quality throughout the product's potential shelf life. None of the three mentioned processes or filling technologies are able to stop the chemical product degradation taking place, a feature mainly affected by time/temperature.

11.7.5 Flavour

Generally the use of an aroma barrier carton will give best protection of the characteristic flavour of a juice product. Mild flavoured products may deteriorate faster than more strongly flavoured variants. Also, when the aroma barrier is not optimal, flavour will change more quickly in small packs than in larger ones due to the greater surface area to volume ratio. Flavour changes due to odour pick-up from the packaging material should not be accepted.

For many fruit juices, particularly citrus juices, the characteristics of the food contact surface of the container are important, however, reports on the loss of flavour into packing material are conflicting. At one time it was believed that polymeric materials such as polyethylene could absorb citrus oils from juice and so reduce flavour intensity. More recently it has been claimed that the absorption of aromatic oils, e.g. d-limonene in the polymer may be beneficial because d-limonene itself does not contribute much to the flavour; also oxidation products from *d*-limonene may create off-flavours. Recently, laminates incorporating hydrophilic barrier materials such as EVOH, have been used in the belief that flavour will be retained and better protected. There is a need for more research in this field, particularly for comparisons to be made in a controlled fashion and with a limited number of variables.

11.8 Packaging of frozen concentrated juices (FCJ)

The United States remains the major leading market for FCJ distributed in retail packs. The traditional and most common pack used is the spiral wound composite can, just below 2 thousand million units being sold per year. Frozen juices in composite cans have, from time to time, been introduced in Europe without securing major sales although these packs are still found in some shops and supermarkets, either imported from the United States or contract packed in own label by a European packer.

The composite can is cylindrical, composed of spiral wound laminated board with metal or plastic ends (Figure 11.10). It has been through various stages of development. Today the two most common models differ in the form of the opening device. In one the pull-tab removes the full lid: this can comprises three parts:

(i) can body;

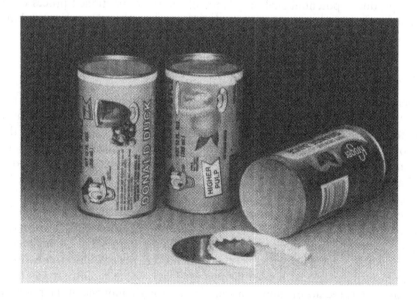

Figure 11.10 Composite containers for frozen concentrated juices.

(ii) top joined to the body by can manufacturer;

(iii) bottom end, applied after filling.

The other, the Mira strip can, comprises four parts:

(i) can body;

(ii) regular end (top lid);

(iii) easy open (Mira strip); these three parts joined together by the can manufacturer;

(iv) bottom end, applied after filling.

A typical US installation for filling and seaming composite cans processes 400–600 cans/min. Composite cans are supplied on pallets and so a large storage space is required for empty cans. From the store, the cans are de-palletised, filled, capped and frozen in a freezing tunnel. The case packer and palletiser are usually located close to the frozen store room.

11.9 Filling in glass containers

Glass bottles, the oldest industrial packaging, still have a high share of the packaged juice product market; they are used mainly for long life shelf stable products in non-returnable bottles. Glass bottles are seen as quality packs and have technical advantages where distribution places special demands on the package, as in long distance travel. Today's bottling operations, based on carousel fillers, can operate up to 80 000–90 000 bottles/h for carbonated products. The average filling equipment used for juice beverages operates at 15 000–30 000 bottles/h. Several medium sized bottling machines offer better flexibility than only one machine. As the filling machines work from 'preformed' packages, they can operate with many times the output of carton-based single serve packages, however, space is required to store a very large number of bottles to meet peak demands.

Once the bottles are properly cleaned and checked for flaws, they enter the filling turntable or carousel. Trouble-free high speed operation requires good evacuation of the air already in the bottles before or during filling. To ensure that the product can enter the bottle in the shortest possible time, without splashing or dripping, air is evacuated by systems with varying degrees of sophistication. In the simplest, air is simply displaced as the product goes in; advanced systems operate with a high degree of underpressure so that bottles can be filled up to the neck at a constant speed. The filler construction can either be based on the bottles being lifted up to the filler nozzles or the nozzles can be lowered down to the bottles. The nozzles either direct product along the sides of the bottle or are inserted to the bottom of the bottle and raised with the liquid as the bottle is filled; the latter type minimises air entrainment.

As the product must be shelf stable, the filling process is either cold aseptic or hot filling. After filling, the bottles go through capping equipment which normally functions on the same carousel principle. In the hot fill operation, the cap will be exposed to heat and then the bottle cooled in a cooling tunnel using water at a series of declining temperatures. The capped bottles from the aseptic machine or the cooling tunnel are labelled, then packed into crates for local distribution or into wraparound corrugated boxes for long distance distribution. Thanks to the strength of the bottles, the boxes can be stacked in warehousing to give a high degree of space utilisation without shelving.

Juice bottles today come in a large variety of shapes and sizes. The system is most efficient around 0.7–1.0 l. Even with single trip bottles, in relatively local distribution re-cycling of the glass allows a satisfactory degree of efficiency.

11.10   Plastic containers and pouches

The poorer technical performance of many plastics materials compared with glass bottles and metal cans does not prevent their satisfactory use for many products, in particular for those with shorter shelf lives. Plastics-based containers and pouches with laminated film or metallic coats are mainly found in cold filled juices and beverages for ambient or refrigerated distribution; plastic containers and pouches for hot filling have not been developed. In recent years developments in co-extrusion technology and the introduction of PET have enabled plastic packs to be produced with improved barrier properties. The principal alternatives are:

   (i) in-house blow moulding on line to feed filler/sealer;
  (ii) use of ready moulded plastic bottles;
 (iii) in-house extrusion of laminate for feeding to form/fill/seal;
 (iv) pre-extruded laminate supplied in rolls;
  (v) filling pre-formed cups.

In-house blow moulding or laminate extrusion requires considerable capital investment and is therefore only practised by major juice or beverage producers. The use of pre-formed plastic bottles, cups and laminates for form/fill/seal operations is common among medium and smaller juice and beverage packers.

Most plastic containers have two components, the container and its closure which can be a screw cap with tamper evident ring or plastic/aluminium laminated heat-sealed membrane. The closure must match the barrier properties required for the container and must meet the requirements for tamper proofing. Selection of a closure also depends on the container size. Portion packs do not require easy re-closure whereas above 500 ml this is a normal consumer requirement.

For a time plastics were considered as 'cheap' alternative packaging materials. However, new materials have provided interesting opportunities in container shaping, size flexibility, barrier properties, transparency/opacity and flexibility. Plastics have now been well accepted both by juice product packers and consumers as attractive and useful alternatives in many packaging applications. Plastics have recently been faced with the environmental issue. Although this issue has many aspects, the environment must be taken into account when selecting materials and investing in new packaging systems. Factors to be considered include the base material composition, degree of re-cycling feasible, volume in relation to total domestic refuse, and environmental effects arising from destruction of used plastic materials. Several manufacturers provide total system technology for plastic packaging; these include Serac, Remy, Romalag, Sidel, Gasti, Bosch, Servac, Formpack and M. B. Freshfill.

References and further reading

Billeruds, A. B., Box 60, S-681 00 Säffle, Sweden.

Dürr, P., Schobinger, V. and Waldrogel, R. (1981) Aroma quality of orange juice after filling and storage in soft packages and glass bottles. *Lebensmittelverpackung* 20, 91–93.

(E B) Eddy Company, Ottawa, Canada 1969. Printing production.

Elopak A/S, P.O. Box 523, N-3412 Lierstranda, Norway. Food Sciences, board and equipment development department, Marketing for special products.

Enso Gutzeit OY, P.O. Box 307, SF-00101 Helsinki, Finland. Food Engineering International May 1987, p. 57, TAB-TOP.

Henriksen, Hinkel, Hopland, Nyronning Posti and Verket, Universitetsforlaget Oslo. Off-set printing.

Iversen, A. (1987) Cartons for Liquids. In: *Modern Processing, Packaging and Distribution Systems for Food*, F. A. Paine (ed.), Blackie, Glasgow.

Lithographers Manual, R. Blair, Graphic Arts Technical Foundation, 4615 Forbes Avenue, Pittsburgh, Pennsylvania, USA, Post Code PA 15213.

Mannheim, G. H., Miltz, J. and Retzler, A. (1987) Interaction between polyethylene laminated cartons and aseptically packed citrus juices. *J. Food Sci.* 52, 737–740.

Marshall, M., Nagy, S. and Rouseff, R. (1985) Factors impacting on the quality of stored citrus fruit beverages. In: *The Shelf Life of Foods and Beverages*, Proceedings of the 4th International Flavor Conference, Rhodes, Greece, G. Charalambous (ed.), Elsevier, Amsterdam.

Moshonas, M. G. and Shaw, P. E. (1987) Flavor evaluation of fresh and aseptically packed orange juices. In: *Frontiers of Flavor*, Proceedings of the 5th International Flavor Conference, Chalkidiki, Greece, G. Charalambous (ed.), Elsevier, Amsterdam.

Pure-Pak Inc.

Svensk Standard S.S. 847002 issue 2.

Teknologisk Forlag, Oslo, Norway 1972. Grafisk teknikk.

Weisenberger, A. (1987) *Confructa Studien*, pp. 143–159, Flüssiges Obst, Schönborn, W. Germany.

12 The formulation of sports drinks

M. A. FORD

12.1 Introduction

Since the earliest days of sporting endeavour, man has always sought a means of achieving success over his fellow competitors. It is said that Charmis of Sparta attributed his success in foot races in the ancient Olympics to consumption of a diet rich in dried figs, an early example of a convenience food rich in carbohydrate! The pursuit of success has continued to the present day but science has played an increasingly important part in understanding the physiological effects of exercise and in promoting the essential role of diet in the achievement of that success. In particular, this knowledge has facilitated the development of suitably formulated sports drinks which have enabled active individuals to capitalise on their talents.

It is now clear that what we eat and drink has a profound effect upon sports performance and this fact was reinforced by the recent publication by Maughan (1993) of a consensus statement agreed by a group of leading academics in sports science which included the following statements: 'Loss of fluid and a reduction of the body's carbohydrate stores are the two major causes of fatigue in prolonged exercise,' and 'The evidence clearly indicates that sports drinks which contain an energy source in the form of carbohydrate together with electrolytes, particularly sodium, are more effective than plain water in improving performance.'

Despite this acceptance and the large numbers of scientific papers published in this area of work, there remains a widespread reluctance to acknowledge the substantial benefits that a correctly formulated sports drink can confer upon those engaged in sporting activities. At the same time, it must be recognised that the advantages that these products offer are less crucial to the armchair athlete, for whom flavour and overall acceptability remain important issues.

In most cases, exhaustion results from a combination of depletion of carbohydrate reserves and dehydration. While rest and the drinking of water can redress the situation, the consumption of a beverage formulated to optimise the delivery of fluid and carbohydrate is a more effective treatment. To prevent this situation from occurring, it is sensible to prepare for a sporting activity by ensuring that carbohydrate reserves and hydration are maintained prior to an event. In many sports it is now well accepted that drinks taken during an event can help to mitigate the effects of fatigue in the later stages and maintain or improve performance. As a result of the increasing interest in sports and exercise generally, the world-wide market for sports drinks has grown and the number of products available to the public has increased rapidly in the last few years.

12.2 Sports drinks

12.2.1 History and background

Christensen and Hansen (1939) reported that endurance in sporting activities was greatly enhanced by a pre-competition diet rich in carbohydrate. In the 1960s a number of pioneering studies emanating from Scandinavia demonstrated the relationship between diet, muscle glycogen storage and exercise potential (Bergström and Hultman, 1966; Hultman and Bergström, 1967). In this period, much attention was directed to augmentation of glycogen reserves prior to competition and this led to the development of a small number of specialised sports products, including a sports drink 'Dynamo' containing a high concentration of carbohydrate. In addition to a mixture of electrolytes intended to aid the replacement of those lost in sweat, this product contained glucose polymers providing 1170 kJ (280 kcal) of carbohydrate energy in a palatable formulation. At this time, the emphasis was on carbohydrate provision. Subsequently, the importance of water as a nutrient in preventing a deterioration in physical performance was perceived and this realisation led to the development of 'isotonic' beverages.

The first isotonic drinks were developed for use by university sports teams. The earliest of these was formulated by Dr Martin Broussard for use by the Louisiana State University football team, the 'Tigers', and named 'Bengal Punch'. The second was developed as a result of the work by Cade et al. (1972) who studied the effect of heat exhaustion on players of the University of Florida football team, the 'Gators'. They found that the loss of volume and compositional changes that occurred in body fluids during vigorous exercise could be prevented or ameliorated by the consumption of a glucose electrolyte drink and use of this product led to immediate beneficial effects for the members of the team. The product received national publicity when the Gators beat their opponents in the prestigious Orange Bowl final. This product, known as 'Gatorade', entered the market in 1969 and became the first to be promoted as a sports drink. With the success of Gatorade, other products appeared on the market and attracted the attention of researchers in sports science with the result that a significant literature has been created, demonstrating the value of these products for people engaged in sporting activities.

12.2.2   The sports drink market

Sports drinks have become an increasingly important segment of the world-wide drinks market. The first 'isotonic' beverages were liquids and this remains the dominant form, packaged in glass and polyethylene terphthalate (PET) bottles, cans and recently, in pouch packs. The second type of sports drink is the powdered product which requires mixing by the individual and avoids the inconvenience of transporting large volumes of liquid. It is frequently used in this form by sports teams where mixing is performed in bulk. Sports drinks are also supplied as liquid concentrates, avoiding the inconvenience of dissolving a powder, while offering the advantage of saving weight and space.

Since entering the sports drink market in 1969, Gatorade has remained the dominant brand in the United States. Gatorade's share of the retail market has continued to grow and now commands an estimated 95% of the sports drink market in North America, worth $1 billion. The market for 1995 is estimated at $1.5 billion. In Europe the growth has been much slower than in the United States and Japan, but is gathering momentum. In 1988, total sports drink consumption was 47 million litres and despite a background of slowing economies and generally poor weather has grown to over 230 million litres with a retail value of some £300 million. It is anticipated that this growth will continue at around 15–20% per annum to reach at least 500 million litres by the year 2000. Even at this level, it is well below the 700 million litres currently consumed in the United States which has less than two-thirds of the population of Europe.

Germany has the largest sports drink market in Europe with 36% of the overall volume, followed by Italy with 19% and the United Kingdom with 11%. In some areas, the markets are dominated by one or two major brands such as Gatorade in Italy and Lucozade Sport in the United Kingdom. Overall in Europe, Wander's Isostar retains a narrow margin over Quaker's Gatorade. Isostar's international expansion in 1984 spawned a host of regional sports drinks brands with very local markets and whose individual sales remain small. It is noteworthy, however, that the sports drink market has attracted a number of the world's largest food and pharmaceutical companies; together twelve of these companies control 76% of the total volume.

The ready-to-drink products (still and carbonated), account for more than three-quarters of the total market, powders 22% and syrups less than 1%. While debate continues about the relative merits of the so-called isotonic and hypotonic drinks, isotonic products dominate the market at present with 68% of the market. Concentrated energy products, powder and liquid comprise a further 9%. There is little doubt that each has a role to play, with hypotonics being the specialised rehydration beverages and isotonic drinks remaining the product of choice when there is a requirement for fluid and carbohydrate provision.

A wide range of flavours is employed by the manufacturers of sports drinks ranging from fruit punch, exotic fruits or grape, favoured by the United States market, to the more traditional citrus flavours popular in Europe. Orange remains the principal flavour, followed closely by lemon and mixed citrus; grapefruit is the only other significant flavour with about 4% share of the market. Recently, milk-based drinks have entered the market with R'Activ being particularly successful in the German market.

The overall appeal of a sports product relies heavily on the flavour impact, particularly when the user has a dry mouth following competition. Most producers rely on the use of added flavours but some products have extended the appeal by the incorporation of fruit materials. It should be noted, however, that while orange remains a popular flavour with the general public, it is not a universally popular flavour when consumed after strenuous exercise.

12.3 The effect of exercise

In order to develop an effective product to meet the requirements of sports people, it is necessary to understand the changes that occur in response to the metabolic stress incurred.

12.3.1 Carbohydrate

Exercise is accomplished by the transformation of chemical energy to mechanical energy in muscle tissue. The energy expenditure incurred by physical activities is dependent on the intensity of the exertion and the duration (Table 12.1). The primary source of fuel for this energy is provided

Table 12.1 Energy expenditure of selected activities (kcal/min)

Jogging	7–8
Cross-country running	11
Rapid walking	5–7.5
Running (1500 m)	16
Running (marathon)	5
Running (sprint)	32
Cycling	5–11
Swimming	5–14
Squash	10–18
Tennis	6–9
Table tennis	4
Football	9
Basketball	8
Rowing	8–15
Golf	2.5–5
Gymnastics	5–7.5

by the oxidation of carbohydrates and fats derived from the food we eat. The relative contribution of each fuel to the total energy consumption during exercise depends on a number of factors, including exercise duration and intensity, and the subject's fitness and nutritional status (Golnick, 1988; Maughan, 1991a).

Carbohydrate is stored in the body in the form of glycogen in the liver and muscles. Liver glycogen reserves are essential for the maintenance of normal blood glucose levels since this is the primary energy source for the brain and central nervous system. In response to a number of metabolic stimuli caused by exercise, there is an increased uptake of blood glucose by the working muscles. To prevent the blood glucose level falling to a point where the nervous system becomes compromised, the liver will be stimulated to supply glucose from the liver glycogen stores. As soon as these stores are emptied and provided that glucose utilisation remains high, blood glucose values decline to hypoglycaemic levels. To protect the central nervous system, glucose uptake by the working muscles decreases significantly, and muscle activity becomes totally dependent on muscle glycogen stores and replenishment from external carbohydrate sources.

Muscular contraction depends on the availability of two energy sources, carbohydrate and fat, but the use of these reserves is never exclusive. For example, in low intensity exercise, fat in the form of fatty acids is the main energy provider but as the intensity of exercise increases the body uses a greater proportion of carbohydrate until in highly intensive activities the percentage ratio of carbohydrate to fat may be 90:10 (Brouns, 1993). The ability to perform repeated high intensity contractions becomes impaired when the glycogen content of the exercising muscles reaches very low levels (Bergström and Hultman, 1967; Costill, 1988).

The quantity of glycogen stored in muscle tissue approximates to 300 g in sedentary individuals but may be increased to amounts in excess of 500 g by a combination of exercise and a high intake of carbohydrate. Bergström and Hultman (1972) have demonstrated a close correlation between the pre-exercise glycogen content in the quadriceps femoris muscle and the work time on a bicycle ergometer, increased stores enabling longer work times to be sustained. While ingestion of carbohydrate during exercise may maintain blood glucose levels and reduce dependency on endogenous reserves, there is a considerable body of evidence to show that glycogen stores are more effectively repleted if the carbohydrate is consumed immediately after exercise when the rate of glycogen resynthesis is at a maximum (Blom et al., 1987; Coyle, 1991). This phenomenon of glycogen supercompensation provides the basis of the dietary technique of 'carbohydrate loading' or 'packing'. Bergström and Hultman (1972) reported that the highest glycogen levels were achieved if, after exercising to deplete the muscles, a carbohydrate-poor diet was consumed for the succeeding two to three days, to ensure that muscle stores were emptied. At the end of this period, the subject enters a final, high

Table 12.2 Proportion of energy derived from nutrients (%)

	Protein	Fat	Carbohydrate
Average 'Western' diet	20	40	40
Healthy diet (non-sports)	15–20	30–35	50
Healthy diet (sports)	15	25	60

carbohydrate 'loading' phase prior to competition. Using this technique it is possible for some individuals to increase the glycogen content of the muscles by more than 100% and it can be demonstrated that there is a linear correlation between muscle glycogen concentration before exercise and the length of time a work load can be sustained.

Notwithstanding the benefits that careful application of this type of dietary regime offers to the athlete, it is not popular because it induces a feeling of 'heaviness' and many subjects do not benefit noticeably from the application of this technique (Wootton *et al.*, 1981), possibly because of a poor understanding of nutrition. The belief in the high protein diet for sports people continues to persist despite recommendations to increase the proportion of carbohydrate in the diet (Table 12.2).

Many workers have studied the provision of carbohydrate in the form of a drink either before or during exercise and an extensive review is outside the scope of the present publication. For a summary, the reader is referred to the paper by Murray (1987). Of interest are recent publications demonstrating the value of carbohydrate electrolyte solutions during exercise. In a study in which a group of runners was required to run the marathon distance in a personal best time, Tsintzas *et al.* (in press) reported that, when given a carbohydrate–electrolyte drink (5.5% carbohydrate), they were able to return performance times that were significantly shorter than their times when provided with water. Notably, the improvement in performance occurred in the second half of the marathon exercise, when they were able to demonstrate an ability to sustain running performance. In a second study (Tsintzas *et al.*, 1993), subjects were required to complete 60 min of intensive exercise on a treadmill, during which the carbohydrate–electrolyte drink (5.5% carbohydrate) was provided in the ratio of 2 ml/kg body weight, following a pre-exercise bolus of 8 ml/kg body weight. The subjects were then required to run to exhaustion, with water only being provided in the second phase. They reported an improvement in endurance capacity and muscle biopsy samples revealed a 28% reduction in glycogen utilisation as a result of carbohydrate ingestion, in the form of the carbohydrate–electrolyte drink, when compared with water.

In summary, the majority of the published studies suggest that carbohydrate supplementation before or during a period of exercise, generally has a beneficial effect on exercise performance.

12.3.2 Fluid

The largest and potentially most serious change in body fluid status during exercise is caused by increased sweat production for temperature regulation. When the body's energy reserves fall to very low levels, the effect is noticeable because of an inability to maintain physical performance. However, the effects of fluid loss on performance are not immediately apparent and fluid status is not generally considered as a factor that is likely to cause an impairment in performance, except in extreme circumstances. Fluid is lost from the gastro-intestinal tract, the respiratory tract and in sweat. During exercise, urine volume is reduced and sweating becomes the major route for fluid loss. At high ambient temperatures, heat is lost from the body by the evaporation of sweat which is governed by two environmental factors, air temperature and the relative humidity. An increase in one or both of these factors results in an increased rate of sweating and a greater need to replace the lost fluid. Although some compensation of the fluid loss is accomplished by the release of water when glycogen is broken down, additional fluid is required to offset the losses occurring and to minimise the effects of dehydration. Unfortunately, thirst is not a good indicator of fluid status and water balance may be impaired before it is realised by the subject.

Sweat production can exceed 1 to 2 l/h, depending on exercise duration, and ambient temperature. Evaporation of 1 l of water from the skin will dissipate 580 kcal (2.4 MJ) of heat from the body. Maughan (1991a) has calculated that for a marathon runner with a body weight of 70 kg and a time of 2 h 30 min, a sweat secretion rate of about 2 l/h is necessary to achieve this rate of evaporative heat loss, resulting in the loss of about 5 l of body water or about 7% of body weight. It is reported that exercise performance is impaired when dehydrated by as little as 2% of body weight. With a loss of water amounting to 4 to 5% of body weight, the capacity for hard muscular work must be expected to decline by 20 to 30%. In extreme circumstances, when exercising in a warm environment, losses equivalent to 10% of body weight can occur, risking circulatory collapse (Maughan, 1991a; Bergström and Hultman, 1972).

Water losses are distributed in varying proportions between the extracellular and intracellular water compartments. For duration events, the principal contribution is from the extracellular water resulting in a reduction in plasma volume and an increase in blood viscosity. Heart rate increases to maintain blood flow to the muscles to supply oxygen and fuel but a high flow to the skin is also necessary to transfer heat to the skin surface where it can be dissipated. In situations where plasma volume is decreased skin blood flow is likely to be reduced thereby compromising the body's mechanisms for losing heat. As a consequence the body temperature rises, cardiac function is impaired and a limitation is placed upon physical performance. It is essential to redress this situation by an intake of fluid, during exercise wherever possible and immediately post exercise when it is not.

Notwithstanding the value of maintaining fluid intake during exercise, the importance of adequate pre-event hydration status was demonstrated in a study by White *et al.* (1992). Prior to an endurance run, subjects were randomly assigned to one of three hydration regimes for five consecutive days, designed to induce states of 'normo-hydration' (normal body weight), 'hypo-hydration' (1.1% reduction in body weight) and 'hyper-hydration' (1.9% increase in body weight). The latter condition was achieved by the consumption of 50 ml/kg body weight per day of a sports drink containing 5.5% carbohydrate in addition to *ad libitum* intake of food. Significant differences were noted in endurance run performance, with the greatest distance achieved with the 'hyper-hydration' treatment. The results demonstrated the necessity of ensuring that adequate levels of hydration and carbohydrate supply are achieved prior to an event, particularly in endurance running over 1 h in duration.

12.3.3 Sodium

Table 12.3 shows the concentration of the major electrolytes in different body fluids; note the differences in the ionic concentrations within the cells compared with the extracellular fluid compartment, plasma.

Sweat is hyposmotic in relation to plasma; that is, water is lost in a greater proportion than sodium or chloride, the two major electrolytes in sweat. Sodium accumulates in the plasma and the resulting increase in osmolality inhibits the sweating response. Thus, correction of the water deficit is of more importance than electrolyte replacement. Serious electrolyte losses are a much less frequent occurrence than dehydration caused by water loss. Nevertheless, compensation of large sweat losses by plain water may induce very low sodium levels in exceptional circumstances. 'Water intoxication', that is, over dilution of the plasma, has been observed in athletes undertaking strenuous exercise in high ambient temperatures (Noakes *et al.*, 1975). For this reason and because of the essential role of sodium in glucose and water absorption (see section 12.4) there are sound reasons for the inclusion of low concentrations of sodium in sports drinks.

Table 12.3 Average electrolyte concentrations in body fluids (mEq/l)

	ICF	Plasma	Sweat	Urine
Sodium	10	132–142	48	128
Potassium	150	3–5	6	60
Magnesium	40	1–2	1–5	15
Calcium	0	4–6		
Chloride	4–8	98–106	40	135

ICF: Intracellular fluid (muscle)

12.3.4 Potassium

At the onset of exercise, the potassium content of the working muscle declines and plasma levels increase slightly, indicating an efflux of potassium from exercising cells. On average 0.47 mEq of potassium are released when 1 g of glycogen is broken down (Bergström and Hultman, 1972), and there will be a slight increase in plasma concentration. However, Costill and his colleagues have demonstrated that there were no significant changes in plasma levels during extended periods of heavy sweating, even on a low potassium intake (Costill, 1977; Costill et al., 1976, 1982).

Under normal circumstances, the need to increase potassium intake is unlikely. Indeed, excessive potassium intake can induce ECG changes and should be discouraged. Although high concentrations of potassium are contraindicated in sports drinks, requirements for a slightly increased potassium intake post-exercise, caused by the enhanced rate of glycogen synthesis in this phase, may be met by the inclusion of fruit in the diet or by the addition of fruit materials to sports drink formulations.

12.3.5 Other ions

Magnesium deficits have been associated with impaired energy metabolism, greater fatigue and the occurrence of muscle cramps (Clarkson, 1991). Low magnesium values have been reported in athletes involved in regular endurance exercise but even with large sweat losses, magnesium losses are small in comparison with dietary intake. Excessive depletion does not occur in well nourished people who have large daily sweat losses (Brotherhood, 1984), and there is no systematic indication to replace this ion other than by normal dietary practices.

Much of the calcium in the body is concentrated in the skeleton and only 1% is present in extracellular fluids and other tissues. During exercise, calcium plays an essential role in initiating muscle contractions and although intense physical effort may provoke a loss of calcium in sweat, these losses can be replaced provided that dairy produce is not excluded from the diet. An exception are female athletes who suffer from 'athletic osteoporosis', characterised by reduced bone density. This condition is associated with depressed hormone levels which are known to regulate calcium metabolism and a reduced calcium intake (Clarkson, 1991). For this reason, the inclusion of calcium in sports drinks should be considered, especially the reduced calorie products aimed at the active female.

The intake of other substances, including iron, zinc and phosphorus is generally adequate in a well balanced diet; in these circumstances, supplementation is unnecessary and will not enhance performance. The importance of trace minerals and vitamins in biological functions, particularly those concerned with energy metabolism, is well understood and for an analysis of their role in the nutrition of the athlete, the reader is referred to the review by Brouns (1993).

12.4 Physiological considerations

Absorption of nutrients, including water, occurs in the small intestine and the first barrier to overcome is the rate at which the contents empty from the stomach. Gastric emptying rate is influenced by a number of factors including volume, caloric content, osmolality and exercise.

Control of gastric emptying is mediated by receptors located in the stomach, duodenum and jejunum, sensitive to changes in volume, osmotic pressure, pH and fat (Minami and McCallum, 1984). The receptors respond to gastric volume and effluent by slowing the gastric emptying rate to prevent the rapid entry of excessively concentrated solutions into the intestine (Brener et al., 1983), thereby ensuring that the effluent becomes isosmotic, facilitating the optimal absorption of the fluids and solutes (Hunt, 1959).

Gastric emptying is stimulated by distension of the stomach and an increase in intragastric pressure. Receptors respond to the increase in pressure by increasing the rate of emptying (Hunt and MacDonald, 1954) and the most rapid rate occurs just after ingestion when the volume is greatest. The rate of emptying follows an exponential time course, decreasing rapidly as the volume remaining in the stomach diminishes. This observation offers the opportunity to maximise the rate at which nutrients are presented to the small intestine by taking a number of small drinks at intervals to maintain a high volume in the stomach, rather than the consumption of one large bolus (Rehrer et al., 1990).

As gastric emptying is stimulated by distension of the stomach, it would be anticipated that carbonation would aid this process due to the release of carbon dioxide. Although Lolli et al. (1952) reported a faster rate of emptying, other workers were unable to confirm this observation (Lambert et al., 1992).

Many workers have studied the influence of carbohydrates on stomach emptying and a comprehensive review is outside the scope of the current discussion. For a more detailed analysis of the literature, the reader is referred to the publication by Murray (1987). In general, concentrated solutions of carbohydrates empty from the stomach more slowly than dilute carbohydrate solutions or water; solutions of glucose containing 2 to 3% glucose exit the stomach almost as fast as water. Maltose, sucrose and lactose are emptied at rates similar to equivalent mixtures of their constituent monosaccharides (Elias et al., 1968).

Osmolality, defined as a measure of the number of osmotically active particles in solution, also has a role to play in gastric emptying. Osmolality is a colligative property and is proportional to the molal concentration of the solutes. Glucose polymers, produced by the controlled hydrolysis of starch, are comprised of polymerised glucose units of varying chain length. Due to their polymerisation, solutions of glucose polymers (glucose syrup, corn syrup, maltodextrin) of equivalent molality will possess identical osmolality values. Using this observation, workers have been able to overcome the concentration limitations imposed when free glucose is used by using solutions of glucose polymers (Wheeler and Banwell, 1986; Neufer et al., 1986). The polymers are hydrolysed by brush border saccharidases and energy content of a drink can be increased without compromising the gastric emptying rate significantly. As with the observations on glucose solutions, the results of studies employing glucose polymers are variable, but substitution of glucose polymers for glucose does not appear to have a deleterious effect on the rate of gastric emptying. It should be noted here that although a high concentration of glucose in the ingested fluid delays stomach emptying, the delivery of glucosyl units to the small intestine by concentrated carbohydrate solutions is enhanced. However, the high carbohydrate concentration causes a net movement of fluid from the tissues into the intestinal lumen, exacerbating any dehydration if present and increasing the risk of gastrointestinal discomfort by increasing the volume of fluid in the organ (Maughan, 1991b).

Notwithstanding the potential slowing effect of carbohydrate on stomach emptying, the presence of glucose is essential for the transport of fluid across the intestinal membranes and thence into the tissues. Water absorption occurs mainly as a result of the osmotic and hydrostatic gradients created by solute absorption. Glucose is actively absorbed, unlike other simple sugars, by a process linked with sodium transport. The active transport of glucose and sodium is believed to occur by the means of a common carrier protein located in the intestinal membrane (Gray, 1970). Although some absorption of sodium will occur in the absence of glucose, in contrast, the absorption of glucose is highly dependent upon an adequate concentration of sodium in the intestinal lumen. In the absence of these two substances, water movement is minimal. As water is absorbed, other solutes present in the intestine are carried along by solvent drag thereby intensifying the osmotic gradient, enabling the process to be likened to a pumping action. It is these observations that have formed the basis of the recommendations for the oral rehydration therapy proposed by the WHO for the treatment of dehydration resulting from acute diarrhoea.

In summary, it is clear that in order to maintain physical performance, there is sound justification for the inclusion of sodium and glucose, either as the monosaccharide or as a polymer, in drinks designed to offset fluid and energy losses occasioned by physical exercise.

12.5 The essentials of a sports drink

Sports drinks are formulated and consumed with the aim of achieving one or more of the following objectives:

—to supply fuel for working muscles
—to maintain or enhance performance
—to provide water to replace that lost in sweat.

Consideration of the effects of physical activity on fluid and energy status and an understanding of the physiology involved in restoring homeostasis lead to the conclusion that essentials of an effective sports drink are water, carbohydrate, sodium ions and isotonicity.

The term isotonic has become almost synonymous with sports drinks. Isotonic drinks are defined broadly as drinks which are in balance with the body's fluids. This slightly imprecise definition can be improved by stating that an isotonic drink contains the same number of osmotically active particles as plasma and therefore has an osmolality value similar to that of plasma, that is 280 to 300 mOsmol/kg.

Although the term isotonic has gained general acceptance as a descriptor of this type of drink, the correct terminology is 'isosmotic' because the products are characterised by a measure of the number of solute particles. The term isotonic should be reserved for solutions which possess identical powers of diffusion; the following examples should clarify the distinction.

When a cell is in contact with a solution that has the same osmotic pressure as the cell contents, there will be no net gain or loss of water by the solution or the cell, provided that the cell membrane is impermeable to all solutes present. As the volume of the cell contents is unaffected, the tone of the cell is unchanged. In this case, the solution is isosmotic and isotonic.

However, if the cell membrane is permeable to one or more of the solutes, for example urea or alcohol, diffusion into the cell contents occurs and the tone of the cell is changed. The solution is isosmotic but not isotonic.

With the reservation that the terminology employed to describe sports drinks is not correct, it is necessary to examine the differences between isotonic, hypotonic and hypertonic products. Drinks that contain more dissolved particles than body fluids are said to be hypertonic and are characterised by osmolality values in excess of 300 mOsmol/kg, typically 600 to 700 mOsmol/kg. Absorption will not occur until the gut contents become isosmotic by withdrawing fluid from body tissues. Although these products may aggravate any existing dehydration, they do have a role to play as a component of the normal pre-event diet, allowing the subject to recharge depleted energy reserves, without the need to consume large quantities of carbohydrate-rich foods.

Drinks that contain fewer solute particles than plasma are described as hypotonic, generally with osmolality values ranging from 50 to 250 mOsmol/kg. Products with an osmolality that is slightly hyposomotic with respect to plasma will maximise the rate of water uptake from the intestinal lumen. Included in this category are the low-calorie and 'light' versions of sports drinks. The latter products are of particular interest to those who are interested in improving their health and who are exercising to control their body weight, allowing them to replace lost fluid without a significant calorie intake.

Isotonic drinks are an acceptable compromise, causing minimal disturbance to the body's equilibrium while providing fluid and carbohydrate in sufficient quantities to restore fluid balance and energy reserves.

In addition to the four essentials detailed previously, the fifth important attribute is acceptability. It is essential that formulators devise means of masking undesirable taste characters because no matter how carefully a drink is designed to meet the physiological needs, the benefits will not be achieved if the consumer is unwilling to drink it.

12.6 Formulation considerations

Commercial isotonic sports drinks are necessarily a compromise. If monosaccharides are employed as the carbohydrate source, the concentration will be low in order to accommodate other essential ingredients that are required to increase effectiveness and achieve palatability, while maintaining isotonicity. However, by using a more complex carbohydrate, such as a partial hydrolysate of starch (glucose polymers, glucose syrup, maltodextrin), it is possible to increase the caloric content of a drink without affecting the osmolality (q.v.).

The presence of sodium is essential for efficient fluid absorption. Urine production is stimulated by drinks that contain very low levels of sodium and much of the ingested fluid is excreted. Most sports drinks are formulated to contain sodium in concentrations approximating to 25 mmol/l, about half that of specialised products designed to combat dehydration. In comparison, most soft drinks of the lemonade or cola type contain very low levels of sodium, 1 to 3 mmol/l (Maughan, 1991a). However, a high sodium content derived from the addition of sodium chloride, tends to make the drink unpalatable and it is important that the drinks have a pleasant taste in order to stimulate consumption. Sports drinks are generally formulated to strike a balance between efficacy and palatability and since the majority of sales are to the less dedicated sports person, it is inevitable that most manufacturers perceive taste as an important criterion. It is here that the skill of the formulator may be employed to optimise the sodium content of a drink by employing a blend of permitted sodium salts to achieve the required sodium content without the concomitant salinity resulting from the use of sodium chloride alone.

To develop a formulation, it is necessary to calculate the proportion that each of the constituents contributes to the total osmolality of the product. Osmolality is related to concentration by the following expression:

$$\text{osmolality (Osmol/kg)} = k\,n\,\text{molality}$$

where k = constant for non-ideality and n = number of particles. For example, for an electrolyte such as sodium chloride dissociating into two ions, $n = 2$. For a non-electrolyte, such as glucose, $n = 1$.

Note the distinction between molar and molal solutions and that the correct terminology is osmolality and not osmolarity as is frequently encountered in publications. The following example illustrates the error for non-dilute solutions when substituting molar for molal concentrations.

A 1 molal solution of sucrose is prepared by the addition of 342 g of sucrose to 1000 g of water. That is, 342 g of sucrose are associated with 1000 g of water. A 1 molar solution of sucrose results from the dissolution of 342 g of sucrose in water and the subsequent dilution of this solution to a total volume of 1000 ml. Since the volume of water displaced by 342 g of sucrose is 212 ml (g) (Wolf et al., 1988), it follows that in the case of a 1 molar solution, 342 g of the carbohydrate are associated with $1000 - 212 = 788$ g of water.

Therefore, g of sucrose associated with 1000 g of water $= (342 \times 1000)/788$

$$= 434\,\text{g}$$

Molality of a 1 molar solution of sucrose $= 434/342$

$$= 1.27$$

Using the expression above,

$$\text{osmolality} = k\,n\,\text{molality}$$

it is possible to calculate the contribution to the osmolality of each of constituents in a formulation. For example the osmolality of a 0.2 molal solution of sodium chloride ($k = 0.93$) is:

$$\text{osmolality} = 0.93 \times 2 \times 0.2$$

$$= 0.372\,\text{Osmol/kg or }372\,\text{mOsmol/kg}$$

Relatively dilute solutions of salts are employed in the formulation of sports drinks and the constant k can be ignored in most cases, reducing the expression to:

$$\text{osmolality} = n\,\text{molality}$$

providing a first approximation to osmolality value. For the above example

$$\text{osmolality} = 2 \times 0.2$$

$$= 0.4\,\text{Osmol/kg or }400\,\text{mOsmol/kg}$$

In practice, electrolyte concentrations in sports drinks do not generally exceed a value of 0.05 molal. At such concentrations, the volume displaced by the solute is negligible and molal concentration equals molar concentration for all practical purposes. Similarly, for dilute solutions, the value of k approaches 1. For dilute solutions of strong electrolytes such as salts, the error resulting from the use of the simplified expression diminishes and the calculated value approximates very closely to the actual value for the osmolality.

For the salt of a polybasic acid, such as sodium orthophosphate (Na_3PO_4), assuming complete dissociation into four particles, the osmotic contribution is much greater than for a solution of sodium chloride of equivalent molality. Thus, the osmolality of a 0.2 molal solution of sodium orthophosphate approximates to:

$$osmolality = 4 \times 0.2$$
$$= 0.8 \, Osmol/kg \, or \, 800 \, mOsmol/kg$$

Non-electrolytes, such as the carbohydrate components of a sports drink, do not dissociate and $n = 1$. Assuming ideal behaviour, and by substituting the appropriate values in the expression above, a 1 molal solution of a carbohydrate will have a value of 1 Osmol/kg or 1000 mOsmol/kg. Consideration of this fact allows the formulator to substitute a carbohydrate of higher molecular weight, thereby increasing the caloric content of the product, without affecting the osmolality of the solution. For example, a 0.2 molal solution of glucose (molecular weight 180) contains 36 g of glucose and will have an osmolality of 0.2 Osmol/kg (200 mOsmol/kg). A solution of a glucose polymer with an average molecular weight of 504 and of equivalent molal concentration will contain 100.8 g of carbohydrate and possess the same osmolality, that is, 200 mOsmol/kg.

It is obvious from the foregoing that carbohydrates are the major contributors to the osmolality of a sports drink but a consideration of the carbohydrate sources available permits the development, if required, of sports drinks of significant energy content, without compromising the osmolality of the product. In practice, mixtures of carbohydrates are employed in commercial products, for example sucrose and glucose polymers or mixtures of monosaccharides and glucose polymers, to achieve a compromise between osmolality, caloric content and taste. The composition of some of the major European sports drinks is shown in Tables 12.4 and 12.5.

Fructose is only occasionally employed despite the report that the addition of small amounts of fructose to solutions containing glucose appears to enhance gastric emptying in comparison with isocaloric glucose solutions (Neufer et al., 1986). Despite the widespread use of sucrose, this sugar slowly inverts in acidic beverages. The increase in monosaccharide concentration resulting from the inversion causes an increase in beverage osmolality, with the consequence that a product may become slightly hyperosmotic during storage.

Flavour is an important determinant of beverage acceptability and Hubbard et al. (1990) reported that an increased fluid intake had been elicited by the addition of flavouring (cherry, raspberry or citrus) to water. When large volumes of liquid had to be consumed during exercise, subjects con-

Table 12.4 Composition of European sports drinks (isotonic)

Product	Carbohydrate sources	Energy (kcal/100 ml)	Na	K	Ca (mg/100 ml)	Mg	P	Cl
Aquarius	S	25	22	2.2	0.8	–	1	24
Athlon	S; MD; Fr	24	27.5	10	9	–	–	–
Enervit Aquasport	S	32	90	26	–	–	–	122
Enervit Tropical								
Enervit Orange	S; Fr	28	52	26	–	7	–	25
Enervit Agrumi								
Enervit G	F; G; S; Fr	37	15.5	15	–	4.5	–	–
Fitgar Mandarin								
Fitgar Lemon	S; F; MD; Fr	24	47	27	–	4	–	40
Fitgar Lemon Tea								
Fitgar Agrumi	S; F	25	38	28	–	10.5	–	58.8
Gatorade	GS; S	24	41	11.7	–	7	–	39
Gatorade powder	S; G	24	41	11.7	–	7	31	39
Isostar Citrus	S; MD	31	55	12	8	4.5	6	50
Isostar Orange	S; MD	29	53	13	8.5	4	7.5	45
Isostar Orange (Germany)	S; IS; Fr	27	45	10	9.6	7.7	–	–
Isostar Lemon (Spain)	S; IS; MD; Fr	26	41	17	12	7	3	39
Isostar Fresh	S; IS; MD; Fr	27	45	10	9.6	7.7	–	–
Isostar powder citrus	S; MD; Fr	35	53	12.6	8	4.5	6.3	59.4
Isostar powder orange	S; MD; Fr	35	54	16.2	9	4.8	7.2	67.5
Lucozade Sport	GS; G; MD; Fr	29	55	20	2	1	–	–
Nesfit	F; MD; Fr	21	11.2	33.2	26	2.8	32.8	16
Pro-long	S; GS; MD; Fr	27	40	22.4	12.5	5.1	–	28.3
Reasport	S; MD; Fr	32.4	58	12	29	5.8	–	63
Red Bull	S; G	40			not specified			
Restore	F; G; MD	30.5	39.5	9.5	–	3.5	–	44.5

Key: S = sucrose, G = glucose, F = fructose, GS = glucose syrup, MD = maltodextrin, IS = invert sugar syrup, Fr = fruit source.

Table 12.5 Composition of European sports drinks (light)

Product	Carbohydrate sources	Energy (kcal/100 ml)	Na	K	Ca (mg/100 ml)	Mg	P	Cl
Enervit Aquasport Light	F; MD	17	45	13	–	4.2	–	61
Isostar Light	G; MD; Fr	16	45	10	9.6	7.7	–	–
Lucozade Sport (low calorie)	MD; Fr	5	30	15	41	1	–	–
Misura Light	F; Fr	18			not specified			
R'Activ Orange	G; SMP; Fr	14.5	–	270	60	20	46	–
R'Activ mixed fruit	G; SMP; Fr	19.7	–	215	54	20	49	–
Replay	F; GS; Fr	19	70	45	16	4	–	40
River Isolight	F; Fr	17.8	–	15	4	1.2	–	–
Sport Activ	F; Fr	17.8	–	30	10	5	–	–

Key: S = sucrose, G = glucose, F = fructose, GS = glucose syrup, MD = maltodextrin, Fr = fruit source, SMP = skimmed milk powder.

sumed more flavoured than plain water. In these studies, consumption of both cool (15°C) and warm (40°C) water was increased by a factor of about 50% when flavouring was introduced.

For products that contain fruit materials or other natural ingredients, an assessment must be made of the effect of such an addition on the osmolality of the product. Finally, the contribution of all other solutes must be taken into account, including those present in the water used in the preparation of the drink. These considerations together with the calculations, offer a means of developing an initial formula that is close to the desired target but the calculated values must be confirmed by instrumental measurements. This is most readily accomplished by instruments that utilise the depression of freezing point as the principle of measurement, offering a rapid and precise evaluation of the osmolality.

12.7 Other nutrients

The desire to maximise sporting prowess, short of employing illegal performance enhancing substances, has resulted in an explosion of interest in many substances which are part of normal dietary intake and may have an influence on physical activity. Although some of these materials are used in sports products, they have not achieved wide acceptance in the main and research continues to establish their value.

12.7.1 Branched chain amino acids

Tryptophan is essential for the formation of 5-hydroxytryptamine (serotonin), a brain neurotransmitter and may be considered to be essential for the efficient transmission of neural impulses; Newsholme (reported by Brouns, 1993) has suggested that serotonin may be involved in the development of fatigue. Research suggests that low levels of branched chain amino acids (BCAA) facilitate the entry of tryptophan into the brain and that serum levels of these amino acids are decreased in the later stages of strenuous exercise. BCAA supplementation should enhance blood levels of these amino acids and reduce the uptake of tryptophan in the brain but in one study employing BCAA, no beneficial effects were observed (Galiano, 1991). The value of BCAA supplementation remains to be confirmed.

12.7.2 Glutamine

Excessively low plasma glutamine levels have been implicated in a weakened immune response, frequently characterised as the 'overtraining' or stress syndrome, resulting in an increased susceptibility to colds and other infections in athletes. A recent patent application (Sandoz Nutrition, 1993) describes the use of the peptide L-alanyl-L-glutamine as a source of glutamine in beverage compositions.

12.7.3 Carnitine

Carnitine is a nitrogenous compound that is obtained from the diet (principally red meats and dairy products) and by endogenous biosynthesis. Fatty acids cannot be utilised for the energy requirements of cells until they have been transported into the mitochondria, a process facilitated by carnitine. It has been suggested that increased carnitine availability may increase the rate of fat utilisation and hence spare carbohydrate reserves during exercise. Despite intensive investigation (Heinonen *et al.*, 1992; Decombaz *et al.*, 1993), no evidence has been found to support the view that energy metabolism in exercise is modified by carnitine administration, to date.

12.7.4 Choline

A reduction in plasma choline concentrations has been observed following extended physical exercise and it has been suggested (Wurtman *et al.*, 1986) that the use of choline or lecithin, the phosphatidyl ester of choline, could sustain muscle performance and delay the onset of fatigue by increasing the rate of acetylcholine synthesis, an important neuromuscular transmitter.

12.7.5 Taurine

This compound, a derivative of the sulphur-containing amino acid, cysteine, is conjugated in the body with cholic acid, forming one of the bile salts which are essential for promoting the absorption of fats. At least one commercial product containing both choline and taurine is available but the advantage that is afforded by the addition of these compounds to a sports drink is not yet clear.

12.7.6 Ubiquinone

Ubiquinone (Coenzyme Q_{10}) is a component of important oxidation–reduction reactions and functions as an electron carrier in aerobic energy production processes in cells. Although ubiquinone supplementation has increased oxygen uptake and exercise performance in cardiac patients, it has offered no significant improvement in cardiovascular functions or endurance performance in normal subjects (Roberts, 1990).

12.7.7 Caffeine

The sparing of carbohydrate reserves in working muscles can be accomplished by an increase in fatty acid metabolism, and techniques that increase the concentration of circulating free fatty acids have been shown to extend exercise time to exhaustion. Caffeine in amounts equivalent to $2\frac{1}{2}$ to 3 cups of coffee has been shown to elevate plasma fatty acid concentrations and improve cycle ergometer performance (Ivy *et al.*, 1979).

A usage level of 150 mg/l in soft drinks has been proposed in Europe but caffeine is on the list of prohibited substances for athletes, with a maximum allowable limit of 12 µg/ml of urine.

12.7.8 Sodium bicarbonate

During high intensity exercise there is a build up of lactic acid in muscle tissue, causing fatigue. Plasma bicarbonate maintains acid–base status by buffering the hydrogen ions released with the lactate. The use of sodium bicarbonate has been proposed to manipulate the pre-exercise acid–base status and has been extensively reviewed (Heigenhauser and Jones, 1991). These workers reported that while bicarbonate supplementation does not have a significant effect on endurance performance, a dose of 0.3 g/kg body weight appears to have a beneficial effect on high intensity exercise of 1 to 7.5 min duration. While bicarbonate supplementation may be useful, it should be noted that several studies have reported unpleasant side effects, including nausea and diarrhoea.

Of importance to beverage formulation is the fact that sodium citrate has a similar effect on buffering capacity, by reason of its oxidation to carbon dioxide, without inducing gastric disturbance (McNaughton, 1990).

In conclusion, the consumption of fluid and carbohydrate is essential to combat the detrimental effect of dehydration and loss of carbohydrate reserves on physical performance. The adoption of principles founded on physiological observations will enable the development of a product that will not only optimise fluid absorption but provide a source of carbohydrate to supplement endogenous reserves and enhance physical performance.

References

Bergström, J. and Hultman, E. (1966) *Nature* 210, 309–310.
Bergström, J. and Hultman, E. (1967) *Scand. J. Clin. Invest.* 18, 16–20.
Bergström, J. and Hultman, E. (1972) *J. Amer. Med. Assoc.* 221, 999–1006.
Blom, P. C. S., Høstmark, A. T., Vaage, O., Kardel, K. and Maehlum, S. (1987) *Med. Sci. Sports Exercise* 19, 491–496.
Brouns, F. (1993) *Nutritional Needs of Athletes*, John Wiley & Sons, Chichester.
Brener, W., Hendrix, T. R. and McHugh, P. R. (1983) *Gastroenterol.* 85, 1592–1610.
Brotherhood, J. R. (1984) *Sports Med.* 1, 350–389.
Cade, R., Spooner, G., Schlein, E., Pickering, M. and Dean, R. (1972) *J. Sports Med.* 12, 150–156.
Christensen, E. H. and Hansen, O. (1939) *Skand Arch. Physiol.* 81, 160–171.
Clarkson, P. M. J. (1991) *Sports Sci.* 20, 91–116.
Costill, D. L. (1977) *Ann. N. Y. Acad. Sci.* 301, 160–174, 183–188.

Costill, D. L. (1988) *Int. J. Sports Med.* 9, 1–18.

Costill, D. L., Cote, R. and Fink, W. J. (1976) *J. Appl. Physiol.* 40, 6–11.

Costill, D. L., Cote, R. and Fink, W. J. (1982) *Amer. J. Clin. Nutr.* 36, 266–275.

Coyle, E. F. (1991) *J. Sports Sci.* 9, 29–52.

Decombaz, J., Deriaz, O., Acheson, K., Gmuender, B. and Jequier, E. (1993) *Med. Sci. Sports Exercise* 25, 733–740.

Elias, E., Gibson, G. J., Greenwood, L. F., Hunt, J. N. and Tripp, J. H. (1968) *J. Physiol.* 194, 317–326.

Galiano, F. (1991) *Med. Sci. Sports Exercise* 23, S14.

Gollnick, P. D. (1988) In *Perspectives in Exercise Science and Sports Medicine*, Vol. 1, D. R. Lamb and R. Murray (eds.), Benchmark Press, Indianapolis, Indiana, pp. 1–37.

Gray, G. M. (1970) *Gastroenterol.* 58, 96–107.

Heigenhauser, G. J. F. and Jones, N. L. (1991) In *Perspectives in Exercise Science and Sports Medicine*, Vol. 4, D. L. Lamb and M. Williams (eds.), Brown and Benchmark, Carmel, Indiana, pp. 183–207.

Heinonen, O. J., Takala, J. and Kvist, M. H. (1992) *Eur. J. Appl. Physiol.* 65, 13–17.

Hubbard, R. W., Szylk, P. C. and Armstrong, L. E. (1990) In *Perspectives in Exercise Science and Sports Medicine*, Vol. 3, C. V. Gisolfi and D. R. Lamb (eds.), Brown and Benchmark, Carmel, Indiana, pp. 39–85.

Hultman, E. and Bergström, J. (1967) *Acta Med. Scand.* 182, 109–117.

Hunt, J. N. (1959) *Physiol. Rev.* 39, 491–533.

Hunt, J. N. and MacDonald, I. (1954) *J. Physiol.* 126, 459–474.

Ivy, J. L., Costill, D. L., Fink, W. J. and Lower, R. N. (1979) *Med. Sci. Sports Exercise* 11, 6–11.

Lambert, C. P., Costill, D. L., McConell, G. K., Benedict, M. A., Lambert, G. P., Robergs, R. A. and Fink, W. (1992) *Int. J. Sports Med.* 13, 285–292.

Lolli, G., Greenberg, L. A. and Lester, D. L. (1952) *New Eng. J. Med.* 27, 490–492.

Maughan, R. J. (1991b) In *Perspectives in Exercise Science and Sports Medicine*, Vol. 4, D. R. Lamb and M. H. Williams (eds.), Brown and Benchmark, Carmel, Indiana, pp. 35–76.

Maughan, R. J. (1991a) *Trends in Food Sci. Technol.* 162–165.

Maughan, R. J. (1993) *Br. J. Sp. Med.* 27, 34.

McNaughton, L. (1990) *Eur. J. Appl. Physiol.* 61, 392–397.

Minami, H. and McCallum, R. W. (1984) *Gastroenterol.* 86, 76–82.

Murray, R. (1987) *Sports Med.* 4, 302–351.

Noakes, T. D., Goodwin, N., Rayner, B. L., Branken, T. and Taylor, R. K. N. (1975) *Med. Sci. Sports Exercise* 17, 370–375.

Neufer, P. D., Costill, D. L., Fink, W. J., Kirwan, J. P., Fielding, R. A. and Flynn, M. C. (1986) *Med. Sci. Sports Exercise* 18, 658–662.

Rehrer, N. J., Brouns, F., Beckers, E. J., ten Hoor, F. and Saris, W. H. M. (1990) *Int. J. Sports Med.* 11, 238–243.

Roberts, J. (1990) *Med. Sci. Sports Exercise* 22, 6–11.

Sandoz Nutrition (1993) *European Patent Application* EP 540,462.

Tsintzas, O. K., Williams, C., Singh, R., Wilson, W. and Burrin, J. (in press) *Eur. J. Appl. Physiol.*

Tsintzas, O. K., Williams, C. and Wilson, W. (1993) *J. Physiol.* 467, 72P.

Wheeler, K. B. and Banwell, J. G. (1986) *Med. Sci. Sports Exercise* 18, 436–439.

White, J. A., Pomfret, D. K., Cosslett, L. and Ford, M. A. (1992) *Excel* 8, 209–214.

Wolf, A. V., Brown, M. G. and Prentiss, P. G. (1988) In *Handbook of Chemistry and Physics*, 68th edn., R. C. Weast, M. J. Astle and W. H. Beyer (eds.), CRC Press, Boca Raton, Florida, p. D-262.

Wootton, S. A., Shorten, M. and Williams, C. (1981) *Proc. Nutr. Soc.* 40, 3A.

Wurtman, R. J., Conlay, I. and Blusztajn, K. (1986) *US Patent* 4,626,527.

13 Nutritional value and safety of processed fruit juices

D. A. T. SOUTHGATE, I. T. JOHNSON
and G. R. FENWICK

13.1 Introduction

The consumption of fruit juices has increased rapidly in the United Kingdom during the last decade, while the consumption of fresh fruit has been relatively stable. In addition to its effects on the statistical importance of fruit and fruit products in the diet, this increased consumption has also been largely responsible for the increased intake of vitamin C over the period in question. In the report of the UK National Food Survey committee for 1986, fruit juices were purchased by 29% of households during the survey period and the consumption was on average 190 g/person per week (UK Ministry of Agriculture, Fisheries and Food, 1987). Since then consumption has increased at a slower rate and has been relatively constant over the past three years; in 1992 it was 222 g per person per week (Ministry of Agriculture, Fisheries and Food, 1993). It is therefore appropriate to consider the nutritional composition of these products in a text on their production. This chapter also

Table 13.1 Fruit juices and nectars: popularity by flavour (% total volume)[a]

	United Kingdom 1987	France 1987	Spain 1987	Netherlands 1986	Italy 1985	United States 1986
Apple	14.2	17.1	4.8	32.9	6	19.0
Citrus						
Orange	66.8	42.1	36.0	55.1	3	54.6
Grapefruit	5.6	5.5	–	1.8	–	3.5
Lime/lemon	–	–	–	–	–	1.8
Grapejuice	–	10.0	–	2.9	–	6.1
Apricot	–	3.7	–	–	30	–
Pear	–	1.8	–	–	22	–
Peach	–	–	15.3	–	25	–
Pineapple	6.7	–	26.2	–	2	2.7
Prune	–	–	–	–	–	0.9
Blends	2.6	–	–	–	–	9.7
Tropical	4.1[b]	19.8[b]	–	7.3[b]	12.0[b]	1.7[b]
Others	–	–	17.7	–	–	–

[a] Adapted from Table 14.5, p. 382, of the first edition of this book.
[b] Different products depending on country; in most European countries this would include tomato.

considers the nutritional role of fruit juices with particular reference to the United Kingdom and issues of food safety that relate to fruit juices.

The pattern of consumption in different countries shows considerable differences in preference, and possibly availability, with the range of fruit juices consumed in the United States being substantially greater than in many European countries (Table 13.1). In the United Kingdom, United States, France and the Netherlands, orange juice is the most popular, whereas in Spain, pineapple, peach and mixed tropical fruits are only slightly less popular than orange, and in Italy, peach, pear and apricot juices are the most popular.

13.2 Composition of fruit juices

In discussing the composition of the various fruit juices, most attention is given to orange, apple, pineapple and grapefruit juices. As a general rule, a freshly prepared fruit juice corresponds in composition to the fruit from which it has been prepared. The major constituent of the expressed pulp is the cell wall material and as this represents a small proportion of the fresh weight of the whole fruit, if the juice extraction has been effective, the correspondence between juice and fruit composition minus the cell wall material of flesh and skin is high.

In practice, expression on the production scale is rarely complete and in many fruits the pectic substances act to prevent complete expression of the juice. Changes in composition occur during the subsequent treatment of the juice and especially when it is concentrated or heat-treated. It therefore follows that fruit juices rarely have a constant composition because of the natural variation in the composition of fruits. The variety used for the juice production and the maturity and cultural variables all produce variations in the composition of the juice.

13.2.1 Proximate composition

Freshly expressed fruit juices have a water content similar to that of the fruits themselves, the removal of the plant cell wall material having only a minor effect overall. Small amounts of nitrogenous material are carried into the expressed juice as free amino acids and the soluble cytoplasmic proteins from the fruit. In general, fruit juices that are marketed as 'cloudy' products contain higher amounts of nitrogenous material because the various clarification procedures are also effective in removing soluble proteins. The amounts of nitrogenous material are too small to have any significant nutritional implications. Very small amounts of essential oils are also expressed into the juice and although these are highly significant as flavour components, fruit juices in a nutritional context are fat-free.

The production of fruit juices has a major effect on the carbohydrates: the extraction removes virtually all the plant cell wall polysaccharides (dietary fibre), proteins, cutins and other associated materials, leaving small amounts of soluble pectin substances in the expressed juice. Juices marketed as clear products usually include a partial pectic hydrolysis in their production since the higher molecular weight pectins often precipitate on standing. Pectic substances in the fruit also prevent effective extraction and recovery of juice and partial hydrolysis is frequently used in the processing. This leads to the fruit juices being very low in non-starch polysaccharides and the carbohydrates present are limited to mixtures of the free sugars in the juices which are characteristic of the fruit and the processing conditions. The high natural acidity and elevated temperatures of pasteurisation will initiate sucrose hydrolysis and the variations in sucrose levels found are largely attributable to hydrolysis of the sugars in the juice leading to a mixture of glucose and fructose whose concentrations further increase after prolonged storage.

The organic acids in the fruit are soluble and extracted into the juice, the pattern of acids being characteristic of the fruit and its metabolic state when harvested and processed. In some fruits the organic acids can be the major components present and, because of this, interpretation of carbohydrate values obtained 'by difference' has long been known to be extremely misleading. Complete information for many products is not available and this lack of information has nutritional consequences.

13.2.2 Inorganic constituents

The inorganic nutrients expressed into the juice show some important differences when comparison is made with the original fruits. Inorganic constituents are associated with the cell wall materials and the concentrations of these in the juices are lower compared to the fruits. Thus potassium, calcium and magnesium concentrations are lower in the juices. The juices are characteristically rich in potassium and have low concentrations of sodium, calcium, magnesium, iron, copper, zinc and phosphorus. The sodium levels are particularly low, being only slightly above the trace concentrations given in nutritional databases.

13.2.3 Vitamins

The vitamins in the fruit are present in the cytosol and are expressed into the juice. The levels of thiamin, riboflavin, and niacin are low, but of the B-group, significant levels of folates are present. Table 13.2 shows some values for vitamins in a selection of fruit juices purchased in the United Kingdom (Holland et al., 1992). The amounts of carotenes are small and very dependent on the type of fruit. Most attention has been focused on vitamin C levels and nutritionally this is of major significance. Vitamin C (L-ascorbic acid) is very

Table 13.2 Vitamins in some fruit juices (per 100 g)[a]

Fruit juice	Type of packaging	Carotene (µg)	Thiamin (mg)	Riboflavin (mg)	Vitamin B6 (mg)	Folates (µg)	Vitamin C (mg)
Apple, unsweetened	bottles, cartons	trace	0.01	0.01	0.02	4	14
Grape, unsweetened		trace	trace	0.01	0.04	1	trace
Grapefruit, unsweetened	cartons, canned, bottled, frozen	1	0.04	0.01	0.02	6	31
Lemon	fresh	12	0.03	0.01	0.05	13	36
Lime	fresh	6	0.02	0.01	0.04	5	38
Orange	freshly squeezed	17	0.08	0.02	0.07	28	48
Orange, unsweetened	cartons, canned, bottled, frozen	17	0.08	0.02	0.07	20	39
Orange, concentrate unsweetened	canned	170	0.31	0.13	0.25	90	210
Pineapple, unsweetened	cartons	8	0.06	0.01	0.05	8	11

[a] From Holland et al., 1992.

Note that the actual concentrations may vary considerably depending on the source of fruit and the precise conditions of processing and storage. These values are given as illustrative ones.

sensitive to oxygen and heat and its oxidation is catalysed by copper ions. In older style processes the losses of vitamin C were substantial, but, since the flavour volatiles are also sensitive to the same factors, conditions that are optimal for flavour retention also favour vitamin C retention and modern processing and packaging practices have reduced the losses. L-Ascorbic acid is a powerful reducing agent and also has value in protecting products against discoloration due to polyphenyloxidases and L-ascorbic acid is used as a processing aid in the production of juices such as apple which are prone to discoloration.

13.3 Composition of individual fruit juice products

Some typical values for the proximate composition of a range of products are given in Table 13.3, compiled from a range of sources, mainly UK and US data (U.S. Department of Agriculture, 1963; Paul and Southgate, 1978; Statens Livsmedelsverk, 1986). As mentioned earlier, fruit juices are a biological product and show natural variations in composition. Commercial production would aim to standardise on the basis of the sugar content and the differences between the data from various sources do not represent significant variations. One important factor to consider when looking at Table 13.3 is the variations in the way sugars have been measured; in the United States values are measured by differences, whereas the UK values are direct analyses. For directly comparable products the difference between the values is a measure of dietary fibre and organic acids. In practice, however, dietary fibre values are usually of the order of 0.1–0.2 g/100 ml and the major cause of discrepancy are the organic acids which are discussed later. In the case of lemon and lime juice the discrepancy is very marked, organic acids accounting for the major part of the carbohydrate by difference.

Concentrated juices are typically about four times as concentrated as the fresh juice. Overall the major citrus juices provide between 9 and 10 g sugar/100 ml, negligible amounts of protein and around 37–43 kcal (154–180 kJ) per 100 ml. Lemon and lime juices have lower true sugar content and lower energy values. The other fruit juices (and nectars which are mixtures of fruit and syrup) provide between 12 and 16 g sugar per 100 ml and between 40 and 63 kcal (197–264 kJ).

13.3.1 Sugars

Although there is considerable variation, both varietal and seasonal, in the amounts and individual species of sugars present in the various juices it is possible to make some generalisations and to give some values that illustrate the types and amounts present (Whiting, 1971; Holland et al., 1992). Table

Table 13.3 Typical values for the composition of fruit juices (per 100 ml)

Fruit and type of juice	Water	Protein	Sugars	Energy		Source
				kcal	kJ	
Citrus						
Orange, freshly pressed	87.7	0.6	8.1	33	140	UK
	88.3	0.7	10.4[a]	42	176	USA
Range of US values	87.2–89.6	0.6–1.0	9.3–11.3[a]			
Frozen concentrate	58.2	2.3	38.0[a]	152	636	USA
Dilute ready to consume	88.1	0.7	10.7[a]	43	180	USA
Canned unsweetened	88.7	0.4	8.8	36	153	UK
	87.4	0.8	11.2[a]	45	188	UK
Canned sweetened	85.8	0.7	12.8	51	213	UK
	86.5	0.7	12.2[a]	49	205	USA
Grapefruit, freshly pressed	90.0	0.5	9.2[a]	37	154	USA
Range of values	89.0–90.4	0.4–0.5	8.8–10.2[a]			
Frozen concentrate						
Unsweetened	59.1	2.1	34.6[a]	139	582	USA
Sweetened	57.0	1.6	40.2[a]	157	657	USA
Canned unsweetened	89.4	0.4	8.3	33	140	UK
	89.2	0.5	9.8[a]	39	163	USA
Sweetened	87.3	0.5	9.7	38	159	UK
	88.7	0.6	10.1	40	167	USA
Lemon, freshly pressed	91.3	0.2	1.6	7	29	UK
	91.0	0.5	8.0[a]	32	133	USA
Frozen concentrate	58.0	2.3	37.4[a]	150	628	USA
Diluted ready to consume	92.0	0.4	7.2[a]	29	121	USA
Canned unsweetened	90.3	0.3	9.0[a]	35	146	USA
Lime, freshly pressed	90.3	0.3	9.0[a]	35	140	USA
			0.7			
Canned unsweetened	90.3	0.3	9.0	35	140	UK
	93.0	0.2	6.7	26	109	USA
Other fruits						
Grapefruit, canned	82.9	0.2	16.6	63	264	Sweden
Apple, canned	88.0	0.2	10.2	39	163	Sweden
Apricot, nectar canned	84.6	0.3	14.6	56	234	Sweden
Pear, nectar canned	86.2	0.3	13.2	51	213	Sweden
Peach, nectar canned	87.2	0.2	12.4	47	197	Sweden
Pineapple, frozen concentrate	53.0	1.3	44.5	172	720	USA
Diluted ready to consume	86.0	0.4	12.8	50	210	USA
Canned unsweetened	86.1	0.3	13.4	51	213	UK
	85.6	0.4	13.5	52	278	USA
Tangerine and other small citrus	89.9	0.5	10.1	40.0	163	USA

[a] Carbohydrate by difference.

13.4 gives some values for the individual sugars in a selection of fruit juices purchased in the United Kingdom.

The major sugars present in citrus products are glucose, fructose, and sucrose. In oranges, sucrose is usually one of the major sugars present, with

Table 13.4 Individual sugars in some fruit juices (g per 100 g)[a]

Fruit juice	Type of packaging	Glucose	Fructose	Sucrose	Total sugars
Apple, unsweetened	bottles, cartons	2.6	6.3	1.1	9.9
Grape, unsweetened	cartons	5.5	6.2	trace	11.7
Grapefruit, unsweetened	cartons–canned, bottled and frozen	3.0	3.3	2.0	8.3
Lemon	fresh	0.5	0.9	0.2	1.6
Lime	fresh	0.6	0.6	0.4	1.6
Orange	freshly squeezed	2.0	2.2	4.0	8.2
Orange, unsweetened	cartons–canned, bottled and frozen	2.8	2.9	3.1	8.8
Orange concentrate, unsweetened	canned	11.7	12.3	20.9	44.9
Pineapple, unsweetened	cartons	2.9	2.9	4.7	10.5

[a] From Holland et al., 1992.

similar concentrations of glucose and fructose; in grapefruit also the amounts of the three sugars are similar but sucrose is predominant. Tangerines contain very much more sucrose; in the more acid lemon and lime juice, glucose and fructose are the major sugars present and in limes the sucrose levels are very low. The sugars in grape juice are usually a mixture of glucose and fructose, but in apple, fructose is usually the major sugar with smaller and similar quantities of sucrose and glucose. Apricots, peach and pineapple are characteristically rich in sucrose. Pears are fructose rich, and sorbitol (the sugar alcohol derived from glucose) has been reported at levels of 20 g/l; much lower levels (4 g/l) have been reported in apple juice (Tanner and Duperrex, 1968).

13.3.2   Organic acids

The most predominant organic acids in most fruits are citric and malic acid; grapes are unusual in having substantial amounts of tartaric acid present. A very wide range of other organic acids can be detected as intermediates of the metabolism of the fruit and organic acid levels are especially dependent on the maturity of the fruit and its storage prior to processing. Ulrich (1971) divides the fruits into three groups:

(i) those where the predominant acid is malic including apple, apricot, peach, pear and some grapes;
(ii) those where citric is the major acid, citrus fruits, where it accounts for 80–90% of the total acidity, and pineapple;
(iii) most grapes which are unusual in having similar concentrations of tartaric and malic acids.

The concentration of organic acids in citrus juices ranges from 4.6 g/100 ml, in lemon, to around 1 g/100 ml in tangerine and orange juices, with grapefruit occupying an intermediary position.

13.3.3 Vitamins

Although fruit juices contain trace amounts of many vitamins, only their vitamin C content is of major nutritional significance (see Table 13.2). The level of vitamin C in a fruit is highly dependent on its maturity and cultural conditions, and the amount present in the juice products is greatly dependent on the processing the product has received. Vitamin C is a powerful reducing agent and is rapidly oxidised at elevated temperatures and, although the natural levels of acidity are protective, some losses occur; further losses occur on storage particularly if the container is open to the atmosphere. Citrus fruits are particularly rich in vitamin C and the freshly pressed juice usually has a similar concentration to the original fruit. Frozen concentrated juices show good retention of vitamin C but canning can produce a reduction to around 70–80% of the original level (Paul and Southgate, 1978). Other fruits used for juice production have lower initial concentrations and therefore have correspondingly lower levels in the juices after processing. The production of apple juice can lead to virtually complete losses of any natural vitamin, although the practice of adding vitamin C as an antioxidant to maintain juice colour results in products that contain the vitamin.

B vitamins. Citrus and other fruit juices contain low concentrations of thiamin, riboflavin and folates (Paul and Southgate, 1978). Nutritionally the contribution to the total intake of these vitamins is of little true significance, with the possible exception of orange juice, both freshly pressed, and in the frozen concentrate, where levels up to 90 µg/100 ml of folates have been reported. At the present time the analysis of folates presents considerable difficulty and there is great uncertainty about the requirement for folates (Bates *et al.*, 1982). It is therefore difficult to make any real judgement on these values.

Carotenoids are also present in many citrus juices and, in principle, make a nutritional contribution. The levels depend on the variety; some tangerines and some grapefruit varieties provide levels that can be considered nutritionally significant. In the context of contributing to the vitamin A activity of the UK diet, the contribution of fruit juices is not highly significant but due to current interest in the role of carotenoids in their own right as antioxidant vitamins, this view may have to be revised in the future.

13.3.4 Inorganic constituents

Some selected values for inorganic constituents are shown in Table 13.5. This shows that the major inorganic constitutent is potassium; most juices contain around 120–180 mg/100 ml. The nectars illustrate the dilution used in their production. The other constituents are present in concentrations that are nutritionally insignificant.

Table 13.5 Some values for inorganic constituents mg/100 ml

Fruit	Type of juice	Na	K	Ca	Mg	P	Fe	Cu	Zn
Orange	Freshly pressed	2	180	12	12	22	0.3	0.05	0.2
	Canned unsweetened	4	130	9	9	15	0.5	–	0.3
Grapefruit	Fresh	1	162	9	12	15	0.2	–	0.05
	Canned	4	129	14	10	10	0.4	–	0.05
Lemon	Fresh	1	124	7	7	6	0.4	–	0.05
	Canned	21	102	11	8	9	0.1	–	–
Grape	Canned	3	132	9	10	11	0.2	–	0.05
Apple	Canned	3	119	7	5	7	0.4	–	0.04
Apricot	Canned nectar	1	82	7	5	9	0.4	–	0.09
Peach	Canned nectar	7	40	5	4	6	0.2	–	0.08
Pear	Nectar	4	33	5	3	3	0.3	–	0.07
Pineapple	Canned	1	140	12	12	10	0.7	0.09	–

13.4 Levels of consumption and nutritional significance

In assessing the nutritional significance of any foodstuff it is necessary to consider its composition and the level of consumption. The measurements of consumption are made at several different levels and each level of measurement provides a different type of estimate and quality of data in the sense of the accuracy of the measurement. Measurements at the food supply level are based on commodities and fruit juices are aggregated into generalised fruit categories. At the household budgetary level, which measures retail food purchases, a general measurement of fruit juices may be provided or the fruit juices may be aggregated into fruit products.

The National Food Survey (UK Ministry of Agriculture, 1987, 1993) in the United Kingdom is a well documented household purchases study and this provides some information on trends and the distribution of purchases. In the early series of the reports, fruit juice consumption was low and actual consumption of juices was not covered in detail. In 1958, for example, consumption of fruit juices was equivalent to 8 ml/head per week, and represented some 11% of the total expenditure on all fruit and fruit products. In 1960 consumption was around 12 ml per head/per week. At this time fruit juices of all kinds were purchased by 8% of all households during the week of the study. Purchases over the next decade rose rapidly from about 15 ml/head per week in 1970, to 38 ml/head per week in 1975 and to 87 ml/head per week in 1980. During this period the households purchasing fruit were 9, 12 and 20%, respectively, of all households. This trend continued until in the latest report (for 1992), consumption was over 222 ml (7.8 oz)/head per week and 29% of households reported purchases.

Thus in the United Kingdom, fruit juices have become a significant component of the diet of many households and this level provides the basis for comments on nutritional significance. The National Food Survey shows that purchase of fruit juice is quite strongly influenced by the income group and size of family, consumption being highest in higher income families of small size. In 1986, fruit juice consumption amounted to 23% of total fruit consumption and exceeded the consumption of fresh oranges. The proportion of the total fruit consumption due to fruit juices varied from 32% in the highest income group to 15% in the unemployed households; the proportion of fruit consumed by pensioners was around 10%. Seasonal consumption showed that consumption of fruit juice as a percentage of total fruit was relatively constant throughout the year, but, absolute consumption was higher during the second half of the year when consumption of fresh oranges was lowest. This indicates that fruit juices tend to supply vitamin C through the year and are especially important during the autumn and early winter.

Apart from the metabolic role of vitamin C, it also has an important function as a promoter of iron absorption. Iron is an essential mineral which is relatively poorly absorbed from the human diet. Its bioavailability is largely determined by the relative proportions of other substances in foods which either bind it in the gut lumen and inhibit absorption, or else form weak but readily absorbable complexes. Ascorbate, and indeed citrate, are both important enhancers of iron absorption which occur abundantly in fruit juices and can significantly increase the bioavailability of iron from any meal which they accompany (Hazel and Johnson, 1987a). Orange juice, which has been investigated by several groups of workers, has been shown to approximately double the absorption of iron from a single meal. For persons at risk from iron deficiency therefore, citrus fruits juices have a considerable advantage over beverages such as tea and coffee which strongly inhibit iron absorption (Hallberg and Rossander, 1982).

The current 222 ml of typical commercial orange juice (UHT or pasteurised carton or from frozen concentrate) would provide around 76 mg of vitamin C per week, that is about 35% of a typical adults' Reference Nutrient Intake (RNI, RDA; Department of Health, 1991), five times the level in 1975. Thus fruit juices are approaching potato in importance as a source of vitamin C for the population as a whole. However, unlike potato, which is eaten by over 90% of consumers, fruit juices, being consumed by less than a third of the population, ensure a full RDA of vitamin C for the average juice consumer. The energy provided by fruit juice is on average about 350 kcal/head per week, that is about 50 kcal/day representing some 2–3% of the RDA for energy for the average consumer, up to 10% of energy for the average fruit juice consumer.

While fruit juices, in many respects, are nutritionally very similar to the fruit itself, this is not true if one considers the dietary fibre provided by the fruit (Southgate et al., 1978); fruit juices contain very small amounts of the soluble components of dietary fibre. The presence of the intact cellular material in fruit has been shown (Haber et al., 1977) to modify the satiating, glycaemic and insulinogenic effects of the sugars in fruit and in this respect the consumption of fruit has quite distinct nutritional properties from the juice. There is no evidence that fruit juice has replaced the consumption of fresh fruit because the position of fruit juices in meal patterns is quite different from fresh fruit.

13.4.1   The glycaemic response to fruit juice

The Committee on Medical Aspects of Food Policy recommended in its report on *Dietary Sugars and Human Disease* (COMA, 1989) that sugars naturally integrated into the cellular structure of a food should be distinguished from those which are free in the food or added to it. Sugars which form an integral part of unprocessed foods such as whole fruits and vegetables were classified by the committee as intrinsic sugars, and those not located within cellular structures were classified as extrinsic sugars. The latter were further subdivided into natural milk sugars and non-milk extrinsic sugars comprising fruit juices, honey, added sugars and table sugars. The principal difference between fruit juices and the intact fruit from which they are derived is the more or less complete loss of cellular structure which occurs as a result of extraction. Thus a comparison of fruit and fruit juices provides an excellent model with which to test the physiological significance of the physical disposition of sugars within foods. Part of the rationale for the concept of extrinsic sugars is based upon the supposed implications for dental health (Edgar, 1993).

It has long been recognised that different foods containing equivalent amounts of total carbohydrate have quite different effects on the appearance of sugar in the circulation following a meal (post-prandial glycaemia), but little serious attention was paid to this issue until the early 1980s. Jenkins *et al.* (1981) introduced the concept of the glycaemic index to enable the blood glucose response to different foods to be compared objectively. The glycaemic index is simply the ratio obtained by dividing the area under the blood glucose curve for a test food by that produced by an equal quantity of a reference food such as glucose or white bread. An average figure is then calculated for a group of subjects. In general the glycaemic effect of fruit is lower than that of an equivalent quantity of carbohydrate from starchy foods.

Haber *et al.* (1977) compared the glucose and insulin response in normal subjects after consumption of equal quantities of carbohydrate from whole apples and apple juice, and observed that the insulin response to whole fruit was lower than that to apple juice. Furthermore consumption of apple juice tended to provoke rebound hypoglycaemia, a reduction in blood glucose concentration below the previous fasting level. Subjective feelings of hunger were also assessed in this study, and whole fruit was found to result in a significantly higher degree of satiety. In a later study the investigation was extended to oranges and grapes (Bolton *et al.*, 1981). The fruit was consumed by healthy subjects over a period of 18–20 min, and was eaten either as whole fruit or as homogenised juice after enzymic depectinisation and removal of macerated fruit tissue. As in the previous study, satiety was consistently judged to be higher after whole fruit than after juice. In the case of oranges, there was a lower insulin response after the fruit and less tendency to rebound hypoglycaemia. Paradoxically however, the insulin response to whole grapes was higher than to the juice, and the authors commented that the glucose content of grapes was somehow more insulinogenic than that of oranges or apples.

In general the faster post-prandial glucose response to fruit juice which has been observed in these studies supports the concept of compartmentalisation of nutrients within intact plant foods, but the physiological significance is uncertain because the magnitude of the difference is small and the glycaemic response to free sugars is blunted when they are consumed in combination with other foods. Insulin-dependent diabetic patients should certainly be aware of the high sugar content and rapid assimilation of fruit juices when considering glycaemic control but otherwise there seem to be few implications for human health.

13.5   Fruit juice and dental caries/erosion

The acidity of most fruit juices coupled with the free sugars raises some questions about the possible links between fruit juice consumption and dental caries. Dental caries is a chronic disease characterised by localised destruction of the dental enamel, which, if untreated, is followed by bacterial infection of the dentine and pulp cavity. There are wide geographical variations in caries incidence, but there is strong epidemiological evidence that the very high rates experienced in the younger age groups in industrialised societies are associated with high intakes of refined carbohydrate foods. Although all carbohydrates are thought to be cariogenic to some extent, sucrose appears to be particularly so.

The primary event in caries formation is acid de-mineralisation of the surface enamel. The principal cause is thought to be the production of acid metabolites by bacteria, the most important of which is *Streptococcus mutans*, localised in dental plaque, a soft layer of bacterial protein and extracellular polysaccharide coating the surfaces of the teeth. During and immediately after the ingestion of readily fermentable carbohydrates, the bacterial activity causes a transient decline in plaque pH, which is associated with a period of net de-mineralisation. Re-mineralisation may subsequently occur, but if dental hygiene is poor and the dietary intake of an appropriate substrate is frequent, the eventual result is the formation of an irreversible lesion.

In addition to this chemoparasitic model of caries formation, it is known that de-calcification and erosion of the tooth enamel can occur as a direct result of reduced oral pH from acid foods. Fruit juices have therefore been suggested as a significant cause of dental decay, both because of their sugar content and their low pH. Isolated clinical examples of dental destruction caused by frequent sucking of lemons and other citrus fruits have been reported for many years but there is some controversy over the significance of such observations in relation to the dental health of the community as a whole.

Touyz and Glassman (1981) have argued that frequent consumption of acidic fruit juices is an important aetiological factor in dental caries, and have stated that a daily intake in excess of 150 ml 'could be considered abusive' (Touyz, 1982). Recent *in vitro* studies indicate that the intrinsic acidity of fruit juices may have a significantly greater de-mineralising action than the acid generated by microbial fermentation of their constituent sugars (Grenby and Wells, 1988). Studies carried out using freshly extracted teeth *in vitro* have shown that citrus and apple juices (citric acid content approx. 1.0%) expose open dentine tubules which are thought to cause pain due to hypersensitivity of the cervical dentine in living teeth (Addy *et al.*, 1987). This did not occur with a fruit beverage (Ribena) containing 0.3% citric acid, although its pH was slightly lower, suggesting that the degree of acid erosion depends upon titratable acidity rather than pH. The same conclusion was reached in a study of the effect of fruit juices and fruit drinks drunk normally by adults (Grobler *et al.*, 1985). This work also showed that a sugar-free drink (Fanta) had no adverse effect of plaque pH, and that the method of drinking can greatly influence the effect.

The views of Touyz (1982) were strongly challenged by Walker (1982) on the grounds that no epidemiological evidence existed to support a positive relationship between fruit juice consumption and caries incidence. Indeed there appeared to be some weak evidence for an inverse correlation (Gedalia, 1981). However a recent study produced some evidence that prolonged high consumption of fruit was associated with an increased incidence of caries in adults (Grobler, 1991), and in a Finnish study, use of juice at night was associated with increased caries in infancy (Paunio *et al.*, 1993). There has been a general decline in the incidence of caries in developed countries over the last two or three decades (Diensendorf, 1986), and this has occurred despite the very large increase in fruit juice consumption which has taken place since the early 1970s. It appears unlikely therefore that consumption of unsweetened fruit juice poses a major risk to the dental health of older children and adults. Nevertheless heavy consumers might take note of the precautions proposed by Touyz (1982), to the effect that the juice should not be drunk too frequently, that it should be taken with meals, rather than as a snack or nightcap, and that it should be followed by an alkaline mouthwash, or a piece of cheese, which is known to have a protective effect against dental de-mineralisation (de Silva *et al.*, 1986).

In the case of infants, however, there is no doubt that inappropriate use of sugary drinks, including fruit juices, in infant feeding can result in caries and dental erosion. Concentrated fruit juices intended to be drunk undiluted by babies have a pH of 3–4, and cause rapid de-mineralisation of dental enamel *in vitro*. When given in a reservoir feeder and used as a nighttime comforter, such juices can remain in the infant's mouth for long periods and give rise to major destruction of the deciduous teeth (Smith and Shaw, 1987). Most consumer products now carry a warning against such practices and some contain buffering salts to increase pH.

13.6 Fruit juice safety

13.6.1 Naturally occurring toxicants

Numerous plants regularly consumed by man contain naturally occurring compounds (natural, or inherent, toxicants) which may pose a risk to health (Cheeke, 1989). Contrary to general public opinion, such risk is considerably greater than that associated with the presence of additives and contaminants (Fenwick, 1986). In comparison with other dietary components (notably vegetables, cereals, nuts and fungi), fruit juices are not generally considered to be a major source of such risk. This is partly due to many biologically active principles being located in the leaves and roots, rather than in the fruits themselves, and partly because when they do occur in the latter, their concentration is greatest in the immature tissue. Even when biologically active compounds do occur in the ripened, harvested fruit they are found mainly in the kernels or outer peel, rather than within the inner, juice-containing tissue.

Where problems are encountered, they are mainly associated with the use of immature, fungally-infected or otherwise low quality raw materials. The major area of concern remains the presence of fungal secondary metabolites produced by infected fruit, most commonly apples, grapes and tomatoes. Whilst steps may be taken to minimise the levels of such compounds, it is essential for the health of the consumer, and the reputation of the producer, that the highest standards of quality control are maintained when selecting raw materials.

The importance of such standards is obvious from the work of Steele *et al.* (1982) who reported an outbreak of haemolytic uremic syndrome amongst children in Canada. This was associated with a locally produced fresh apple juice and was attributed to microbial spoilage and concomitant toxin formation. Such instances are, fortunately, extremely rare and whilst fungal metabolites are certainly the most probable cause of health risk, consideration will initially be given to other classes of natural toxicant.

Cyanogenic glycosides, which readily yield hydrogen cyanide on hydrolysis (Montgomery, 1980), are found in the kernels of lemon, lime, apple, pear, cherry, plum and apricot. A wide range of biologically active principles has been identified in the outer layers of the peel of citrus, perhaps the most significant being the flavonoid glycosides which possess a range of phar-macological properties including hypotensive (blood-pressure-lowering) activity (Kumamoto et al., 1985; Matsubara, 1986). A great many fruits, and their processed juices, contain flavonones and there has been much discussion about their biological effects (Robbins, 1980; Regnault-Roger, 1988); thus both mutagenic and antimutagenic properties have been attributed to quercetin and structurally related compounds (Stavric, 1984). Flavonoids have a complex effect on cell metabolism, may induce coronary vasodilation and spasmolysis (Jongebreur, 1952), inhibit lipoxygenase activity (Sekiya and Okuda, 1982) and leucocyte function (Long et al., 1981) and affect platelet aggregation (Beretz et al., 1982) and adhesion (Sempinska et al., 1977). If, on balance, the presence of quercetin in the diet were to be judged undesirable, given its ubiquity in dietary constituents it would be difficult to attain a significant reduction in human intake. However, it should be emphasised that fruit juices in the normal diet contribute only a small part of the total flavanol intake.

Persimmon fruit and juice is intensely astringent as a consequence of the presence of tannins (Salunke and Desai, 1984); whilst tannins are known to bind strongly to protein, inactivate enzymes and reduce the availability of essential minerals, it is probable that the astringency of the compounds in persimmon will act as an effective deterrent to excessive human exposure. The toxic alkaloid, tomatine, is present in tomato plants and fruit, but is rapidly metabolised as the fruit matures (Jadhav et al., 1981) and so should not pose a problem in tomato products provided that the quality of the raw material is maintained.

Nitrate, from ground water or green vegetables, may be converted in the human body to nitrite which may subsequently react with secondary or tertiary amines to form mutagenic nitrosamines. The presence in fruit juices of ascorbate effectively inhibits the formation of such undesirable species. Other nitrite-scavenging products may also be present, thus 3-hydroxy-2-pyranone has been found in the juice of kiwi fruit (Chinese wild plum) and has been suggested to be derived from ascorbate during processing and preparation (Normington et al., 1986). Sato et al. (1986) have examined the effects of various fruit and vegetable juices on diethylnitrosamine formation. Orange and komatsuma juices were most inhibitory, apple and plum juices showing least activity. The effects of some of the samples varied widely, suggesting geographical, seasonal and agronomic factors were important modulators of chemical composition. Levels of ascorbate were apparently unrelated to nitrosamine formation, emphasising the importance of other constituents.

Patulin, 4-hydroxy-4H-furo[3,2-c]pyran-2(6H)-one, (I) (Figure 13.1) is a toxic antibiotic produced by certain species of Aspergillus, Penicillium and Byssochlamys (Stott and Bullerman, 1975; Doores, 1983). Whilst moderately to highly toxic to cells, the significance of patulin as a natural toxicant

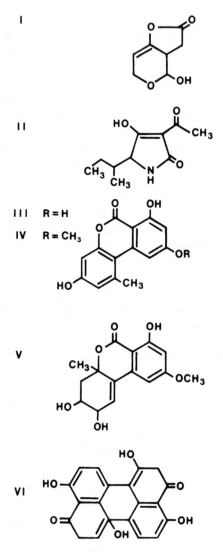

Figure 13.1 Metabolites isolated from *Alternaria*-infected tissue; I, patulin; II, tenuazonic acid; III, alternariol; IV, alternariol monomethyl ether; V, altenuene; VI, altertoxin 1.

remains unclear (Friedman, 1990). Strains of patulin-producing moulds have been isolated from a variety of food products and animal feedingstuffs, although natural patulin production is mainly observed in fungally infected apples and the juices and other products prepared therefrom. Although there have been suggestions that patulin may be found in visually acceptable, undamaged produce, this has yet to be proved unambiguously. Such claims, if confirmed, would obviously have significant consequences for apple juice production. There have also been claims that certain apple varieties may be particularly susceptible to patulin production. One such variety is Bramley, but once again unequivocal data in support of these claims is still awaited. Other fruits have been found to contain patulin, including pears, grapes, peaches, apricots, bananas, blueberries, plums and tomatoes (Akerstrand *et al.*, 1976; Frank *et al.*, 1976, 1977; Andersson *et al.*, 1977; Scott *et al.*, 1977; Thurm *et al.*, 1979). Thurm *et al.* (1979) also identified the presence of this fungal metabolite in sour cherry, currant and sea buckthorn juices, suggesting a relatively wide distribution.

Patulin was discovered over 40 years ago at a time of great interest in naturally occurring antibiotics, and many of the moulds producing this metabolite were identified at that time (Stott and Bullerman, 1975; Doores, 1983). Because of its broad spectrum biological activity and potency, patulin was tested in humans, but this was discontinued once its toxicity became apparent. Dickens and Jones (1961) obtained evidence for the carcinogenicity of patulin following subcutaneous injection in rats. Subsequently, however, the validity of this type of carcinogenicity testing has been questioned (Grasso, 1970). In marked contrast to the above, neither Osswald *et al.* (1978) nor Becci *et al.* (1981) obtained any evidence for the carcinogenicity of patulin in long term studies in which the toxin was administered by intubation.

Patulin possesses a hemiacetal lactone structure (Stott and Bullerman, 1975), is relatively stable under acid conditions (half-life at pH 6, 1310 h) but breaks down much more readily in an alkaline environment (half-life at pH 8, 62 h) (Brackett and Marth, 1979a). In apple juice, the compound is stable for up to 3 weeks (Scott and Somers, 1968), this being a consequence not only of the low pH of the product, but also of its lower concentration of SH-containing compounds as compared with, for example, orange juice.

The reaction of patulin with cysteine and cysteine-containing proteins has been offered as an explanation of the limited occurrence of the compound in a wider range of fungally contaminated foodstuffs (Hofmann *et al.*, 1971). Whilst detailed information regarding the structures and toxicity of such patulin-SH adducts is difficult to obtain, it has been suggested that these do retain some measure of biological activity, albeit much reduced (Lieu and Bullerman, 1978). Friedman (1990) has comprehensively reviewed the *in vivo* and *in vitro* toxicity of patulin. Primary toxic effects appear to involve the plasma membrane, and effects on key enzymes, including muscle aldolase, RNA polymerase, aminoacyl-tRNA synthetases have been described. Patulin has also been observed to interfere with transcription and translation as a result of its interaction with DNA and RNA. The FAO/WHO Joint Expert Committee of Food Contaminants (JECFA, 1990) recommended a Provisional Tolerable Weekly Intake (PTWI) level for patulin of 7 µg/kg body weight/week (this should be compared with estimations of the dietary intake which ranged between 0.21 and 1.82 µg/kg body weight/week for different populations, including children). Calculations based upon UK surveillance data, indicate that in extreme circumstances the intake of patulin from cloudy juices could range up to 0.9 and 1.5 µg/kg body weight/week for children and adults, respectively (MAFF, 1993). Intakes from clear juices were calculated to be about five-fold lower.

The UK Committees on Toxicity, Mutagenicity and Carcinogenicity of Chemicals in Food, Consumer Products and the Environment (1991) have advised that patulin be classified as an *in vivo* mutagen. Whilst the available data are incomplete, patulin has been shown to exhibit adverse effects on the developing foetus in rodents, and has moreover been shown to provoke adverse gastrointestinal, neurological and immunomodulatory effects. The background to consumer concerns over the presence of patulin in apple products in the UK has been summarised by Hopkins (1993). Nevertheless, as is almost always the case with assessing the risk associated with human consumption of a natural toxicant, the available data is insufficient to draw any firm conclusions. The importance of assessing the level of any risk associated with patulin consumption may be seen in the fact that in the USA, apple juice ranks second only to milk as a natural beverage.

There is an emerging consensus, within the International Association of Fruit Juice and Nectar Producers (AIJN, 1991) and individual national governments that apple products conform to a maximum patulin level of 50 µg/kg. This was established in the UK as a provisional advisory level for apple juice in 1993. Figures in Table 13.4 should be considered from this perspective.

Methods of analysis for patulin include reverse-phase high performance liquid chromatography (HPLC) and gas liquid chromatography (GLC). As in the case of analysis of other fungal metabolites, the extraction procedures are important especially where rapid and reliable methods are sought for the monitoring of quality by surveillance agencies and processors. A rapid HPLC method has been reported (Kubacki and Goszcz, 1988), although the stability of the analyte in the basic extraction medium has been questioned. Rovira *et al.* (1993) have recently described an improved method, and the procedure of Forbito and Babsky (1985) has also been widely adopted. Tarter and Scott (1991) have summarised GLC methods involving a variety of derivatisation (and detection) techniques. The ISO method (ISO, 1993) has been extensively used for quality assurance, although modifications have been made to increase speed of analysis without reducing accuracy and reproducability (for example, see British Soft Drinks Association, 1993).

The use of semi-quantitative thin layer chromatographic (TLC) methods has been made more rapid by the application of diphasic extraction (Prieta *et al.*, 1992), whilst Taniwaki *et al.* (1992) have used this method to examine migration of patulin in experimentally infected apples. The results confirmed that the toxin does not diffuse through the whole fruit and that trimming can significantly reduce patulin from infected apples. There was no correlation found between the weight of the rotten parts removed and the amount of

Table 13.6 Patulin content of apple juices and ciders

Products	Contamination (%)	Range (μg/l)	Reference
Roadside stand juice	58	10–350	Brackett and Marth (1979b)
Home-made juice	40	6–16400	Lindroth and Niskanen (1978)
Juice concentrate	20	50–690	Lindroth and Niskanen (1978)
Juice, commercial	27	<39	Lindroth (1980)
home-made	30	<14400	Lindroth (1980)
concentrate	30	5–1478	Lindroth (1980)
Juice	3	106–216	Wilson (1981)
Juice concentrate	100	55–610	Bohuon and Drilleau (1980)
Juices and concentrates	0	–	Czerwiecki (1980)
Juices	21	5–15	Cavallaro and Carreri (1982)
Juice	17	10–130	Woller and Majerus (1982)
Juice	30	200–1200	Jacquet et al. (1983)
Juice	42	5–56	Mortimer et al. (1985)
Juice	37	40–440	Stoloff (1976)
Juice	82	2–73	Josefsson and Andersson (1976)
Juice	40	20–200	Thurm et al. (1979)
Juice	0	–	Eyrich (1975)
Juice	84	20–7400	Meyer (1978)
Juice	100	2–60	Ruggieri et al. (1982)
Juice	86	<402	Ehlers (1986)
Cider	21	250–2000	Jacquet et al. (1983)
Cider	62	44–309	Ware et al. (1974)
Cider	100	244–3993	Wheeler et al. (1987)
Juice	65	5–629	Watkins et al. (1990)
Juice	57	5–646	Burda (1992)
Cloudy juice	72	10–434	MAFF (1993)
	26	50–434	MAFF (1993)
Clear juice	42	10–118	MAFF (1993)
	35	10–50	MAFF (1993)

patulin found in the apples. The authors concluded that trimming 1 cm around the rotten tissue would be satisfactory.

Levels of patulin in apple products are listed in Table 13.6 whilst Table 13.7 indicates levels in other fruits. Patulin is found both in clear juices (which are prepared by the dilution of concentrates) and in cloudy juices. The conclusion of the JECFA Committee (1990) was that implementation of good manufacturing practices should be the main way to reduce patulin exposure below the Provisional Tolerable Weekly Intake. According to the British Soft Drinks Association (1993), a number of factors, economic and social, may have contributed to an increase in patulin contamination from both fresh juice production and out-of-season clear juice concentration preparation. It was against this background that a code of practice for apple juice production has been recently published.

Doores (1983) has described the control of patulin in apple products via the addition of chemicals such as ascorbic acid and sulphur dioxide, treatment with charcoal (Van, 1989), fermentation (see above) or trimming. Avoidance of the problem is, however, the most effective remedy and to

Table 13.7 Patulin content of other fruit juices

Product	Contamination (%)	Range (μg/l)	Reference
Pear juice	83	2–25	Ruggieri *et al.* (1982)
Pear juice	100	<20	Ehlers (1986)
Pear juice, concentrate	0	–	Lindroth (1980)
Pear juice	100 (6/6)	<3	Möller (1986)
Lemon juice, concentrate	0	–	Lindroth (1980)
Orange juice	0	–	Lindroth (1980)
Pineapple juice	0	–	Lindroth (1980)
Pineapple juice	100 (1/1)	<25	Ruggieri *et al.* (1982)
Blueberry juice, concentrate	100 (1/1)	15	Lindroth (1980)
Blackcurrant juice	66 (2/3)	8–10	Lindroth (1980)
Blackcurrant juice, concentrate	100 (2/2)	<5	Lindroth (1980)
Redcurrant juice	0	–	Lindroth (1980)
Redcurrant juice, concentrate	100 (2/2)	5–361	Lindroth (1980)
Golden currant juice	0	–	Lindroth (1980)
Rhubarb juice	0	–	Lindroth (1980)
Peach juice	33 (2/6)	<25	Ruggieri *et al.* (1982)
Apricot juice	33 (1/3)	<25	Ruggieri *et al.* (1982)
Grape juice	22	<50	Altmeyer *et al.* (1982)
Grape juice	16	50–230	Altmeyer *et al.* (1982)
Grape juice	0	–	Lindroth (1980)
Grape juice, concentrate	33	10	Lindroth (1980)
Grape juice	0	–	Mortimer *et al.* (1985)

this end the occurrence of patulin in apple, and other, produce may be minimised by application of appropriate pre- and post-harvest fungicides, storage under low temperature conditions, rejection or removal of spoiled tissue and effective in-plant sanitation and quality control. Given that the present consensus of information suggests that patulin is neither carcinogenic nor acutely toxic to man, its presence in fruit and fruit juices is more an indicator of poor quality standards than overt toxicity of the product *per se*. It has been suggested that patulin in frozen blueberries was the cause of diarrhoea in children (Akerstrand *et al.*, 1976). It must be emphasised that, as with most other natural toxicants, the effects of chronic exposure are unknown. In this context it should be remembered that in most developed countries the scientific basis of the risk associated with even potent mycotoxins (such as aflatoxins) is non-existent or very weak. Generally this is based upon anecdotal, unsupported statements of carcinogenic risk for humans. It will not then be surprising that the paucity of toxicological and survey data (Blunden *et al.*, 1991) makes it impossible to assess the hazard, if any, associated with the consumption of these particular secondary metabolites.

Moulds of the genus *Alternaria* are ubiquitous, being found in decaying organic material and soils. Many *Alternaria* species are plant pathogens, frequently being associated with post-harvest decay of fruits and vegetables. According to King and Schade (1984), over 30 different metabolites having varied chemical structures (Figure 13.1) have been isolated from *Alternaria*-infected tissue, with tenuazonic acid (II), alternariol and its monomethyl ether (III, IV), altenuene (V) and altertoxin 1 (VI) of particular importance in relation to food plants (Watson, 1984; Stinson, 1985).

In contrast to the detailed information available on patulin, there are few reports dealing with the natural occurrence of *Alternaria* toxins in fruit. Stinson *et al.* (1981) have reported tenuazonic acid (100–1400 µg/kg) in tomatoes and levels of > 1000 µg/kg in apples. Scott and Kanhere (1980) have found both tenuazonic acid (10–100 µg/kg) and an isomer (probably D-allo-tenuazonic acid) in tomato paste, whilst in a particularly extensive study Stack and co-workers (Stack *et al.*, 1985; Mislivec *et al.*, 1987) analysed 142 samples of mouldy tomato from commercial processing lines and found just over half (73) samples to contain 0.4–70 mg tenuazonic acid/kg. Evidence was also found for the sporadic occurrence of low levels of dibenzopyrone metabolites (see below), but these could not be quantified. Although the potential for tenuazonic acid contamination of soft fruit and citrus has been established, levels produced by inoculation are much higher than those found naturally (e.g. 106 mg/kg in ripe tomatoes; Stinson *et al.*, 1980; 139 mg/kg Stinson *et al.*, 1981). For this reason it is inappropriate to estimate dietary exposure to fungal toxins via artificial (laboratory) conditions, since the latter rarely reproduce the complex integration of environmental conditions affecting fungal proliferation and toxin production in nature (Watson, 1984).

Alternariol (III), and its monomethyl ether (IV), are dibenzopyrones which have been found (Stinson *et al.*, 1981) in apples (0.7–59 mg/kg and 0.2–2.3 mg/kg, respectively) and in tomatoes (3–50 µg/kg and 3–8 µg/kg, respectively). The structurally related altenuene (V) has been identified in one sample (out of nineteen) of infected tomatoes (10 µg/kg) and at rather higher levels (500 µg/kg) in apples (Stinson *et al.*, 1981). Altertoxin 1 (VI) is apparently of much more limited occurrence having been found in apples but apparently not in tomatoes (Stinson *et al.*, 1981).

The potential for the wider contamination of fruit by *Alternaria* species has been demonstrated by Stinson *et al.* (1980, 1981), who found fungal metabolites in oranges, lemons and blueberries. So far as is known, there have been no investigations of *Alternaria* toxins passing into tomato and other juices through the use of infected raw materials, but this must be considered a distinct possibility. King and Schade (1984) have summarised data on the toxicity of alternariol toxins. Tenuazonic acid, the dibenzopyrone toxins and altertoxin I all possess toxicity towards mammalian cells (Pero *et al.*, 1973), with the first and last of these toxins being more toxic to mice than alternariol or its monomethyl ether. A combination of these latter compounds was, however, shown to possess greater toxicity towards *Bacillus mycoides* than that expected from the sum of their individual activities, and a similar synergism was found in a study of foetotoxicity. It has generally been considered that tenuazonic acid was the most important of the *Alternaria* food toxins, not the least because it has been implicated in onyalai, a haemorrhagic disease in Africa (Steyn and Rabie, 1976), but an additional concern is the mutagenic or carcinogenic activity of altertoxin I (Pollock *et al.*, 1982; King and Schade, 1984). Recently Younis and Al-Rawi (1988) have described the foetotoxic effects of altenuene, altenariol monomethyl ether and altertoxin 1.

It is thus important to investigate the chronic toxicity of the various *Alternaria* toxins, individually and in combination; more extensive analytical surveys are also necessary to establish likely human exposure (Schade and King, 1984). Whilst GC (Kellert *et al.*, 1984) and HPLC (Wittkowski *et al.*, 1983; Stack *et al.*, 1985; Frisvad, 1987) methods have been reported for the analysis of *Alternaria* toxins, it is probable that both surveillance exercises and toxicity studies would be better served by the availability of immunological assays. Whilst it is probable that apple and tomato products will contribute most to the total, other processed fruit products should not be excluded from the survey. Reduction of these fungal metabolites may be achieved, as with patulin, by a combination of appropriate pre- and post-harvest treatments, good storage conditions, selection of undamaged raw materials, trimming if necessary and attention to processing plant sanitation and quality control.

13.6.2   Metal contamination

The concentration of potentially toxic inorganic constituents is low in fresh fruit juices under normal circumstances (Nagy, 1977) and only rises to undesirably high levels as a result of prolonged storage in unsuitable processing equipment or packaging. The type of contamination most likely to occur is the introduction of iron, tin and lead as a result of the deterioration of soldered steel cans. Not surprisingly, levels of all three elements have been found to be significantly higher in commercially canned juices compared to juices packed in cardboard cartons or laminated pouches (Mesallam, 1987). The chromium level of acidic fruit juices in contact with steel surfaces also tends to rise (Kumpulainen, 1992). The mechanisms and incidence of metal contamination in commercially canned fruit juices are discussed below, but it should be noted that the recent near-universal introduction of stainless steel processing equipment and non-ferrous packaging materials has largely eradicated the problem.

The traditional can used extensively by the food industry is made of carbon steel, coated inside and out with a thin coating of tin. The tin is relatively corrosion resistant, and acts as a protective layer between the steel can and its contents. However, in the presence of an acidic electrolyte, an electrochemical circuit is created with the tin layer acting as a sacrificial anode. This leads to the release of tin into the contents of the can, coupled with the liberation of hydrogen. Breakdown of the underlying steel occurs more rapidly if the protective layer of tin becomes exhausted at any point. Contamination of any food by heavy metals is greatly increased by the unwise practice of storing foods in open cans in the refrigerator. For obvious reasons it is particularly important to avoid this in the case of infant foods.

Iron.   Iron levels in fresh citrus juices have been reported to range between 0.2 and 5 mg/kg for orange juice (Birdsall, 1961; Nagy, 1977) and from 0.6 to 1.9 mg/kg for grapefruit juice (Nagy, 1977). Iron is an important nutrient which is incompletely absorbed from human foods. However, its bioavailability from citrus juices, and from foods eaten with them, is likely to be particularly good because of the high levels of both citrate and ascorbate which are known to act as enhancers of intestinal absorption in man (Hazel and Johnson, 1987b). Fruit juices therefore have a dual role to play in improving iron nutrition, both as a source of enhancers which increase the absorption of non-haem iron from the general diet, and as moderate sources of iron in their own right. Nevertheless, there are some in the population for whom iron overload in middle life is a clinically significant problem. The possible adverse effects of iron contamination are shown by the prevalence of iron overload in some parts of Africa caused by locally brewed beers containing high levels of iron from brewing utensils. It is important therefore that iron contamination is minimised. The current FAD/WHO Codex Alimentarius limit for iron in orange juice is 15 mg/kg.

Nagy *et al.* (1980) investigated the release of tin and iron into commercially filled, unlacquered cans of unconcentrated orange juice over a period of 12 weeks at a variety of temperatures. Although tin accumulated steadily over the experimental period in a temperature dependent manner, there was relatively little release of iron. However, in a recent survey of the iron and tin contents of 122 cans of mainly imported fruit juices and nectars purchased in Penang, Malaysia, 18.9% contained iron in excess of 15 mg/kg, and 34.4% exceeded 10 mg/kg (Seow *et al.*, 1984). Of the various juices sampled, 66.7% of the cans of apple juice and 75% of the tomato juices exceeded 15 mg/kg.

Seow *et al.* (1984) observed that the iron content of completely unlacquered and end-lacquered cans was significantly lower than that of fully lacquered products. The authors explained this seeming paradox in terms of the behaviour of tinned steel undergoing corrosion. In unlacquered cans, the very high ratio of tin to steel in the exposed surfaces provides cathodic protection against electrolytic dissolution of the steel plate. In lacquered cans, the tin is protected, but this allows corrosion of the base steel to occur at discontinuities in the lacquer coating.

Tin.   In contrast to iron which poses a theoretical risk of chronic iron overload to a comparatively small number of individuals, tin can cause acute food poisoning. However, tin is poorly absorbed and rapidly excreted by man, and therefore can be tolerated at surprisingly high levels in foods (Reilly, 1980). Nevertheless significant outbreaks of human intoxication, with severe gastrointestinal symptoms, have occurred as a result of ingesting fruit juices contaminated with tin (Horio *et al.*, 1967; Benoy *et al.*, 1971). There is also some evidence from studies with animals to suggest that lower levels of tin consumption can interfere with the metabolism of inorganic nutrients (deGroot, 1973).

As has been discussed in the previous section, tin is released steadily from the surface of unlacquered cans containing citrus fruit juices. Nagy *et al.* (1980) observed that the reaction was strongly dependent upon temperature but concluded that storage below 43°C should ensure that the tin content of single strength orange juice does not reach the Codex Alimentarius tolerance limit of 250 ppm within 12 weeks. In the survey of 122 cans carried out by Seow *et al.* (1984) however, 13.1% of the cans contained tin in excess of this limit.

The breakdown of tin-plate is greatly accelerated by the presence of nitrite, which can be derived from nitrate inside the can. The level of nitrate in the canned juice, or in any water used in its preparation from concentrate, is therefore an extremely important determinant of the ultimate level of tin contamination in unlacquered cans (Horio *et al.*, 1967; Hall *et al.*, 1982). Certain constituents of fruits can also increase the rate of dissolution of the tin. The anthocyanins found in coloured fruits such as blackcurrants act in this way and such fruits must be stored in fully lacquered cans.

Lead. Lead is a widespread contaminant in the environment which is capable of causing both acute and chronic poisoning in man. A significant proportion of human lead intakes are derived from the diet. The lead content of fresh fruit juices will depend upon environmental levels in the area of origin, but can be assumed to be relatively low. In 1937, Roberts and Gaddum (1937) reported a range of 0.03–0.14 ppm for blood oranges in the United States; more recently Rouseff and Ting (1980) reported a figure of 0.06 ppm for fresh Florida grapefruit juice. The Codex Alimentarius limit for lead in canned fruit juices intended for consumption by infants is 0.3 ppm.

Lead levels in commercially canned fruit juice are significantly higher than those of the unprocessed product. In the recent survey of packaged orange juice undertaken by Mesallam (1987), canned juices tended to have higher lead levels than those packed in cardboard cartons or laminated pouches. However, the highest average concentration obtained in the study was 0.188 ppm from canned juice, and the highest single observation was 0.322 ppm this being the only sample which exceeded the Codex Alimentarius limit. These figures are similar to those obtained by Nagy and Rouseff (1981) who examined 168 samples of commercially canned single strength Florida orange juice and found four which exceeded a lead concentration of 0.3 ppm; the observed range was 0.02 to 0.32 ppm. A recent study of commercial infant foods yielded figures in the middle of this range for a variety of juices packed in glass jars (Dabeka and McKenzie, 1988).

The major source of lead in fruit juices packed in conventional steel cans is the solder, which contains approximately 98% lead and 2% tin. The extent of lead contamination of citrus juices is increased to some extent by high acidity (Roussef and Ting, 1980) but temperature is less important than in the case of tin (Nagy and Roussef, 1981). The main variation between cans appears to be due to differences in the exposed area of tin within the seams. Stringent manufacturing standards to reduce solder area within the can, or better still the use of fully welded cans (Jorhem and Slorach, 1987), appear to be the best methods of minimising lead contamination, as lacquering is apparently ineffective. According to the American Can Manufacturers Institute, lead-soldered cans were last manufactured in the USA in 1991 and any currently available must have been imported. In 1993 The Food and Drug Administration was proposing to impose a final ban on the commercial use of any lead-soldered food cans in the USA.

Aluminium.   Aluminium is the most abundant element in the earth's crust; the metal is widely used in cooking utensils and as a packaging material, and significant quantities inevitably enter the human food chain. Metallic aluminium is not acutely toxic to man, but in recent years much interest has been focused on its possible role in the aetiology of Alzheimer's disease (Candy, 1986); this remains highly controversial however, and is by no means universally accepted (Hughes, 1989).

Information on the aluminium content of fruit juices is limited; Greger (1988) quotes a figure of 0.4 mg/kg for orange juice but cautions that because of methodological difficulties any estimate for food-borne aluminium needs to be treated with caution. The most likely source of aluminium in fruit juice products is contamination from processing equipment, although the use of aluminium for this purpose or for packaging is now rare. Like tin, aluminium cans are vulnerable to corrosion in the presence of acidic contents, but this is mainly prevented by the use of vinyl epoxy resin or similar lacquers. The quantity of aluminium leaching from processing equipment depends upon a variety of factors such as the residence time, temperature and the acidity of the liquid. In one comparative study, boiling water in an aluminium pan led to an aluminium concentration of 0.54–4.3 mg/l, the concentration in milk rose to 0.2–0.8 mg/kg, but in foods containing acidic fruit juice the concentration rose to 2.9–35 mg/kg (Liukkonen and Piepponen, 1992).

A recent UK Food Surveillance Paper (Ministry of Agriculture, Fisheries and Food, 1985) reported that a mixed pineapple and grapefruit juice packaged in aluminium contained 1.2 mg/kg of aluminium. This content was 2–3 times higher than that of several canned beers measured on the same survey (presumably due to the higher acidity). Fruit juice, however, makes only a relatively small contribution to the total intake of dietary aluminium in the UK population, estimated to be 6 mg/day.

References

Addy, M., Absi, E. G. and Adams, D. (1987) *J. Clin. Peridontol.* 14, 274.

AIJN (1991) *AIJN Code of Practice–General, physical, chemical and microbiological criteria for fruit and vegetable juices and nectars in the European Community*, Confruct-Studien 35, 10–36.

Akerstrand, K., Molander, A., Andersson, A. and Nilsson, G. (1976) *Var foda* 28, 197.

Altmeyer, B., Eichhorn, K. W. and Plapp, R. (1982) *Z. Lebensm.-Unters. Forsch.* 175, 172.

Andersson, A., Josefsson, E., Nilsson, G. and Akerstrand, K. (1977) *Var foda* 29, 292.

Bates, C. J., Black, A. E., Phillips, D. R., Wright, A. J. A. and Southgate, D. A. T. (1982) *Human Nutrition Appl. Nutrition* 36A, 667–669.

Becci, P. J., Hess, F. G., Johnson, W. D., Gallo, M. A., Babish, J. G., Dailey, R. E. and Parent, R. A. (1981) *J. Appl. Toxicol.* 1, 256–261.

Benoy, C. J., Hooper, P. A. and Schneider, R. (1971) *Food Cosmet. Toxicol.* 9, 645.

Beretz, A., Cazenave, V. P. and Anton, R. (1982) *Agents Actions* 12, 382.

Birdsall, J. J., Derse, P. H. and Teply, L. J. (1961) *J. Am. Diet. Assoc.* 38, 555.

Blunden, G., Roch, O. G., Rogers, D. J., Coker, R. D., Bradburn, N. and John, A. E. (1991) *Med. Laboratory Sci.* 48, 271–282.

Bohuon, G. and Drilleau, J.-F. (1980) *Ann. Fals. Expert. Chem.* 73, 153.

Bolton, R. P., Heaton, K. W. and Burroughs, L. F. (1981) *Am. J. Clin. Nutr.* 34, 211.

Brackett, R. E. and Marth, E. H. (1979a) *J. Food Protect.* 42, 862.

Brackett, R. E. and Marth, E. H. (1979b) *Z. Lebensm.-Unters. Forsch.* 169, 92

British Soft Drinks Association (1993) *Code of Practice for Apple Juice Production*, The British Soft Drinks Association Limited, London.

Candy, J. M., Klinowski, J., Perry, R. H., Perry, E. K., Fairbairn, A., Oakley, A. E., Carpenter, T. A., Atack, J. R., Blessed, G. and Edwardson, J. A. (1986) *Lancet* i, 354.

Cavallaro, A. and Carreri, D. (1982) *Boll. Chem. Unione Ital. Labor. Provinciali* 33, 527.

Cheeke, P. R. (ed.) (1989) In: *Toxicants of Plant Origin*, CRC Press, Boca Raton, FL.

COMA (1989) Committee on Medical Aspects of Food Policy. *Dietary Sugars and Human Disease*, DoH Report on Health and Social Subjects, 37, HMSO, London, 1989.

Committees on Toxicity, Mutagenicity and Carcinogenicity of Chemicals in Food, Consumer Products and the Environment (1991) Her Majesty's Stationery Office, London.

Czerwiecki, L. (1980) *Rocz. Panst. Zaklado Hig.* 31, 271.

Dabeka, R. W. and McKenzie, A. D. (1988) *Food Addit. Contam.* 5, 333.

deGroot, A. P. (1973) *Food Cosmet. Toxicol.* 11, 19.

Department of Health and Social Security (1979) Recommended daily amounts of food energy and nutrients for groups of people in the United Kingdom. Reports on Health and Social Security Subjects, No 15, HMSO, London.

Department of Health (1991) Dietary reference values for food energy and nutrients for the United Kingdom. Report on Health and Social Security Subjects, No 41, HMSO, London.

Dickens, F. and Jones, H. E. (1961) *Br. J. Cancer* 15, 85.

Diesendorf, M. (1986) *Nature* 322, 125.

Doores, S. (1983) *CRC Crit. Rev. Food Sci. Nutrition* 19, 133.

Edgar, W. M. (1993) *Caries Res.* 27, 64.

Ehlers, D. (1986) *Lebensm.-Gerichtl. Chem.* 40, 2.

Eyrich, W. (1975) *Chem. Mikrobiol. Technol. Lebensm.* 4, 17.

Fenwick, G. R. (1986) *Proc. Nutr. Soc. Aust.* 11, 11.

Forbito, P. R. and Babsky, N. E. (1985) *J. Assoc. Off. Anal. Chem.* 68, 950–951.

Frank, H. K., Orth, R. and Figge, A. (1977) *Z. Lebensm.-Unters. Forsch.* 163, 111.

Frank, H. K., Orth, R. and Hermann, R. (1976) *Z. Lebensm.-Unters. Forsch.* 162, 149.

Friedman, M. (1990) In: *Biodeterioration Research Vol. III*, G. Llewellyn and C. O'Rear (eds.), Plenum, New York, pp. 21–54.

Frisvad, V. C. (1987) *J. Chromatogr.* 392, 333.

Gedalia, I. (1981) *J. Am. Dent. Assoc.* 102, 306.

Grasso, P. (1970) *Chem. Br.* 6, 17.

Greger, J. L. (1988) In: *Trace Minerals in Foods*, K. T. Smith (ed.), Marcel Dekker, New York, Chap. 9.

Grenby, T. H. and Wells, J. C. (1988) *Lancet* i, 992.

Grobler, S. R. (1991) *Clin. Prev. Dent.* 13, 13.

Grobler, S. R., Jenkins, G. N. and Kotze, D. (1985) *Br. Dent. J.* 158, 293.

Haber, G. B., Heaton, K. W., Murphy, D. and Burroughs, L. F. (1977) *Lancet* ii, 679.

Hall, M. V., Jewell, K. and Henshall, J. D. (1982) *Shelf-Life of Canned Fruits and Vegetables;* CEPRA Technical Memorandum No. 267.

Hallberg, L. and Rossander, L. (1982) *Human Nutr. Appl. Nutr.* 36A, 116.

Harwig, J., Scott, P. M., Stolz, D. R. and Blanchfield, B. J. (1979) *Appl. Environ. Microbiol.* 38, 267.

Hazel, T. and Johnson, I. T. (1987a) *Br. J. Nutr.* 57, 223.

Hazel, T. and Johnson, I. T. (1987b) *J. Sci. Food Agric.* 38, 73.

Hofmann, K., Mintzlaff, H. J., Alperden, I. and Leistner, L. (1971) *Fleischwirtschaft* 51, 1534.

Holland, B., Unwin, I. D. and Buss, D. H. (1992) Fruits and nuts. First supplement to McCance and Widdowson's *The Composition of Foods*, 5th edn, Royal Society of Chemistry, Cambridge.

Hopkins, J. (1993) *Food Chem. Toxic.* 31, 455–456.

Horio, T., Iwamoto, Y. and Shiga, I. (1967) In: *Proceedings of 5th International Congress on Canned Foods.*

Hughes, J. T. (1989) *Lancet* i, 490.

ISO, International Standards Organisation (1993) *Apple juice, apple juice concentrates and drinks containing apple juice. Determination of patulin content–Part 1: Methods using high performance liquid chromatography,* ISO 8128-1:1993 (E).

Jacquet, L., Lafont, J. and Vilette, O. (1983) *Microbiol. Aliment. Nutr.* 1, 127.

Jadhav, S. J., Sharma, R. P. and Salunke, D. K. (1981) *CRC Crit. Rev. Toxicol.* 11, 21.

JECFA (1990) *Toxicological Evaluation of Certain Food Additives and Contaminants,* World Health Organisation, Geneva, pp. 143–165.

Jenkins, D. J. A., Wolever, T. M. S., Taylor, R. H., Fielden, H., Baldwin, J. M., Bowling, A. C., Newman, H. C., Jenkins, A. L. and Goff, D. V. (1981) Glycaemic index of foods: a physiological basis for carbohydrate exchange. *Am. J. Clin. Nutr.* 34, 362–366.

Jongebreur, G. (1952) *Arch. Int. Pharmacodyn. Ther.* 90, 384.

Jorhem, L. and Slorach, S. (1987) *Food Addit. Contam.* 4, 309.

Josefsson, E. and Andersson, A. (1976) *Var foda* 28, 189.

Kellert, M., Blaas, W. and Wittkowski, M. (1984) *Fresenius' Z. Anal. Chem.* 318, 419.

King, Jr., A. D. and Schade, J. E. (1984) *J. Food Protect* 47, 886.

Kubacki, S. J. and Goszcz, H. (1988) *Pure Appl. Chem.* 60, 871.

Kumamoto, H., Matsubara, Y., Iizuka, Y., Okamoto, K. and Yokoi, K. (1985) *J. Agric. Chem. Soc. Jpn.* 59, 677.

Kumpulainen, J. T. (1992) *Biol. Trace Elem. Res.* 32, 9.

Lieu, F. Y. and Bullerman, L. B. (1978) *Milchwissenschaft* 33, 16.

Lindroth, S. (1980) *Technical Research Centre of Finland,* Espoo, Publication 24.

Lindroth, S. and Niskanen, A. (1978) *J. Food Sci.* 43, 446.

Liukkonen, L. H. and Piepponen, S. (1992) *Food Addit. Contam.* 9, 213.

Long, G. D., DeChatelet, L. R., O'Flaherty, J. I., McCall, C. E., Boss, D. A., Shirley, P. S. and Parce, J. W. (1981) *Blood* 57, 561.

Matsubara, Y., Yonemoto, H., Kumamoto, H., Iizuka, Y., Okamoto, K. and Yokoi, K. (1986) *Yakugaku* 35, 435.

Mesallam, A. S. (1987) *Food Chem.* 26, 47.

Meyer, R. A. (1978) *Lebensmittelindustrie* 25, 224.

Ministry of Agriculture, Fisheries and Food (1985) *Survey of Aluminium, Antimony, Chromium, Cobalt, Indium, Nickel, Thallium and Tin in Foods,* Food Surveillance Paper No 15, HMSO, London.

Ministry of Agriculture, Fisheries and Food (MAFF) (1987; 1993) *Household Consumption and Expenditure 1986.* Report of the National Food Survey Committee and National Food Survey, 1992, HMSO, London.

MAFF (1993) *Mycotoxins: third report,* Ministry of Agriculture, Fisheries and Food, Steering Group on Chemical Aspects of Food Surveillance, Her Majesty's Stationery Office, London, pp. 46–50.

Mislivec, P. B., Bruce, V. R., Stack, M. E. and Bandler, R. (1987) *J. Food Protect* 50, 38.

Möller, T. (1986) *Var foda* 38, 404.

Montgomery, R. D. (1980) In: *Toxic Constituents of Plant Foodstuffs,* 2nd edn. Academic Press, New York.

Mortimer, D. N., Parker, I., Shepherd, M. J. and Gilbert, J. (1985) *Food Addit. Contam.* 2, 165.

Nagy, S. (1977) In: *Citrus Science and Technology*, Vol. 1, S. Nagy, P. E. Shaw and M. K. Veldhuis (eds.), AVI, New York, Chap. 13.

Nagy, S. and Rousseff, R. L. (1981) *J. Agric. Food Chem.* 29, 890.

Nagy, S., Rouseff, R. and Ting, S.-V. (1980) *J. Agric. Food Chem.* 28, 1166.

Normington, K. W., Baker, I., Molina, M., Wishnok, J. S., Tannenbaum, S. R. and Song, P. (1986) *J. Agric. Food Chem.* 34, 215.

Osswald, H., Frank, H. K., Komitowski, D. and Winter, H. (1978) *Food Cosmet. Toxicol.* 16, 243.

Paul, A. A. and Southgate, D. A. T. (1978) McCance and Widdowson's *The Composition of Food*, 42nd edn., HMSO, London.

Paunio, P., Rautava, P., Helenius, H., Alanen, P. and Sillanpaa, M. (1993) *Caries Res.* 27, 154.

Pero, R. W., Posner, H., Blois, M., Harvan, D. and Spalding, J. W. (1973) *Environ. Health Perspect.* 87.

Pollock, G. A., BiSabatino, C. E., Heimsch, R. C. and Hilbelink, D. R. (1982) *Food Chem. Toxicol.* 20, 899.

Prieta, J., Moreno, M. A., Blanco, J. L., Suarez, G. and Dominguez, L. (1992) *J. Food Protection* 55, 1001–1002.

Regnault-Roger, C. (1988) *Experientia* 44, 725.

Reilly, C. (1980) In: *Metal Contamination of Food*, Elsevier Applied Science, London, Chap. 8.

Robbins, R. C. (1980) In: *Citrus Nutrition and Quality*, American Chemical Society, Washington, Chap. 3.

Roberts, J. A. and Gaddum, L. W. (1937) *Ind. Eng. Chem.* 29, 574.

Rouseff, R. L. and Ting, S. V. (1980) *J. Food Sci.* 45, 965.

Rovira, R., Ribera, F., Sanchis, V. and Canela, R. (1993) *J. Agric. Food Chem.* 41, 214–216.

Ruggieri, G., Valletrisco, M., Ruggieri, P. and Nicola, I. (1982) *Ind. Bevande*, 270.

Salunke, D. K. and Desai, B. B. (1984) In: *Postharvest Biotechnology of Fruits*, Vol. II, CRC Press, Boca Raton, FL, Chap. 15.

Sato, K., Yamada, T., Yoshihira, K. and Tanimura, A. (1986) *Shokuhin Eiseigaku Zasshi* 27, 619.

Schade, J. E. and King, Jr., A. D. (1984) *J. Food Protect* 47, 978.

Scott, P. M. and Kanhere, S. R. (1980). *J. Assoc. Off. Anal. Chem.* 63, 612.

Scott, P. M. and Somers, E. (1968) *J. Agric. Food Chem.* 16, 483.

Scott, P. M., Fuleki, T. and Harwig, J. (1977) *J. Agric. Food Chem.* 25, 434.

Sekiya, K. and Okuda, H. (1982) *Biochem. Biophys. Res. Commun.* 105, 1090.

Sempinska, E., Kostka, B., Krolikowska, M. and Kalisiak, E. (1977) *Pol. J. Pharmacol. Pharm.* 29, 7.

Seow, C. C., Abdul Rahman, Z. and Abdul Aziz, N. A. (1984) *Food Chem.* 14, 125.

Silva, M. F. de A., Jenkins, G. N., Burgess, R. C. and Sandham, H. J. (1986) *Caries Res.* 20, 263.

Smith, A. J. and Shaw, L. (1987) *Br. Dent. J.* 162, 65.

Southgate, D. A. T., Bingham, S. and Robertson, J. (1978) *Nature* 274, 51–52.

Stack, M. E. and Prival, M. J. (1986) *Appl. Environ. Microbiol.* 52, 718.

Stack, M. E., Mislivec, P. B., Roach, J. A. G. and Pohland, A. E. (1985) *J. Assoc. Off. Anal. Chem.* 68, 640.

Statens Livsmedelsverk (1986) *Livsmedlstabller*, Liber Tryck AB Stockholm, Sweden.

Stavric, B. (ed.) (1984) *Fed. Proc.* 43, 2454.

Steele, B. T., Murphy, N. and Rance, C. P. (1982) *J. Paediatr.* 101, 963.

Steyn, P. S. and Rabie, C. J. (1976) *Phytochemistry* 15, 1977.

Stinson, E. E. (1985) *J. Food Protect* 48, 80.

Stinson, E. E., Osman, S. F., Heisler, E. G., Siciliano, J. and Bills, D. D. (1981) *J. Agric. Food Chem.* 29, 790.

Stinson, E. E., Bills, D. D., Osman, S. J., Siciliano, J., Ceponis, M. J. and Heisler, E. G. (1980) *J. Agric. Food Chem.* 28, 960.

Stott, W. T. and Bullerman, L. B. (1975) *J. Milk Food Technol.* 38, 695.

Stoloff, L. J. (1976) *J. Assoc. Off. Anal. Chem.* 59, 317.

Taniwaki, M. H., Hoenderboom, C. J. M., Vitali, A. de A. and Eiroa, M. N. U. (1992) *J. Food Protection* 55, 902–904.

Tanner, H. and Duperrex, M. (1968) *Schweiz. Z. Obst.-Weinbau* 104, 508.

Tarter, E. J. and Scott, P. M. (1991) *J. Chromatogr.* 538, 441–446.

Touyz, L. Z. G., Glassman, R. M. and Naidu, S. G. (1981) *South Afr. J. Sci.* 77, 423.

Touyz, L. Z. G. (1982) *Tydskrif Tandheelkd South Afr.* 37, 663.

Touyz, L. Z. G. and Glassman, R. M. (1981) *J. Dent. Assoc. South Afr.* 36, 195.

Thurm, V., Paul, P. and Koch, C. E. (1979) *Die Nahrung* 23, 131.

U.K. Ministry of Agriculture, Fisheries and Food (1987) *Household Food Consumption and Expenditure 1986*, Annual Report of the National Food Survey Committee, HMSO, London.

Ulrich, R. (1971) In: *The Biochemistry of Fruits and their Products*, A. C. Hulme (ed.), Academic Press, New York and London, Vol. 1, 89–118.

U.S. Department of Agriculture (1963) *Composition of Foods Handbook* 8, US Department of Agriculture, Washington, DC.

Van, J.-A. R. (1989) *Diss. Abstr.* 49, 3527.

Walker, A. P. R. (1982) *Tydsrif Tandheelkd South Afr.* 37, 663.

Ware, G. W., Thorpe, C. W. and Pohland, A. E. (1974) *J. Assoc. Off. Anal. Chem.* 57, 1111.

Watkins, K. L., Fazekas, G. and Palmer, M. V. (1990) *Food Australia* 42, 438–439.

Watson, D. H. (1984) *J. Food Protect* 47, 485.

Wheeler, J. L., Harrison, M. A. and Koehler, P. E. (1987) *J. Food Sci.* 52, 479.

Whiting, G. C. (1971) In: *The Biochemistry of Fruits and Their Products*, A. C. Hulme (ed.), Academic Press, New York and London, Vols. 1–31.

Wilson, R. D. (1981) *Food Technol. (New Zealand)* 16, 27.

Wittkowski, M., Baltes, W., Kronert, W. and Weber, R. (1983) *Z. Lebensm.-Unters. Forsch.* 177, 447.

Woller, R. and Majerus, P. (1982) *Flüss. Obst.* 49, 564.

Younis, S. A. and Al-Rawi, F. I. (1988) *J. Biol. Sci. Res. (Baghdad)* 19, 245.

14 Legislation controlling production, labelling and marketing of fruit juices and fruit beverages
J. S. DRANSFIELD

14.1 Fruit juices, concentrated fruit juices and fruit nectars

14.1.1 Introduction

Most countries have a general regulation in some form, often an Act, which controls the overall quality of food. Under the UK Food Safety Act (1990), food must be of the nature, substance or quality demanded by the purchaser, must not be harmful to health and must comply with food safety requirements. It is an offence to sell food that does not comply with these requirements. General legislation, similar in nature to the Food Safety Act, also exists in other countries. The Food Safety Act provides, in addition, that further regulations can be made in certain areas where necessary. These regulations may cover standards of composition, permitted ingredients and additives, labelling, hygiene and quantity control. A standard of composition on a particular food type may include a number of these aspects.

The scope of legislation on fruit juices and beverages varies throughout the world, despite efforts to improve the ease of international trade via harmonised legislation. Raw materials used must be of suitable quality and in some countries agricultural and grading standards are also specified. Most countries in addition have a form of enforcement whereby products that do not meet the required quality can be removed from the market, and have an authority that can act to respond to any complaints on product acceptability. The stringency of this enforcement varies in different countries, and sometimes within a country. Most major countries either publish regulations in consolidated form or have volumes that discuss the regulations and/or case law within that country, and which can be used for reference.

14.1.2 Fruit juice regulations in European Community (EC) countries

One of the aims of the member states forming the Community was to achieve harmonisation of laws in a number of important areas so that barriers to trade could no longer be justified, the result being a true common market. As the size of the Community increased, this was found to be an increasingly difficult task. A new initiative was introduced by the European Commission

(1985) with a view to ensuring that the internal market was completed by the end of 1992. In the earlier periods of discussion many EC draft Directives on specific subjects such as soft drinks, bread, chocolate and beer were considered. Many of these did not progress beyond the discussion and draft stage. However, some were passed to become Directives, including one on fruit juices, fruit nectars and similar products. With the later change of emphasis to 'horizontal' legislation on subjects affecting many foods, such as additives and labelling, the idea of specific commodity or 'vertical' Directives lapsed. However, those that were passed have been adopted in most cases by the member states within the required time span, and these form the basis of the member states' legislation for the particular commodity. Hence, the basis of the legislation on fruit juices and fruit nectars is Council Directive (1993) 93/77/EEC relating to fruit juices and certain similar products. This Directive consolidates the provisions of former Directive 75/726/EEC and its amendments. This action was necessary for reasons of clarity and rationality, as Directive 75/726/EEC had been substantially amended. The new Directive revokes Directive 75/726/EEC and its amendments.

The main areas covered by the Directive are definitions of the various products covered (fruit juice, concentrated fruit juice, dried fruit juice and fruit nectar); permitted additional ingredients and additives; labelling requirements for the final products, and authorised processes and treatments during manufacture. Directive 75/726/EEC contained a number of derogations, which meant that, despite the existence of the Directive, significant differences existed in the laws in the individual member states. Some specific derogations enabling national rules to be followed still exist in the consolidated Directive, 93/77/EEC. These include the addition of dimethylpolysiloxane to pineapple juice and the use of diffusion processes for the manufacture of certain fruit juices intended for the manufacture of concentrated fruit juices. The derogations regarding additives will cease once appropriate Community provisions become applicable.

Fruit juice is defined in the Directive as the juice obtained from fruit by mechanical processes, fermentable but unfermented, having the characteristic colour, odour and flavour typical of the fruit from which it comes. This definition extends to include the product obtained from concentrated fruit juice having sensory and analytical characteristics equivalent to those of juice obtained directly from the fruit. Fruit must be in good condition and free from deterioration and contain all the essential constituents needed for the production of fruit juices and fruit nectars. Tomatoes are not considered as fruit for the purposes of the Directive; therefore, the member states' regulations on fruit juices do not cover tomato juice unless they have a separate national regulation.

Only certain ingredients and additives may be added to fruit juices. Ascorbic acid is permitted in an amount required to provide an antioxidant effect; member states may decide their own rules for the use of vitamins, including vitamin C, in these products. Use of other additives is strictly controlled and only certain substances may be added to the specified levels of use. The Directive authorises the addition of sulphur dioxide to grape juice and other juices and citric acid to pineapple juice. Addition of both sugars and acid to the same fruit juice is prohibited.

Fruit juices other than pear and grape may contain added sugar to a maximum of 15 g/l of juice, for correction of acidity, without its having to be declared on the label. If sugar is added to juices for sweetening purposes, the amount permitted to be added depends on the type of juice. In this case, the name of the product must be accompanied by the term 'sweetened' followed by an indication of the maximum quantity of sugar added. The Directive does not specify Brix values for particular products.

With regard to the labelling of fruit juices, the provisions of the Council Directive (Labelling) (1978) 79/112/EEC apply; in addition, specific labelling requirements are detailed in the fruit juices Directive. The labelling of these products in the EC member states is therefore controlled both by the appropriate regulations on fruit juices and fruit nectars and by the general food labelling regulations.

In the UK, the composition of fruit juices and fruit nectars is controlled by the UK Fruit Juices and Fruit Nectars Regulations (1977), as amended. Labelling is covered by these regulations and by the UK Food Labelling Regulations (1984), as amended. The Fruit Juices and Fruit Nectars Regulations require a statement 'made with concentrated X juice', where X is the name of the fruit, if a juice has been manufactured using a concentrate, and specify the precise product name to be used in each case. Specific labelling requirements are also given for sweetened juices, reconstituted, dried and concentrated juices and carbonated juices.

The Directive also covers composition of fruit nectars. These are the unfermented but fermentable products that are obtained by the addition of water and sugars to fruit juice, concentrated fruit juice, fruit purée or concentrated purée or a mixture of these that comply with the specifications given. Sugars can be added to maximum 20% by weight of the product; specific forms of sugar that can be used are listed. Minimum contents of juice and/or purée as percentage by weight of the finished product and minimum total acid contents expressed as g/l tartaric acid are specified (Table 14.1). With regard to the addition of sugars to nectars, Commission Directive (1993) 93/45/EEC provides for the production of nectars without the addition of sugars or honey under certain conditions. This applies to fruit listed in points II and III in Table 14.1 and apricots, which can be used individually or mixed together to manufacture nectars without the addition of sugars or honey where their naturally high sugar content warrants this.

Labelling requirements for nectars include a declaration of the minimum content of fruit purée, or fruit juice or a mixture of these in the form 'fruit content: x% minimum'. A number of different types of nectars are charac-

Table 14.1 EC minimum requirements for fruit content and acid content of fruit nectars

Fruit nectars made from	Minimum total acid content expressed as tartaric acid (g/l of finished product)	Minimum juice and/or purée content (% by weight of finished product)
I. Fruits with acid juice unpalatable in the natural state		
Passion fruit (*Passiflora edulis*)	8	25
Quito naranjillos (*Solanum quitoense*)	5	25
Blackcurrants	8	25
Whitecurrants	8	25
Redcurrants	8	25
Gooseberries	9	30
Sallow-thorn berries (*Hippophae*)	9	25
Sloes	8	30
Plums	6	30
Quetsches	6	30
Rowanberries	8	30
Rose hips (fruits of *Rosa* sp.)	8	40
Sour cherries	8	35
Other cherries	6[a]	40
Bilberries	4	40
Elderberries	7	50
Raspberries	7	40
Apricots	3[a]	40
Strawberries	5[a]	40
Mulberries/blackberries	6	40
Cranberries	9	30
Quinces	7	50
Lemons and limes	–	25
Other fruits belonging to this category	–	25
II. Low-acid, pulpy or highly flavoured fruits with juice unpalatable in the natural state		
Mangos	–	35
Bananas	–	25
Guavas	–	25
Papayas	–	25
Lychees	–	25
Azeroles (Neapolitan medlars)	–	25
Soursop (*Annona muricata*)	–	25
Bullock's heart or custard apple (*Annona reticulata*)	–	25
Sugar apples	–	25
Pomegranates	–	25
Cashew fruits	–	25
Spanish plums (*Spondia purpurea*)	–	25
Umbu (*Spondia tuberosa aroda*)	–	30
Other fruits belonging to this category	–	25
III. Fruits with juice palatable in the natural state		
Apples	3[a]	50
Pears	3[a]	50
Peaches	3[a]	45
Citrus fruits except lemons and limes	5	50
Pineapples	4	50
Other fruits belonging to this category	–	50

[a] Limit not applicable for products described as 'succo e polpa' and 'sumo e polpa'.

teristic of particular member states. The Directive was worded so that these individual products could retain their identity, for example 'succo e polpa' for fruit nectars obtained exclusively from fruit purée and/or concentrated fruit purée in Italy, and 'sumo e polpa' for fruit nectars obtained exclusively from fruit juice and fruit purée and/or concentrated fruit purée in Portugal.

Hence, the basis of the legislation for these products is the Directive and as the rules have been adopted by all of the member states, the regulations on these products are similar in content.

Greek regulations are part of the Food Code; the Directive has been adopted and, in addition, the regulations control fruit juice and nectar manufacture to include manufacturing and packing equipment, chilling installations, packing machinery and pasteurisation units. Emphasis is specifically on hygiene. Tomato juice is covered in a separate regulation.

Italy has adopted the Directive and has detailed regulations on similar products manufactured from almonds and orgeat.

In Denmark, Belgium and France the regulations cover vegetable juices in a detail similar to that for fruit juices.

The Dutch regulations on fruit juices contain additional criteria in the form of detailed specifications of purity; these are described as authenticity criteria and include such detail as levels for sugar-free extracts, ash, minerals, acids, sugars and amino acids. A Brix level is given for extracts of the unsweetened juice after correction for acidity. These authenticity criteria are specified for orange, apple, grape and grapefruit juices.

The Directive has been adopted into German food law by two regulations, one on fruit juices and another on fruit nectars and fruit syrups. In addition, there are further 'guidelines' on fruit juices, which, while having no legal force, are an indication of trade and consumer expectations. For fruit juices, the guidelines include a description of Good Manufacturing Practice with regard to physical techniques used in processing, the quality of water used for the restoration of fruit juice concentrates, the quality of fruit juices manufactured from concentrates, label declarations and additional specifications for fruit juices.

From this it can be seen that despite the adoption of the Directive by the member states, legislation for these products is not identical in all member states. However, the basic composition and labelling requirements stipulated do enable industry to have a slightly easier task when processing fruit juice products for export to more than one country in the EC.

Although the fruit juices Directive was consolidated in 1993, a review of this and other vertical food composition Directives was already underway. At the EC Summit meeting in Edinburgh in December 1992, it was agreed that the Commission should carry out a review of these vertical Directives. It was felt that the legislation on these products was excessively detailed and could be streamlined and replaced, under the new approach to harmonisation, by minimum requirements to be met by products circulating freely within the Community.

In August 1993, the Commission produced a first draft of a possible revised Directive on fruit juices. This revised text was not a formal proposal from the Commission, but a working draft that the Commission intended to discuss briefly with the member states. The various draft Directives have been discussed in detail, and the Commission has since suggested possible options for the way in which the review of the Directives should progress. The majority of the member states favour simplification, involving removal of provisions covered by horizontal legislation, removal or harmonisation of derogations and changes to take account of technological progress. The Commission has indicated that consultation with the member states and the industries concerned on the simplification of the vertical Directives has begun, and its objective is to present appropriate proposals during 1994.

14.1.3   Fruit juice regulations in EFTA countries

Switzerland.   Although Switzerland is not, at the time of publication, a member of the EC, its intention of harmonising its legislation with that of the EC and its close proximity in geographical terms play an important role in trade. The Swiss regulations include detailed requirements for composition and labelling of fruit juices, vegetable juices and fruit nectars. For mixtures of juices, the proportion of fruits represented pictorially must be equivalent to that of the fruits used in the product. If less than 5% fruit is used it is not permitted to use a pictorial representation.

Fruit nectars are covered in similar depth and include specified fruit contents. Generally, the Swiss regulations are very strict; if an ingredient is not permitted by a regulation it is necessary to contact the Swiss authorities regarding its use.

Austria.   There are two chapters of the Austrian Food Code concerning juices and similar products, one on 'Obstsirupe' (formerly 'Obstrohsäfte' and 'Obstsirupe'), which is made from juices of berries and soft fruit, and a second covering 'Fruchtsäfte'. The term 'natural' is not permitted for 'Obstsirupe'. The standard for 'Obstsirupe' covers permitted added ingredients and labelling requirements.

The chapter on 'Fruchtsäfte' covers juices from other types of fruit, and also dried and concentrated products and fruit nectars. Detailed requirements for labelling and permitted ingredients are specified. Claims for the addition of vitamin C are covered. Permitted processes and treatments for product manufacture are also included.

Scandinavia.   Legislation of the Scandinavian countries, Norway, Sweden and Finland, emphasises use of additives, permitted maximum levels and the specific foods in which these additives are permitted. Mostly, the national standards of composition are sketchy and tend to include some specific labelling requirements but without the detail characteristic of the standards required by a number of the other European countries. For many foods there are no standards of composition in terms of minimum levels of ingredients, but only those additives included in the positive lists may be used. In the case of fruit juices and nectars, however, Norway and Sweden have now adopted the EC rules on fruit juices and certain similar products. This is as a result of the Agreement forming the European Economic Area (EEA) (1994), which entered into force at the beginning of 1994, effectively establishing one of the largest free trade areas in the world having harmonised technical rules. The compositional and specific labelling requirements for fruit juices and nectars in these countries now reflect the provisions of the EC in this area, although Norway has retained its previous provisions for additives in these products whereas Sweden has adapted its additives rules to reflect more closely those of the EC. Finland has issued a notice to the effect that it will accept such products complying with the EC provisions.

14.1.4   Fruit juice regulations in USA and Canada

USA.   Standards of composition for foods, labelling and use of additives are covered by the Food and Drug Administration (FDA) and the United States Department of Agriculture (USDA). The FDA regulations are entitled the Code of Federal Regulations, Title 21; standards of composition for a range of juices and products are included. Among these standardised products are canned fruit nectars, lemon juice, grapefruit, prune and pineapple juice and a number of different types of orange juice, namely canned, frozen, pasteurised, frozen concentrated, orange juices for manufacturing, and orange juice with preservative. Tomato juices are also covered. Labelling requirements that are specific to these products include the need to declare any sweeteners added and the addition of orange concentrate to adjust the solids content as permitted by the standard. Juice may be heat-treated to reduce enzyme or microbial activity.

   These types of orange juices are for direct sale to the consumer; in addition, the standards for orange juice for manufacturing and orange juice with preservative both cover products intended for further processing. Orange juice for manufacturing is prepared from unfermented oranges as for retail orange juice, except that the oranges may be below the Brix and Brix/acid ratio specified for oranges used to produce juice for retail sale. Orange juice with preservative contains preservative in order to inhibit spoilage organisms and reduce microbial activity so that the product retains its quality for

Table 14.2 USA Brix values for single-strength fruit ingredients for fruit nectars

Name of fruit	Brix value
Apple	13.3
Apricot	14.3
Blackberry	10.0
Boysenberry	10.0
Cherry	14.3
Guava	7.7
Loganberry	10.5
Mango	13.0
Nectarine	11.8
Papaya	11.5
Passionfruit	14.5
Peach	11.8
Pear	15.4
Pineapple	13.0
Plum	14.3

processing. Thus the emphasis on fruit juice quality, in terms of both composition and hygiene, is ensured by compliance with these regulations. The standards for other juices are of a similar nature.

In Title 7 of the Code of Federal Regulations, the USDA publishes grade standards for a range of fruit juices. These are standards of quality, and should be consulted by anyone intending to export these products to the United States. Labels for fruit juices that meet the minimum quality standards and those that have no standards of quality need not make reference to quality; however, if reference is made, the product must correspond to the usual understanding of the grade. The terms 'Fancy' or 'Grade A' may only be used on products meeting the specifications established for such grades by the USDA.

Canned fruit nectars are defined as pulpy, liquid foods consisting of specified amounts of fruit ingredients (depending on the fruit type), water and the optional sweetening ingredients listed. Brix values are specified for the different fruits (Table 14.2).

For all of these products the permitted additional ingredients are listed. A method of analysis of the Association of Official Analytical Chemists (AOAC) is used to determine the consistency of the finished product and whether it is of suitable quality.

Canada. The Canadian Food and Drugs Regulations include standards for a number of juices, specifically lemon, lime, grapefruit, apple, grape, orange and pineapple. Provision is made within the regulations for the production of other juices, but without specific criteria. In addition, there are a number of national standards for internal and international trade; these grade standards have been published for a number of fruit juices. Standards are also specified for apricot, peach and pear nectar. Both labelling requirements and permitted additional ingredients are included in the standard; additives permitted are included in the additives regulations. Thus, the composition of the juices most commonly used is strictly controlled, as in the USA.

14.1.5   Fruit juice regulations in other major countries

Australia.   Traditionally, food legislation in Australia was the responsibility of the individual States or Territories, with each authority producing its own regulations. In 1987, the National Health and Medical Research Council (NHMRC) initiated the move towards a harmonised system of food legislation across Australia, producing a Food Standards Code that has gradually been adopted by all States and Territories. In August 1991, the National Food Authority (NFA) was established, which is now responsible for setting all food standards in Australia. Under agreement between Commonwealth, States and Territories, the States/Territories will adopt all new food standards and amendments to the existing food standards by reference. Any existing anomalies between previous NHMRC food standards and State or Territory food standards will remain until the whole standard is reviewed and regazetted by the NFA; after this review process the regazetted standard will be picked up by States by reference under the new agreement.

The Australian Food Standards Code includes a detailed standard of composition for fruit juice and a separate one for orange juice. A maximum of 25 g/kg added sugar is permitted to be added without the product being considered sweetened, except in the case of pineapple, lemon or lime juice, where a maximum of 40 g/kg of sugar may be added. If more sugar is added, the name of the product must indicate this as 'sweetened (name of fruit) juice'. Minimum levels of vitamin C that must be present in different juice types are also specified.

Orange juice is defined as that juice coming from the endocarp of one particular species, *Citrus sinesis* (L) Osbeck. In contrast to other fruit juices, sweetened orange juice must declare the percentage of added sugar in the form 'contains not more than (%) added sugar'. From this the consumer can clearly see the product type. As in EC legislation, there is a requirement to indicate when the product has been produced from concentrate. Over twenty purity criteria that must be met are listed for orange juice.

A further indication of the trend towards controlling use of terms that imply quality is the fact that orange juice can only be described as fresh if it is expressed from the fruit in front of the purchaser and does not contain any food additives. Products can only be described as pure if they are composed of one ingredient and do not contain any additives. Fruit nectars are not standardised in Australia; they would have to comply with the labelling requirements for all foods, and provisions concerning non-standardised products in terms of additives.

Table 14.3 Specifications for certain fruit juices in New Zealand

Type of juice	% Soluble solids[a]	Essential oil measured in ml/kg[b]	Total titratable acidity per 100 ml[c]	
			Minimum (g)	Maximum (g)
Orange, mandarin				
or tangelo	10	0.4	0.65	1.5
Grapefruit	9	0.3	1.0	2.0
Lemon	6	0.5	4.5	–
Apple	10	–	0.3	0.8
Grape	15	–	–	–
Pineapple	10	–	–	–
Blackcurrant	11	–	–	–
Lime	8	–	6.0	–

[a] As determined by refractometer at 20°C uncorrected for acidity and read as °Brix on the International Sucrose Scales.

[b] As determined by the AOAC (1980) method (Official Methods of Analysis of the AOAC (1980) 22.088-22.089 and 19.127 Essential Oil – Official First Action).

[c] Measured at 20°C, calculated as anhydrous citric acid.

New Zealand. The regulations are similar in format to those of Australia, with general requirements and provisions in areas such as labelling and contaminants in a consolidated volume with detailed standards of composition, which include permitted ingredients and additives and additional labelling requirements. Standards of composition are laid down for fruit juice, concentrated fruit juice and fruit nectar. Added carbohydrate sweetener in dry form is permitted to maximum 20% in fruit juice; if more than 5% carbohydrate sweetener has been added the term 'sweetened' must be included as part of the name of the food. Specifications are given for particular types of fruit juice (Table 14.3).

Fruit nectars are treated in a similar way; minimum fruit ingredient contents are specified, depending on the fruit used, and the minimum soluble solids content of a nectar is set at 13% or 12% for citrus fruit nectars. The claims 'pure' or 'real' or terms having a similar meaning may not be used on the packages of fruit nectars.

Latin America. In most countries of the region the control of foodstuffs is the responsibility of the Ministry of Health or an equivalent institution. In some countries, standards organisations produce a wide variety of food-related standards, which, in the absence of specific national legislation, are applicable. In all countries of the region, except Mexico, registration of food products is a general requirement; in certain countries, the registration number must appear on the labelling of the product. In addition, in most countries, imported foodstuffs must be accompanied by a health certificate and require authorisation from the authorities prior to circulation.

Although national food law in certain countries of Latin America is not keeping pace with modern technological developments, food legislation in many countries is not only highly evolved and complex but also undergoing radical changes. In addition, there is a general trend towards harmonisation with an important impact on the food legislation in the area. In particular, MERCOSUR, the Southern Common Market consisting of Brazil, Argentina, Uruguay and Paraguay, has developed harmonised food labelling regulations to be followed in the above countries. ICAITI, a standards organisation based in Guatemala, has published a wide variety of food-related standards that are voluntarily applicable in the countries grouped in the 'Central American Common Market', i.e. Guatemala, Costa Rica, Honduras, Nicaragua and El Salvador.

Comprehensive standards on fruit juices are laid down in Argentina, Brazil, Mexico, Colombia, Dominican Republic, Ecuador, Uruguay, Venezuela and countries of the Central American Common Market. These standards vary in scope and detail; in general, they include compositional requirements, provisions on permitted ingredients and additives and specific labelling requirements for specified types of fruit juices. Other countries have no standards of composition for fruit juices, for example Cuba, Chile, Panama and Peru, but the use of additives in these products is covered by the relevant positive additives lists.

Middle East. Middle Eastern countries control the products being imported and produced in their territory largely by means of standards. Some of these have the force of law; others are not mandatory. Countries with mandatory standards for fruit juices include The Lebanon, Saudi Arabia, Jordan, Kuwait and Israel; the last three countries have also issued mandatory requirements for fruit nectars. The range of products covered by these standards varies widely.

Harmonisation of food law by the issuing of Gulf Standards on diverse commodities is one of the main objectives of the Gulf Cooperation Council (GCC), comprising Saudi Arabia, Kuwait, United Arab Emirates, Oman, Bahrain and Qatar. However, Gulf Standards for fruit juices and nectars have yet to be published. In the absence of national standards for these products, Middle Eastern countries will normally accept products complying with standards produced by the Codex Alimentarius Commission (CAC), an international body under the auspices of the World Health Organization (WHO) and the Food and Agriculture Organization (FAO) of the United Nations. The CAC has produced a number of standards for fruit juices and nectars (see below).

Far East. An area of the world that is increasingly producing a greater amount of legislation relating to food is the Far East. Often countries are not satisfied to accept products from other areas of the world without having assurances of quality and suitability by means of their own standards or regulations. Sometimes the standards do not have the force of law and rather act as a guide, but notice should still be taken of these by exporters to the area.

One of the largest areas of opportunity for export is Japan. There are some standards for products, a positive additives list, labelling requirements and often purity criteria for products as well as testing methods. The additives list permits the use of sulphur dioxide and L-cysteine hydrochloride in fruit juice and sodium benzoate and potassium sorbate in fruit juice and concentrated fruit juice. There is no standard of composition, so in theory imports into Japan are not restricted by minimum fruit content levels and similar requirements. In practice, there is a means of ensuring quality of both imported and home-produced products and this is by means of using Japanese Agricultural Standards. This system guarantees the quality of processed foods that adhere to these standards. The standards are set by the Japanese Ministry of Agriculture, Forestry and Fisheries; if products pass through a series of qualifying grading procedures they are entitled to bear the Japanese Agricultural Standard or JAS mark. Only Registered Grading organisations are able to verify the tests on the authority of the Ministry. The Japanese Agricultural Standard for Fruit Drinks includes products such as fruit juices, concentrated fruit juices and fruit purée. Details are given of compositional requirements and labelling, including misuse of claims.

Other countries that have regulations on fruit juices and similar products are Malaysia, the Philippines, Pakistan, India, Indonesia, South Korea and Singapore. These vary in scope and detail, but essentially act to ensure product quality. Other countries have detailed additives regulations, for example Taiwan, Thailand and China, but have no standards of composition for these products. The countries of the region do not appear to have a common link in their legislation.

14.1.6 Fruit juice standards produced by Codex Alimentarius

The Codex Alimentarius Commission is a joint body set up by WHO and FAO in 1962 to implement the Joint FAO/WHO Food Standards Programme. The main considerations that initiated this programme were the need to protect the health of the consumer, to ensure fair practices in the food trade, and to facilitate international trade in food. Its role is to examine the need for and produce, where necessary, standards for foods in order to encourage harmonisation of food legislation throughout the world. Membership of the Commission comprises member nations and associate members of FAO and/or WHO that have notified these organisations of their wish to be considered as members.

Where it is felt necessary to produce specific recommendations on a food commodity, this is usually done in the form of standards. Codex standards are produced by a formal procedure. Member governments are involved throughout by attendance of delegates at the meetings and by making formal comments. Different countries host the various Codex Committees relating to particular food commodities, subject matters or regional matters. As the Commission works to establish a standard, a system built into the working procedure ensures that member governments have the opportunity to comment on the proposed standard.

A number of countries, for example some in the Middle East and Latin America, may accept Codex standards as authoritative legislation where they have no national regulations, or in addition to national law. A Codex standard may be formally accepted by a country; these Codex acceptances and other responses from member governments are published periodically by the Commission, so enabling the path of acceptance and harmonisation to be traced. Bodies such as the European Community also consider Codex standards when drafting their own legislation. The Codex Alimentarius Commission is, therefore, a body with a very important role to play in facilitating world trade and helping to achieve a greater degree of international harmonisation.

A number of Codex standards have been published on fruit juices and concentrated fruit juices, including tomato juice, and a number of nectars, including apricot, pear and peach. Most of these standards date from 1981; standards issued since that date include the general standards for fruit juices and fruit nectars not covered by the individual standards (1989). At the time of writing, the Codex Alimentarius is being re-issued in a new format, so that standards relating to a particular food group or subject are included in a single volume. In 1992, the various standards for fruit juices and nectars were re-issued in volume 6 (Codex Alimentarius, 1992) of this new series. The standards for fruit juices and nectars are detailed and include definitions of the product, essential composition and quality factors, provisions relating to food additives, contaminants, hygiene, weights of fill of containers, and specific labelling requirements; methods of analysis and sampling are also included in volume 6.

14.2 Non-carbonated fruit drinks and beverages

14.2.1 Introduction

Soft drinks can be divided into two main types, carbonated and non-carbonated. Legislation allows for the production and manufacture of many different types of products. Food legislation in a large number of countries covers soft drinks; since they are a popular product with all age groups legislation is considered desirable to ensure quality. Some countries have quality specifications, which include microbiological standards for the water that is used in soft drink manufacture, as this is considered an important aspect of formulation. Regulations differ with respect to standards of composition, permitted additives, for example preservatives and clouding agents, and labelling. Some countries have legislation on packaging for containers for soft drinks and may include provisions for returnable bottles; this area is not covered here. Some of the differences in the regulations are examined in the following sections.

14.2.2   Fruit drink regulations in EC countries

In contrast to fruit juices, there is no EC Directive on soft drinks, although an EC draft directive on this subject was considered in the earlier periods of discussion of the vertical directives. As the emphasis in European Community food law is now on harmonisation of areas affecting the health of the consumers, for example, additives and provision of more information on products that they purchase, for example labelling, it is now unlikely that standards of composition for soft drinks will be established at Community level. The Commission considers that differing national standards in terms of product quality and composition should not be a barrier to trade under the Treaty of Rome and that therefore use of these alone in preventing trade throughout member states is not justified. The end of 1992 had been set by the Commission as the date for completion of harmonisation of food legislation. At the time of writing, however, there are certain areas, for example additives, which are not fully harmonised at EC level. Hence, national rules continue to apply with respect to the use of additives in soft drinks in the member states. Although national composition standards may differ this should not prevent trade within the EC.

At the time of writing, EC member states do have their own standards of composition for soft drinks and it is necessary to consider these when exporting.

In the United Kingdom the composition of soft drinks is controlled by the UK Soft Drinks Regulations (1964), as amended. These regulations specify minimum potable fruit or fruit juice contents, maximum permitted saccharin contents and product descriptions for a number of types of sweetened and semi-sweet soft drinks. Squashes, comminuted fruit and barley drinks and waters are amongst the products included. The definition of soft drink in the regulations includes fruit drinks, fruit juice squashes, crushes or cordials but excludes fruit juices and nectars. Use of additives is permitted as determined by the appropriate additives regulations, including those on preservatives, emulsifiers and stabilisers and sweeteners, except that addition of saccharin and acids is controlled by the Soft Drinks Regulations themselves. Any additive used to give a clouding effect must be one listed in the regulations; clouding agents such as sucrose acetate isobutyrate are not permitted.

Maximum calorie contents of 7.5 calories per fluid ounce for products consumed after dilution and 1.5 calories per fluid ounce for products consumed without dilution are specified for low-calorie soft drinks. The maximum saccharin contents do not apply to low-calorie or diabetic products; drinks specially made for diabetics may not contain added sugar.

All soft drinks must comply with the provisions of the UK Food Labelling Regulations (1984), as amended, in addition to the specific requirements given in the Soft Drinks Regulations. The labelling regulations require that products indicate a suitably descriptive product name (where this is not prescribed in the regulations), a full ingredients list, an appropriate indication of durability (i.e. a date mark), a declaration of origin if absence could be misleading, details of usage, storage instructions and the name and address of the manufacturer, packer or EC seller. These regulations also cover requirements for using a number of restricted claims, including vitamin claims, on food labels. Requirements for the net quantity declaration are specified in the appropriate weights and measures regulations, and include provisions for the size and format of this declaration. Separate regulations require products to be marked with a lot or batch identification mark unless specifically exempt.

The Soft Drinks Regulations specify the names to be used for certain standardised products; for example, a citrus crush meeting minimum juice standards must be described as 'X crush', where X is the name of the citrus fruit or fruits used, and crushes prepared from citrus juice and barley water must be described as 'X barley crush' where X is the citrus juice used. Semi-sweet drinks must have the words 'semi-sweet' immediately preceding the name of the product.

Products that do not comply with requirements for fruit juice or potable fruit content must not show any pictorial device on the label suggestive of fruit or fruits. The label may not use any word suggestive of fruit or fruits not represented in the flavouring of that drink unless the following rules are observed:

(i) the suffix '-ade' is added to the name of the fruit or to the last-named fruit if there is more than one; or

(ii) the word 'flavour' is used, immediately preceded by the names of the fruit or fruits suggested.

Hence, the names 'lemonade' and 'orange flavour drink' indicate that the products being described do not meet the minimum compositional requirements with respect to juice content.

Claims relating to the presence of natural, or absence of artificial, ingredients or additives are often included in the labelling of foods for retail sale. Recommended conditions have been laid down by the Food Advisory Committee (revised 1993) with regard to use of these natural claims, with the aim of standardising the basis of their use so as not to cause confusion in the retail market. It is recommended that permitted food additives may be described as 'natural' only if they are obtained from recognised food sources by appropriate physical processing (including distillation and solvent extraction) or traditional food preparation processes. Flavourings should only be so described when they comply with the appropriate UK and EC legislation on flavourings.

Certain nutrition claims, for example energy, minerals and vitamins, are already comprehensively covered by the UK Food Labelling Regulations. In addition, misleading claims are prohibited under the UK Food Safety Act (1990), which requires that labelling must not be false or likely to mislead as to the nature, substance or quality of the product. However, the strong consumer interest in diet and health and the response of many manufacturers and retailers have resulted in a large increase in the number of nutrition claims appearing on the labels. The UK Food Advisory Committee (1989) considered that some of these claims were providing misleading information, and produced recommendations setting out conditions for the use of certain nutrition claims, to aid manufacturers in their use. These recommendations include use of claims such as 'low in sugar' and 'sugar-free', which are often used for soft drinks. Draft guidelines on nutrition claims, which are based on the FAC's recommendations, have been issued for comment (MAFF, 1993).

Owing to the increased awareness of the consumer with respect to diet and health and the increased demand for nutrition information to be given on food product labels, nutrition labelling is often seen on the labels of soft drinks. In the United Kingdom, nutrition labelling is now subject to the provisions of the UK Food Labelling (Amendment) Regulations (1994), which implement the provisions of the EC Council Directive (1990) on nutrition labelling (90/496/EEC). Under the Regulations, where a nutrition claim is made nutrition labelling must be given in accordance with the prescribed format. Otherwise, nutrition labelling may be given voluntarily, in which case it must be given in the same format as prescribed nutrition labelling, subject to minor qualifications. Before 6 October 1995, where nutrition labelling is provided, the information must consist of at least energy and the amounts of protein, carbohydrate and fat (referred to as 'Group 1' in the Directive). Other nutrients may also be declared if the manufacturer so wishes. From that date, nutrition labelling may alternatively include, as a minimum, energy and the amounts of protein, carbohydrate, sugars, fat, saturates, fibre and sodium (i.e. the 'Group 2' format in the Directive), provided that where a claim is made in respect of sugars, saturates, fibre or sodium this format must be given. A further optional list of nutrients that may also be declared is given. Requirements relating to the derivation of declared values, energy conversion factors and the manner of marking the nutrition information are prescribed. The vitamins and minerals that may be declared and their Recommended Daily Allowances (RDAs) are listed in the Regulations.

The general trend in UK legislation, in line with European Community thinking, is away from detailed standards of composition and towards more informative labelling on the product. Before the Soft Drinks Regulations were amended in 1993, soft drinks were required to contain a minimum quantity of sugar; this requirement has now been removed. It had been proposed to introduce a maximum caffeine limit for soft drinks. At this stage, however, MAFF has decided not to introduce a maximum caffeine limit and the FAC has been asked to review its 1988 recommendation about the need for such a limit.

As part of its 'Deregulation Initiative' the Government is reviewing all aspects of food legislation to identify areas that can be simplified and consolidated. In September 1993, MAFF issued a food law deregulation plan, which had been drawn up in consultation with the Department of Health; this document sets out a plan of action. In the area of food composition, MAFF intends to review certain regulations, including those on soft drinks, to determine whether they can be simplified or revoked in whole or in part. The regulations on soft drinks are timed for review in 1994.

The situation concerning other EC member states varies. The Irish Republic has no standards of composition; uses of some additives are specifically controlled, for example preservatives, but otherwise restrictions are few. The Netherlands has detailed standards of composition for a number of types of soft drink; minimum sugar and fruit content levels are specified, permitted ingredients listed and labelling requirements additional to those of the general labelling regulations detailed. Belgium and Luxembourg have similar regulations on soft drinks as a result of the Benelux Economic Union Decision on which these countries' regulations are based.

Germany is unusual among the member states in having guidelines rather than regulations on soft drinks. These guidelines are an indication of trade and consumer expectation and should be followed as appropriate. Labelling and permitted additives requirements additional to the requirements of the additives list and the general labelling regulations are also included in the guidelines. According to the guidelines a distinction is made between fruit juice drinks and 'Limonaden'. The former contain fruit juice, juice concentrate, fruit pulp, pulp concentrate or a mixture of these to specified levels depending on the fruit. 'Limonaden' contain essences with natural flavours, acids and their salts and sugar so that the minimum sum of sugars is 7%. 'Limonaden' are also usually carbonated; if no carbon dioxide is present then this should be declared. Fruit juices and fruit pulp may be used as optional ingredients; where they are used, 'Limonaden' should contain a minimum of half the fruit juice content customary in fruit juice drinks. According to the guidelines, a clouding effect may only originate from fruit constituents used in these products. In soft drinks based on citrus fruits it may also originate from preparations of the non-preserved flavedo layer and in quinine-containing soft drinks from preparations of the entire peel.

Italy has regulations covering (i) fruit juice drinks, which must contain natural juice or fruit concentrate to a minimum of 12 g/100 ml and have a minimum dry residue of 10%, and (ii) soft drinks sold using non-juice fruit names, e.g. 'Chinotto' (sour orange) or those under a fantasy name, which must have a dry residue of a minimum of 8%. Fruit extract, natural essences and acids may also be present in these soft drinks.

France has a Code of Practice for fruit juice drinks, fruit drinks and fruit pulp drinks; the minimum fruit content is 10%. If a blend of fruits is used and these do not each exceed 2% in the product, the description 'other fruits' can be listed in the ingredients list without the need to specify the individual types.

Part of the Greek Food Code is concerned with non-alcoholic beverages of natural fruit juices. Beverages with the name of a particular fruit may contain only this type. Where mixtures of juices are contained in a product, the total fruit juice content must not be less than 20%.

Denmark has no specific standard of composition for soft drinks, but very detailed additives regulations are in force. Only the additives that are positively listed may be used. The list is updated and amended regularly.

Spanish regulations include standards for flavoured soft drinks, soft drinks with extracts, soft drinks with fruit juices, soft drinks with crushed fruits, mixed soft drinks, soft drinks for dilution and solid products for the preparation of soft drinks. These regulations include specific provisions on permitted optional ingredients, the use of additives and artificial sweeteners, and packaging and labelling requirements.

Portuguese regulations on soft drinks ('refrigerantes') include compositional, labelling and packaging requirements for different types of soft drinks, including those containing fruit juice. Provisions on the use of additives in soft drinks are included in the Portuguese positive additives list.

Member states still differ in their attitudes towards use of artificial sweeteners in these products. Use of these additives is strictly controlled in each member state and they may be used only where specified by the regulations. The use of artificial sweeteners and sugar in the same product is permitted (and practised widely) in the United Kingdom. In general, acceptability of intense sweeteners and bulk sweeteners is increasing and becoming more widely permitted.

In summary, EC member states have their own national regulations on composition, each country giving consumers a range of options with respect to product type. Labelling of the products generally enables consumers to judge the nature of the product being purchased. All of the member states have adopted the EC Labelling Directive (Council Directive, 1978); hence the basic labelling requirements such as a product name, ingredient list and date mark must be given in all cases. Any additional labelling specific to soft drinks is controlled by the individual countries' regulations. With respect to additives, it is not possible to produce a summary of additives currently permitted as these are constantly changing. However, at EC level common positions have been reached on the proposals for directives on colours, sweeteners, and the so-called 'miscellaneous additives' (i.e. the remaining categories of additives). When finalised, these Directives will eliminate differences relating to the use of these additives in soft drinks in the member states.

In addition to the harmonisation of nutrition labelling at EC level, there have been developments with regard to nutrition claims. The Consumer Policy Service of the Commission has produced a draft proposal (1993) for a Directive on the use of claims. This draft, now in its third revision, requires that claims are not false, misleading, or liable to be misleading, and covers the use of claims such as 'reduced' and 'without added' in respect of energy, nutrients, ingredients or non-nutritional substances.

Another area that is under discussion at EC level, which will have an impact on the soft drinks industry, includes quantitative ingredient declarations, or 'QUID'. The QUID proposal (Proposal for Council Directive, 1992) would require the mandatory quantitative declaration of certain ingredients or groups of ingredients used in the manufacture or preparation of a foodstuff. Under the proposal, the quantity of an ingredient must be declared when (i) it appears in the name under which a foodstuff is sold or can be derived from it; (ii) it is emphasised in the labelling; or (iii) it is essential to characterise the foodstuff and distinguish it from products with which it could be confused. Hence, in the case of soft drinks, the quantity of fruit or fruit juice would be required to be stated on the label when these are referred to in the product name, or are emphasised in the labelling, for example by the use of pictorial illustrations of fruit on the label. There have been detailed discussions on the QUID proposal and Presidency papers containing suggested changes have been issued. The discussions in this area are continuing.

14.2.3 Fruit drink regulations in EFTA countries

Switzerland. Switzerland has very detailed regulations on labelling and composition of fruit drinks including the use of additives. The regulations define fruit juice table drinks; these contain fruit juice, concentrated fruit juice or fruit syrup, diluted with water or natural mineral waters, with possible addition of sugar. A minimum of 10% fruit juice is required (6% for fruit juice table drinks manufactured only with lemon juice) and the percentage in the final product must be declared. In fruit juice table drinks based on more than one fruit, specific reference to fruit type in the name is only allowed if more than 2% juice of that type is present in the final product. Pictures of fruit or other types of plants are prohibited on the label of fruit juice table drinks.

Austria. Austrian regulations contain standards of composition for fruit juice lemonades, lemonades and artificial lemonades; these may all be carbonated or non-carbonated. Carbonated and non-carbonated fruit juice lemonades must comply with minimum specified fruit juice contents depending on the type of fruit, and a minimum sugar content of 8% (maximum 4% if the product is described as low in sugar). The minimum juice content must be declared. Whey is a permitted ingredient (to maximum 49%); if special emphasis is put on the presence of whey, then a minimum of 40% of the product must be whey. Fruit juice lemonades are considered high quality, and no colours may be added. The terms 'natural' or 'special' may not be used; pictures of sliced fruit or fruit that is dripping juice may be depicted on the label.

Lemonades have no minimum fruit juice levels. Herb extracts and natural flavours are permitted but artificial flavours are not acceptable in this type of product. In compiling these regulations the Austrian authorities considered the type of claims that might be made regarding the product and the impression the product label would give to the customer. For lemonade products pictures of whole fruits may be used but not pictures of dripping fruits. Products with less than 4% sugar are described as 'low-sugar'; products with no added sugar can declare 'only with sugar intrinsic to the fruit'. Claims using the term 'natural' are not acceptable.

Artificial lemonades contain artificial flavours and may be coloured. No pictures or references to fruit and herbs may be given.

Scandinavia.   The trend in the Scandinavian countries is for limited standards of composition for food. General labelling requirements apply to all foods and these countries each have a very strict positive additives list, arranged by food type rather than additives group. If a particular additive is not listed for a specific product, the authorities will consider a request for use, depending on technical need, and, if it is accepted, they will add it to the additives list at the next amendment.

Sweden, as a result of its commitment under the EEA Agreement (EEA, 1994), no longer operates specific compositional regulations on soft drinks. Norway still has standards for squashes and soft drinks; if a product is described as a fruit squash, a minimum of 8% w/w juice (as a derogation from the previously required 20% juice) must be present and the juice content must be declared. Finland allows the term 'juice drink'/'squash' to be used for a juice product to which water or sweetening agents have been added and which has a minimum of 35% w/w fruit/berry/vegetable juice present. 'Juice drink'/'squash drink' is reserved for the product containing a minimum of 10% w/w juice. When these products are put up for retail sale, the minimum juice contents must be declared.

14.2.4 Fruit drink regulations in USA and Canada

USA.   In contrast to the more detailed standards that exist for some other foods, the USA has only limited regulations for soft drinks. Standards are restricted to products such as frozen concentrates for lemonades and the artificially sweetened equivalent product, or the coloured equivalent product. Other non-carbonated soft drinks have no standard of composition and may contain additives and ingredients that are generally recognised as safe (GRAS). Substances considered as GRAS are listed in the Code of Federal Regulations, Title 21, Parts 182 and 184. Additives on the GRAS list must have been shown to be safe by scientific procedures carried out by qualified experts, or by experience based on common use prior to 1958. Some additives affirmed as GRAS have specific usage restrictions. Carbonated and non-carbonated beverages that do not have a standard of composition but contain any amount of fruit or vegetable juice must bear a suitably descriptive name. If the common or usual name uses the word 'juice', a qualifying term such as 'beverage' should be used, for example 'diluted grape juice beverage'. In addition, there are also provisions for 'multiple juice beverages', 'blends of single strength juices' and 'modified juices'.

Percentage juice declarations are required for foods purporting to be beverages that contain fruit or vegetable juice. If the beverage contains fruit or vegetable juice, the percentage juice declaration must be given, for example 'Contains 50% apple juice', unless exempt from this requirement.

For those products that do not contain fruit or vegetable juice, the declaration 'containing (or contains) no X juice' or 'does not contain X juice' or 'no X juice' where the X is filled with the name of the fruit or vegetable that is represented, suggested or implied must be given. If a non-specific fruit or vegetable juice is represented, the word 'vegetable' or 'fruit' must be inserted. Alternatively, the label may declare 'contains zero (0) percent (or %) juice'.

In addition to these requirements for beverages, there are a number of additional provisions in the Code of Federal Regulations describing uses of terms such as 'artificially flavored' and 'natural flavor' for describing foods. All of these labelling requirements are designed to make clear to the consumer the nature of product that he is purchasing. With effect from 8 May 1994, nutrition information must be provided for all products intended for human consumption and offered for sale, unless specifically exempted. The nutrition labelling declaration must contain information regarding the levels of specific mandatory nutrients per serving in the order and format described in the regulations, and certain nutrients may be given voluntarily unless a claim is made about them. In order to provide consumers with information regarding the contribution of the food components to their overall diet, Percentage Daily Values (PDVs) must be given for the nutrient components.

Canada. Canadian regulations comprise a number of standards of composition for foods, generally applicable labelling regulations and detailed additives provisions. A standard of composition is set for fruit-flavoured drinks; these have minimum vitamin and mineral contents specified, but must not give the impression that the vitamin content or nutritional value is similar to that of fruit juices unless they are sold as a fruit juice substitute or a breakfast drink, and are not represented as being soft drinks.

Other soft drinks are not standardised and have no specific composition or labelling requirements. Product labelling is covered by the Consumer Packaging and Labelling Regulations. Canada has also produced guidelines that cover use of descriptions such as use of the term 'natural', and specify the conditions under which these descriptions may be used to describe food or food ingredients. Use of additives is controlled by the Food and Drug Regulations; those additives that are permitted in non-standardised foods are accepted for use in soft drinks. Maximum levels of use are set where necessary.

14.2.5   Fruit drink regulations in other major countries

Australia. The Australian States have comprehensive standards for a number of different types of soft drinks. Fruit juice drinks must have a minimum 350 ml/l juice or 250 ml/l juice in the case of lemon juice, blackcurrant juice or guava juice, or minimum 500 ml/l juice for pineapple juice, pear juice, apple juice or a mixture of these and may contain permitted artificial sweeteners. Fruit drinks require a minimum 50 ml/l of juice or 35 ml/l juice in the case of passion fruit juice. Soft drinks prepared from water and flavourings may contain fruit juice, fruit pulp and orange peel extract. In addition, the standards specify which ingredients may be used and maximum levels.

Labelling requirements are given so that the consumer can ascertain the nature of the product being bought, and not be misled or confused by illustrations implying that the product is of a different composition than it really is. Therefore pictorial representations, designs, expressions or illustrations of fruit or fruit juice that imply the presence of fruit in the drink may not be used if the product is a fruit-flavoured drink or other flavoured soft drink and contains no fruit juice. Products can only be described as 'pure' if they are single-ingredient foods and if they contain no additives.

Fruit juice cordials are not classified as soft drinks but are included with fruit juice syrups and fruit juice toppings. Fruit juice cordials are products prepared from fruit juice or comminuted fruits or both, water and sugars.

New Zealand. The pattern for New Zealand is similar to that of Australia in that there are detailed standards of composition for many products published in one consolidated text, which also includes labelling requirements and general provisions. Products described as fruit drink must contain a minimum of 5% fruit juice; in addition, the declaration 'contains not less than (%) of (name of fruit) juice' or, where more than one fruit is used, the percentage of each fruit juice in descending order of the proportions present, must be stated. This indicates clearly to the consumer the percentage of fruit juice present in the product. Other soft drinks are fruit-flavoured drinks (containing fruit extracts, possibly with cereals or vegetables, and water) and artificial drinks, which need only be flavoured. Use of claims on labels is as commonplace as in

Europe; fruit drinks, fruit-flavoured drinks and artificial drinks may not be described as 'pure' or 'real' or by any other word of similar connotation. In the case of artificial drinks, there must be no expressions or other devices that suggest or imply the presence of fruit or fruit juice in the drink. Syrups or cordials are covered by separate standards, which specify composition and labelling requirements. Fruit and other cordials may not be described as 'pure' or 'real'.

New Zealand and Australia are endeavouring to harmonise their food regulations and the health ministers have supported the development of a joint Australasian food regulatory system. Harmonisation of the New Zealand and Australian regulations would simplify the situation for a manufacturer wishing to export similar products to both markets.

Latin America. Most countries in Latin America have comprehensive legislation covering composition, additives and labelling requirements for carbonated and non-carbonated beverages. The Argentinian Food Code includes requirements for different types of soft drinks, soft drinks concentrates and powders and syrups for soft drinks; in addition, requirements for dietetic soft drinks are laid down. Brazilian food legislation includes general standards of quality and identity for different types of carbonated and non-carbonated soft drinks and specific provisions for dietetic soft drinks. The General Health Regulations in Mexico lay down requirements for soft drinks and powdered preparations, syrups and concentrates for soft drinks. In the countries of the Central American Common Market, the provisions of ICAITI standard 34154/1986 on soft drinks are applicable.

Specific legislation on soft drinks is also laid down in other countries of Latin America, including Chile, Dominican Republic, Ecuador, Uruguay and Venezuela. In some countries where specific national regulations on soft drinks are not laid down, for example in Panama, the authorities recommend following the provisions of Codex Alimentarius.

Middle East. A number of Middle Eastern countries have mandatory standards for fruit drinks, including Saudi Arabia, Jordan, Iraq and Israel.

Although a Gulf Standard for carbonated non-alcoholic beverages has been issued, there is currently no similar standard for non-carbonated products. In the absence of national provisions, the Gulf States tend to accept products complying with Saudi Arabian rules, or with Codex Alimentarius Standards.

The Saudi Arabian standard for orange drink requires a juice content of 10–50% w/w and minimum 11.5% w/w total soluble solids. In the case of all orange drinks, the percentage of juice or purée used must be declared in a prominent position on the label. If the words 'orange juice' appear in the product name, the product must contain not less than 50% w/w orange juice.

In Jordan there are standards for natural drinks and fruit squashes.

Natural drinks are defined as products based on natural fruit juices; a declaration of the juice content must be included in the labelling of these products. The Jordanian authorities strictly monitor products in their market and the use of all additives is under their direct control.

Israel has detailed standards of composition for non-carbonated soft drinks based on fruits of citrus and other fruit varieties. Certain artificial sweeteners are permitted to specified levels in low-calorie products but their addition triggers specific labelling requirements. Citrus-based products should contain a minimum of 15% citrus ingredients (minimum 10% for lemon soft drinks).

There are certain common prohibitions in the countries of the region that are relevant to soft drink manufacture. Most countries either ban the use of artificial sweeteners completely (e.g. Abu Dhabi) or restrict their use to dietetic products only (e.g. Saudi Arabia). Also, the Muslim prohibition on alcohol consumption is significant; many countries require products to be totally free from ethanol and, in more fundamentalist countries, such as Saudi Arabia, even the use of ethanol in a solvent function for food flavourings is prohibited.

Many countries restrict the proportion of shelf life that may have elapsed before a product is imported. Saudi Arabia, Bahrain, United Arab Emirates and a number of other countries have set mandatory shelf life periods that should be used in the calculation of expiry dates. One Gulf Standard on this subject has already been published and a number of others are being developed.

In order to obtain permission for export to the Middle East, it is often necessary to contact the appropriate authorities in the country concerned or to export through an agent in the country. When labelling products, it is important not to use any illustrations or symbols that would cause offence to the Muslim religion, for example a cross or Star of David.

Far East. A number of Far Eastern countries have legislation covering composition, additives and labelling of soft drinks. Export of soft drinks to this region involves checking the extent of the legislation in each of the countries. No two Far Eastern countries have the same requirements; the regulations differ in each case.

Japanese legislation includes standards of manufacture for soft drinks including purity specifications with regard to contaminants, heavy metals, storage conditions and appearance. These are not standards of composition in the sense that ingredients are specified, and no minimum fruit juice or sugar levels are specified. However, a quality control system is in operation by means of the Japanese Agricultural Standard for Fruit Drinks, which includes specifications to enable the use of the JAS symbol for fruit juice drinks, fruit pulp drinks and other soft drinks containing fruit juices. Japan has a positive list of additives but this only lists synthetic substances. It is a general understanding that additives considered as 'natural' are generally permitted in foods, although this is not actually stated in the regulations; it is likely that these additives would be acceptable for use in soft drinks.

Outside Japan, where countries have standards of composition for foods, soft drinks are usually included. Malaysia, Singapore, India, Indonesia and Bangladesh all have standards on various types of squashes and non-carbonated beverages. The Malaysian regulations are similar in detail and format to those of Australia and New Zealand. Where there are no or limited standards of composition, for example Taiwan and Thailand, specific additives regulations list those additives that may be used in soft drinks and assign permitted maximum levels as appropriate.

14.2.6 Fruit drink standards produced by Codex Alimentarius

No Codex standards have been published on soft drinks and non-carbonated beverages; hence, for countries that tend to follow Codex recommendations it is necessary to formulate their own standards. However, a number of aspects relating to food manufacture in general are covered by the Codex; for example food hygiene, an important aspect of soft drink manufacture. Codes of hygienic practice have been published in a number of areas where it is deemed necessary owing to particular problems; this is not the case for beverages. The Code of Practice relating to general principles of food hygiene concerns all aspects of the food industry (Codex Alimentarius, 1985).

References

Agreement on the European Economic Area (EEA) (1994) Protocol adjusting the Agreement on the European Economic Area and certain Decisions and Acts relating thereto. *Off. J. European Communities* L1 (3/1/94), 1–606.

Codex Alimentarius (1985) *Recommended International Code of Practice General Principles of Food Hygiene*, CAC/RCP 1-1969, Rev. 2.

Codex Alimentarius (1992) *Fruit Juices and Related Products*, Vol. 6 second edition.

Commission Directive (1993) 93/45/EEC of 17 June 1993 concerning the manufacture of nectars without the addition of sugars or honey. *Off. J. European Communities* L159 (1/7/93), 133.

Council Directive (Labelling) (1978) (18 December) on the approximation of the laws of the member states relating to the labelling, presentation and advertising of foodstuffs, 79/112/EEC. *Off. J. European Communities* L33 (8/2/79), 1–14, as amended.

Council Directive (1990) (24 September) on nutrition labelling for foodstuffs, 90/496/EEC. *Off. J. European Communities* L276 (6/10/90), 40–4.

Council Directive (1993) 93/77/EEC of 21 September 1993 relating to fruit juices and certain similar products. *Off. J. European Communities* L244 (30/9/93), 23–31.

Draft proposal (1993) for a Council Directive on the use of claims relating to foodstuffs, document SPA/62/Rev. 3 Orig. FR, 15/7/93.

European Commission (1985) *Completing the Internal Market*, White Paper from the European Commission to the European Council, June 1985.

MAFF (1993) Draft guidelines for the use of certain nutrition claims in food labelling and advertising.

Proposal for a Council Directive (1992) amending Directive 79/112/EEC on the approximation of the laws of the member states relating to the labelling, presentation and advertising of foodstuffs, COM(91) 536 final - SYN 380, *Off. J. European Communities* C122 (14/5/92), 12–4.

UK Food Advisory Committee (1989) Recommended conditions for making particular nutrition claims.

UK Food Advisory Committee (1993) Recommended conditions for the use of the term 'natural' in food labelling and advertising.

UK Food Labelling Regulations (1984) SI 1984 No. 1305, as amended, HMSO, London.

UK Food Labelling (Amendment) Regulations (1994) SI 1994 No. 804, HMSO, London.

UK Food Safety Act (1990) Elizabeth II, Chapter 16, HMSO, London.

UK Fruit Juices and Fruit Nectars Regulations (1977) SI 1977 No. 927, as amended, HMSO, London.

UK Soft Drinks Regulations (1964) SI 1964 No. 760, as amended, HMSO, London.

15 Water and effluent treatment in juice processing

I. PATERSON and P. J. COOKE

15.1 Water treatment

15.1.1 Introduction

Water is colourless, odourless, tasteless and is essential for the continued existence of the human race. The water that is available in the world is continually recycled by means of the hydrological cycle. During this process water from the sea, rivers and lakes is evaporated by the sun and forms clouds in the atmosphere. The changing atmospheric conditions then cause the water to descend to the earth in the form of rain, sleet, hail or snow. Initially, the water released from the clouds is very pure, but as it falls it picks up impurities such as carbon dioxide, dust, smoke, nitrogen, oxygen and other atmospheric gases. When it reaches the earth, the water collects small amounts of soil particles, inorganic and organic matter, bacteria, algae and other foreign materials, dependent on the environment in which it falls. Ordinary water, therefore, contains various impurities in variable amounts. These impurities will, to a greater or lesser degree, adversely affect the flavour of the water.

15.1.2 Water quality for juices and beverages

In non-carbonated fruit juices and fruit beverages, the main ingredient is water. Although mains water from the public supply is satisfactory for potable uses, it will only partially meet the specifications of the soft drinks manufacturer, and will almost certainly require further treatment in order to meet the necessary specifications.

Table 15.1 Current UK legislation for drinking water: microbiological standards

Parameters	Max. number of organisms per sample	Volume of sample (ml)
Total coliforms	0	100
Faecal coliforms	0	100
Faecal streptococci	0	100
Sulphite-reducing clostridia	0	20
For samples taken at the tap:		
Total bacteria counts at 37°C	10	1
at 22°C	100	1

Table 15.2 Current UK legislation for drinking water: chemical standards

Parameters	Units of measurement	Maximum concentration
Arsenic	$\mu g\,As/l$	50
Cadmium	$\mu g\,Cd/l$	5
Cyanide	$\mu g\,Cn/l$	50
Chromium	$\mu g\,Cr/l$	50
Mercury	$\mu g\,Hg/l$	1
Nickel	$\mu g\,Ni/l$	50
Lead	$\mu g\,Pb/l$	50
Antimony	$\mu g\,Sb/l$	10
Selenium	$\mu g\,Se/l$	10
Pesticides and related products:		
(a) individual substances	$\mu g/l$	0.1
(b) total substances	$\mu g/l$	0.5
Polycyclic aromatic hydrocarbons	$\mu g/l$	0.2
Chloride	$mg\,Cl/l$	400
Calcium	$mg\,Ca/l$	250
Boron	$\mu g\,B/l$	2000
Barium	$\mu g\,Ba/l$	1000
Total hardness	$mg\,Ca/l$	60
Alkalinity	$mg\,HCO_3$	30

Table 15.1 summarises the current UK legislation for drinking water supplies, as far as microbiological standards are concerned. Furthermore, water intended for human consumption in the UK must not contain pathogenic organisms such as *Salmonella*, pathogenic staphylococci, faecal bacteriophages or enteroviruses.

Table 15.2 summarises the current UK legislation for drinking water supplies, as far as chemical constituents are concerned.

Although other countries may well have different standards, the UK standards serve as a good guideline. The water authorities in the UK are bound by law to supply water that is 'pure and wholesome', but that does not mean that it is clear and colourless, and fit for use in the juice and beverage industry. Water from the public supply will, therefore, vary in quality from time to time. Water from private wells or boreholes, will vary to a greater degree, since it will not have undergone any treatment at all. This variability creates the need for further treatment, in order to produce a water suitable for the manufacture of juices and beverages of consistent quality.

15.1.2.1 Assessment of water quality. The main parameters to be evaluated in the assessment of water quality are as follows:

Appearance and taste. Good quality water is clear, colourless, tasteless and free from turbidity and insoluble solids. Poor quality water may be yellow or slightly brown, or slightly turbid with a deposit which may contain inorganic hydroxides and silicates, small biological organisms and organic debris.

pH.   Most water supplies have a pH value of between 6.5 and 8.5. Any exceptions that occur may be due to the result of particular chemical composition, e.g. water containing free carbon dioxide will have a pH lower than 6.5 and water containing carbonates and bicarbonates will have a pH higher than 8.5.

Total dissolved solids.   When good quality water is evaporated and dried at 180°C, any residue found should not exceed 500 mg/litre.

Total hardness.   Hardness is due to the presence of calcium and magnesium salts in solution. Temporary hardness is due to calcium and magnesium bicarbonates and permanent hardness to calcium and magnesium chlorides, sulphates and nitrates. Total hardness is expressed as the sum of temporary and permanent hardness (Table 15.3).

Water to be used in the manufacture of non-carbonated fruit juices and beverages should either be soft or moderately soft in character.

Table 15.3 Classification of hardness in water

Classification	$CaCO_3$ (mg/l)
Soft	< 50
Moderately soft	50–100
Hard	100–200
Very hard	200–300

Alkalinity.   The alkalinity of water is due mainly to the presence of bicarbonates, carbonates and hydroxides of calcium, magnesium, sodium and potassium. If the water has a high alkalinity, the acidity of the beverage is neutralised, and this can cause a bland taste. Most soft drink manufacturers reduce the alkalinity of the water to below 50 mg/l as $CaCO_3$.

Nitrogenous substances.   Sewage and decaying vegetable matter are significant pollutants in the water system. Both of these pollutants contain ammonia in different amounts, thus, a measurement of the free ammonia content of the water can indicate how badly the water has been polluted. Soil bacteria produce nitrites as they degrade nitrogenous material. Nitrites are also found in sewage and manure, and should, therefore, be absent from potable waters. Fertilisers and decaying animal matter can cause nitrates to be present in water. Nitrates should be kept to a minimum in potable waters.

Chloride.   High levels of chloride in water, i.e. 100 mg/l, may be an indication of pollution by sewage, or contamination by sea water.

Organic content.   Various acids, e.g. humic acid and fulvic acid, polysaccharides and microbial contaminants, can be present in water in very small amounts. The organic content of water is usually determined by measuring the permanganate value (PV), which is a measure of the oxidisable organic substances, or by measuring the total organic carbon value (TOC), which is a measure of both oxidisable and non-oxidisable organic substances. In soft drinks manufacture, the water to be used should have a PV level of less than 1 mg/l as oxygen, or a TOC value of less than 2 mg/l as carbon.

Microorganisms.   Microorganisms of all types can be found in water from all natural sources. These must be removed from the water or controlled prior to use. High bacterial counts indicate polluted water, and the presence of coliform bacteria indicate faecal contamination.

Phosphates and silicates.   The phosphate (PO_4) of water is representative of the residues of the decomposition of sewage or vegetation, and should not be present in good quality water at levels greater than 0.05 mg PO_4/l. Water that has percolated through sand, rock and clay, picks up silicates as it does so. Levels are normally between 5 and 20 mg/l as silica (SiO_2). Levels higher than this can cause floc problems in soft drinks.

Trace metals.   Water that is polluted by industrial effluents may contain small amounts of various metals, e.g. zinc, copper, iron, cadmium and lead. Although they may not be toxic at the levels found, they can cause problems for the soft drink manufacturer, e.g. brown deposits.

Chlorinated substances.   Water from the mains supply is generally treated with chlorine prior to entering the supply system, as a means of maintaining sterility. Ammonia or ammonia compounds in the water, will react with any free chlorine and form chloramines. Chloramines give water a medicated taste. In the manufacture of soft drinks, the total chlorine must be reduced to as low a level as possible, i.e. 0.05 mg/l Cl.

Pollution indicators.   It is totally impractical to check water for every potential pollutant. However, the presence of certain parameters can give an indication of the type and sources of pollution present. Table 15.4 summarises these.

Table 15.4 Indicators for pollutants in water

Indicator	Pollutant
Coliform bacteria especially *E. Coli*	Intestinal pathogens due to faecal contamination
Yellow/brown colour	Inorganic chemicals, organic debris
pH lower than 6.5	Free CO_2
pH higher than 8.5	carbonates and bicarbonates
Free ammonia	Sewage and/or decaying vegetable matter
Nitrites	Soil bacteria, sewage or manure
Nitrates	Fertilisers, decaying animal matter
High levels of chloride e.g. 100 mg/l	Sewage or sea water
Trace metals	Industrial effluent

15.1.3   Treatment methods

As mentioned above, raw water can contain a variety of unwanted materials in various concentrations. These materials must be removed or reduced to an acceptably low level before the water can be used in the beverage industry. There are a number of treatment methods that can be used. The choice of method or combination of methods to be used, is dependent upon the source of raw water and the treatment it has already undergone. Further treatment is necessary in order to:

   (i) remove all colour and suspended particles,
  (ii) remove all substances in the water that are likely to have an adverse effect on the appearance, taste, odour and stability of the finished product,
 (iii) adjust the pH to a pre-determined level,
 (iv) ensure that all microorganisms and bacteria are killed and removed,
  (v) provide water of a consistent quality throughout the year.

15.1.3.1   Chemical methods.

Coagulation and flocculation. This method involves the removal of insoluble and semi-soluble impurities in a reaction vessel by the addition of chemicals to the incoming raw water. These impurities are the main causes of colour and off tastes being present in the water.

The suspended particles are destabilised by chemical coagulation, and the particles agglomerate to form flocs which precipitate to the bottom of the reaction vessel. This coagulation is normally achieved by the addition of aluminium or iron salts, usually aluminium sulphate or iron sulphate. When these salts are added to water they produce highly charged hydrolysed ions, e.g.

$$Al_2(SO_4)_3 + 6H_2O \rightarrow 2Al(OH)_3 + 6H^+ + 3SO_4^{2-}$$

The agglomeration of the particles in this process is referred to as flocculation, although in practice, the whole process is referred to as coagulation. The various mechanisms involved in this process are complex, and depend on the nature of the raw water, the type of insoluble impurities, the temperature and pH.

The most widely used coagulant is aluminium sulphate, as it is very effective in the treatment of coloured water. Recently however, the aluminium content in water and foods generally has been linked to Alzheimer's disease, and this has caused increased use of other coagulants, e.g. ferric sulphate and alternative treatment methods. The pH of water in this reaction must be controlled so that the impurities remain insoluble in the floc and can be removed. If aluminium is used the pH must be within the range pH 5.5–7.0. If iron is used then the range must be pH 9–10. Normally the dosage rate is between 2 and 5 mg of iron or aluminium per litre of water. A simple coagulation plant is shown in Figure 15.1. The raw water and the coagulating chemicals are pumped into the reaction chamber (1). These chemicals must be mixed rapidly and thoroughly with the water to ensure maximum contact with the insoluble impurities. Coagulation takes place in section (2) and flocculation begins. Floc particles are then formed in section (3), and as they increase in density, they settle out and form a bed in section (4). The blanket formed in section (4) must be monitored carefully, and drawn off as necessary. All water must then flow through this 'blanket zone' as it is called, on its way to the top of the vessel. The flocculation takes place in an inverted cone-shaped area.

This means, that as the water flows upwards, its velocity continually decreases. This helps the flocculated particles to settle at the bottom of the tank. As the water nears the top, it is moving so slowly that the suspended flocs cannot be taken to the top. The water above this area is reasonably clear,

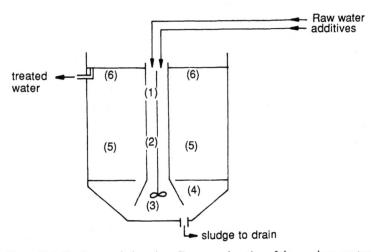

Figure 15.1 Simple coagulation plant. For an explanation of the numbers, see text.

section (5). Clear water, at the end of the reaction is drawn off in section (6).

In this process, various chemical reactions take place. We have given below those that take place when the coagulating chemicals are aluminium sulphate and ferrous sulphate.

Aluminium sulphate.   Aluminium sulphate is best used when the raw water is coloured or has a high level of semi-soluble impurities giving high turbidity.

$$Al_2(SO_4)_3 . 14H_2O + 3Ca(OH)_2 = 3CaSO_4 + 2Al(OH)_3 + 14H_2O$$

| alum | lime | calcium sulphate | aluminium hydroxide | |

If no lime is added, then the acidity of the system will increase

$$Al_2(SO_4)_3 + 6H_2O = 2Al(OH)_3 + 3H_2SO_4$$

| alum | | aluminium hydroxide floc | sulphuric acid |

This reaction must therefore be controlled so that the pH is kept between 5.5 and 7.0. Aluminium will dissolve if the pH is lower than 5.5 or above 7.0. This will then cause a heavy deposit in the final product, and difficulties with filling.

Ferrous sulphate.   In this process, several reactions occur. Lime reacts with the bicarbonates to produce insoluble carbonates and hydroxides. Ferrous sulphate reacts with the lime to produce ferrous hydroxide which is soluble. This in turn reacts with chlorine to produce ferric hydroxide. Ferric hydroxide is insoluble, forms a floc and the insoluble carbonates and hydroxides are then taken out of solution.

$$Ca(HCO_3)_2 + Ca(OH)_2 = 2CaCO_3 + 2H_2O$$

calcium bicarbonate (soluble)　　calcium hydroxide (lime)　　calcium carbonate (insoluble)

$$Mg(HCO_3)_2 + Ca(OH)_2 = 2CaCO_3 + Mg(OH)_2 + 2H_2O$$

magnesium bicarbonate (soluble)　　calcium hydroxide (lime)　　calcium carbonate (insoluble)　　magnesium hydroxide (insoluble)

$$FeSO_4 + Ca(OH)_2 = Fe(OH)_2 + CaSO_4$$

ferrous sulphate　　lime　　ferrous hydroxide (soluble)　　calcium sulphate

$$Fe(OH)_2 \xrightarrow[\text{chlorine}]{\text{oxidised by}} Fe_2O_3 . nH_2O$$

ferric hydroxide (floc)

De-alkalisation.   In this process, acidic cation exchange resins are used to reduce the alkalinity of the water. This is achieved by the resin taking up the calcium and magnesium responsible for the alkaline hardness and producing water and carbon dioxide. The water produced will have very low dissolved solids, and no alkalinity. This means, that in order to have an alkalinity of 50 mg/l, the treated water must be blended with an amount of raw water to raise the alkalinity to this level. The final water from this process must then be filtered before use.

15.1.3.2   Physical methods.

Filtration.   Coagulation and flocculation will not remove all the insoluble solids. The remaining solids will be removed by passing the water through a sand filter. Such a filter is basically a fine layer of sand on the top of various layers of gravel, each layer increasing in coarseness towards the bottom of the tank.

The water that has undergone coagulation and filtration is dispersed evenly over the surface of the sand. As the water flows down through the various layers in the filter vessel, more and more insolubles are removed. The filtered water is then removed from the bottom of the tank.

Taste and odour control.   Although some treatment methods can significantly reduce unwanted tastes and odours, the most effective way to do this is to pass the water through an activated carbon filter. Activated carbon is produced by burning coal or coconut shells to produce a very porous material. This material has a high absorptive capacity, due to its large internal surface area. As the water passes through the filter, impurities are either absorbed at the activated sites or are catalysed by the carbon to form chemicals that are more easily removed or controlled.

Membrane filtration.   A number of filtration methods exist, each with a limit as to the type/size of particles they can remove. In standard filtration systems, the liquid is passed through at right angles to the filter and the unwanted particles are retained by the filter as the liquid passes through. These systems do not, however, remove very small particles or unwanted dissolved solids.

These very small particles can be removed by membrane filtration techniques. In these systems, the liquid is under pressure and flows parallel to the membranes. A proportion of the liquid passes through the membrane leaving behind the unwanted particles. The continuous flow means that the membrane filter produces two liquids, a filtered product and a liquid containing a high concentration of unwanted particles. Membrane filtration can be described by various names, dependent upon the size of the particles the system can remove.

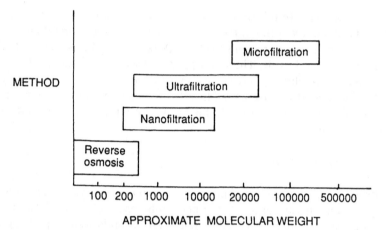

Figure 15.2 Molecular weights of particles removed from water by various membrane filtration techniques.

There are four main types of membrane filtration:

1. *Microfiltration.* This method can remove suspended solids, bacteria and precipitated materials, generally between 0.1 and 1.0 μm. Dissolved solids will pass through this system.
2. *Ultrafiltration.* This system will remove large molecules, e.g. proteins and some microbiological contamination up to 0.1 μm. It will not remove dissolved salts or very small molecules with a molecular weight less than 1000.
3. *Nanofiltration.* As the name suggests, nanofiltration removes particles

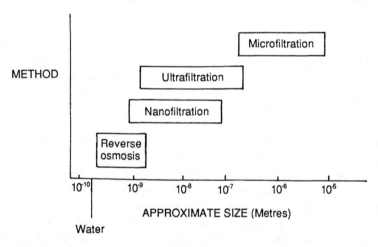

Figure 15.3 Sizes of particles removed from water by various membrane filtration techniques.

greater than 1 nm, i.e. 10 Å (1×10^{-9} mm). Hardness and colour are removed using this method, and the total dissolved solids are greatly reduced. Molecules with a molecular weight less than 200 cannot be removed by this system.

4. *Reverse osmosis.* At present this is the best system available. The membranes used in this system can remove all dissolved salts, including inorganic and organic molecules with a molecular weight greater than about 100. Water molecules pass through the membranes freely. Using this method, virtually pure water can be produced.

The sizes and molecular weights of particles that can be removed by the various membrane filtration techniques are summarised in Figures 15.2 and 15.3.

15.1.3.3 *Methods for microbiological control/sterilisation.*

Chlorination.   Chlorine is the most generally used disinfectant and sterilant for water supplies. Its effectiveness is dependent on various factors, pH, temperature, concentration and contact time. Chlorine reacts with various substances in the water, e.g. ammonia, nitrites and organic water. There must be enough chlorine added to ensure that these reactions are completed and that there is enough free chlorine for bactericidal action.

In order to achieve the best effect, chlorination chemicals should be added immediately after the treatment process (if aluminium sulphate has been used) or added to the coagulation tank (if ferrous sulphate has been used).

If chlorine is added directly to raw water containing humic organic matter, trihalomethanes (THMs) are produced. THMs should be absent from treated water, as their presence in quantities greater than 100 mg/l can cause unpleasant taints. THMs have also been suspected as being carcinogens. Used at the correct levels, chlorine will remove tastes and odours by oxidising the impurities.

When chlorine is added to water, it produces hydrochloric and hypochlorous acids.

$$Cl_2 + H_2O = HCl + HOCl$$

Hypochlorous acid dissociates to give the hypochlorite ion.

$$HOCl = H^+ + OCl^-$$

At pH 10, all available chlorine is present as the OCl^- form. All the available chlorine at pH 5.5 is present as active HOCl. Chlorine does not react at pH 4, and so at this level or lower, it is present in its molecular form, Cl_2. Active HOCl and molecular chlorine are much more efficient disinfectants than the hypochlorite ion, and so disinfection with chlorine is much more efficient at low pH.

However, when canned products are cooled with water, chlorinated water *must not* be used. Using chlorinated water in this way would cause the hydrochloric acid formed to react with the cans, causing pitting, and the cans would eventually leak.

Ultraviolet light.   This process involves the use of high intensity ultraviolet (UV) light to inactivate microorganisms by destroying their nucleic acids. A wide range of organisms is affected by UV radiation. These include protozoa, bacteria, yeasts, moulds, viruses and fungi, and are inactivated in a relatively short time. This process stops the microorganisms reproducing, and at higher doses could kill them. The wavelength of the light to be used is chosen such that it will eliminate bacteria and other organic matter, usually around 254 nm.

Poor quality water will reduce the effectiveness of the UV radiation as soluble solids will absorb it, and insoluble solids will scatter it. Best results are obtained if the feed water to the unit has been clarified. The ultraviolet units should be installed close to the point of use.

Ozone treatment.   Ozone gas is a powerful disinfectant, but can be used to remove colour, control taste and precipitate metal ions. It is produced by creating a silent discharge across a current of oxygen or dry air. The ozonised air is then dispersed in the water in a suitable tank by porous diffusers, water spraying or a venturi arrangement. Any ozone that is not used must be destroyed, since it is very toxic (the maximum amount permitted in a working environment is 0.1 ppm in air). Ozone is destroyed by passing it through a furnace at 300°C, or by reaction with metal oxides. Any plant using this method must, therefore, install monitors to measure the amount of ozone in the air. This method of disinfection is, therefore, much more expensive than chlorination. Ozone is difficult to disperse in the water as it is only slightly soluble. Nevertheless, it has the advantages of being unaffected by pH and removes unwanted tastes very quickly by oxidising the organic matter present.

Numerous combinations of the above methods of water treatment can be used. The methods chosen will depend on the quality of the raw water available, and the standard and quality of water required.

15.2   Effluent management and treatment

15.2.1   Introduction

The answer to many effluent treatment and disposal problems has as much to do with management as technology or engineering. In particular, those aspects of management concerned with planning and performance measurement are of special significance. Within a single chapter it is impossible to discuss the technology and engineering in detail; numerous texts are available (Metcalfe and Eddy, 1991; Horan, 1990; Tebbutt, 1992).

No two effluents are the same so the technology required and its application will vary, even for sites within the same process industry. For this reason, no information is given here on average or expected effluent strengths and volumes; this would merely be repeating what is already widely available. It is important to quantify the effluent under consideration by positive measurement, however for those who wish to effect comparisons perhaps for benchmarking purposes, the publication by Barnes *et al.* (1984) is still a good source of information.

This chapter describes procedures that can be used when dealing with effluent issues, and should enable a decision to be made as to whether treatment is really needed at all. They also provide a basis for ensuring that any treatment processes are designed on proper quantitative criteria.

A major trend in recent years has been the move towards quality management systems but sometimes only the narrower aspects of manufacturing are considered. The wider aspects of a manufacturing operation, including the utility services, are often ignored with no single point of accountability for performance or cost. For example, the production departments that produce effluent may not be responsible for its disposal, hence there is little or no incentive to give it any attention. It is only the recent drastic increases in disposal costs, and to some extent increased environmental awareness, that are at last forcing companies to start the process of managing effluent.

Traditionally many companies have discharged effluent with little further thought. Treatment is only carried out if the company is forced to do so by the need to meet discharge conditions. Isolating effluent from other factory operations in this way can be the first step on the road to high discharge costs, treatment plant failures and even prosecution. An integrated process of managing effluent must begin before even a single product is made as only by considering the factory layout and the manufacturing processes can a start be made to control the way in which effluent is produced. Treatment and discharge processes and their costs can then be controlled.

The traditional approach referred to above may be compared with an output orientated production process, that is one in which there is little or no control over the inputs to the process. A result of such an operation would be a process out of control where the management cannot plan and only reactive controls can be applied. In effluent disposal terms the result is the so called 'end-of-pipe' solution where what flows out of the drain is simply accepted and dealt with as resources and regulations permit.

Manufacturing managers will generally accept that to achieve a product of given specification an input orientated process must be used where the quality of the product is to some extent maintained by controlling the inputs to the process. In principle, effluent treatment processes are no different from manufacturing processes. There is an input, the untreated effluent, and this is treated or converted in one or more ways to products, the treated effluent, plus any residues. As with the manufactured product of the factory, quality specifications can be applied to the product, i.e. the treated effluent. Applying the same logic as above it follows that if there is no control over the input to an effluent treatment process the output will be variable and may not meet the required quality criteria. This is probably the major reason why industrial effluent treatment processes so commonly fail to meet their discharge consent requirements. With insufficient control over what is discharged to drain it will be difficult, if not impossible, to develop design criteria for any treatment process.

The usual solution is to conduct a detailed analysis of the manufacturing process. This should begin by drawing up an accurate mass balance of inputs and outputs. A 1% error may be small in manufacturing terms but could be equivalent in effluent treatment terms to releasing the untreated sewage load from a small to moderately sized town. Points at which process materials can be lost to drain should be identified and losses fully quantified. For an existing process these losses can be measured directly but it will be necessary to make estimates where a new process is being planned. Usually, it will be possible to use information from similar operations elsewhere.

15.2.2   Measures of effluent strength

The mass load of substances in the effluent stream, needs to be measured as one or other of the parameters used to quantify effluent. Effluent strength is generally assessed by one of two tests both of which provide measures of the amount of oxygen required to oxidise the organic substances present.

The biochemical oxygen demand (BOD) measures the amount of oxygen used over a specified time period by microorganisms inoculated into a sample of effluent and using the organic substances within it as substrate. The amount of oxygen used, which is called the oxygen demand, is proportional to the quantity of biochemically oxidisable substrate in the effluent.

The other principal test is that of chemical oxygen demand (COD). The sample is boiled with a strong acid solution of potassium dichromate which oxidises the sample. The extent of oxidation, which is proportional to the quantity of oxidisable substances in the effluent, is usually measured by back titration of excess dichromate.

These two tests do not measure specific substances in the effluent. They provide a measure of the oxygen needed for oxidation of a broad range of substances and hence the load that an effluent will place upon a treatment plant or river. The COD will always give a higher result than the standard five-day BOD since bacteria will not generally complete the oxidative processes in this time period. The COD test will also measure substances that may not be oxidised by bacteria under the conditions of the BOD test.

Generally, for the types of effluent generated in the juice and beverage industry, which are usually carbohydrate based, BOD can be expected to reach 50 to 70% of the COD. It should be appreciated that measurement errors on a BOD test can be of the order of 20%; this is one of the reasons why COD is more frequently used–its accuracy should normally be around 5% or better.

Other analytical techniques that may be applied to evaluate effluents include determinations of suspended solids and fats, oils and greases. The test for oils and fats is rather non-specific and really only measures the substances that can be extracted from a sample by shaking it with a solvent, such as petroleum ether. The test may be referred to as for 'fats, oils and grease', 'petroleum ether extract', 'gravimetric oil' or 'extractable non-volatile matter'.

Standardised methods should be used for all analyses such as those published by the Water Research Centre. There are also a number of books available giving summaries or detailed methods of analysis (Dean Adams, 1990).

15.2.3   Mass load determination

If the quantity of a process material from a manufacturing operation lost to drain is measured and its COD is known, then the COD mass load can be calculated. For example, suppose 100 l of concentrate are lost from the bottom of a blending tank when it is washed and the concentrate is found to have a COD of 700 000 mg/l. Multiplying these two figures and converting to kilograms it will be seen that the total COD load will be 70 kg.

By measuring the COD of all process materials it should then be possible to generate a basic but effective model of effluent production for any manufacturing process.

15.2.4   Volume assessment

Volume is more difficult to assess than strength as it is very much a function of operator practice except where automated cleaning routines are used. For an existing process, estimates can be made from tank volumes, hosepipe flow rates and time in use etc. though these techniques have a tendency to underestimate the volume. For a planned new process an estimate should be made which is based on the most likely methods of cleaning. Reference should also be made to similar operations elsewhere. It is worthwhile metering as much water as possible; the cost is relatively low in most cases and the information generated is a valuable management tool for encouraging accountability.

15.2.5   The total picture

We have considered losses from the process itself, but in most factories there will be other contributions to effluent some of which could match losses from the main processing operation. Trade effluent is usually defined as 'any liquid either with or without particles of matter in suspension therein, which is wholly or in part produced in the course of any trade or industry carried on at trade premises'. Some of the more obvious contributions that fall within this legal definition of trade effluent are as follows.

15.2.5.1 *Cooling water.*   This will generally comprise a bleed from a recirculating system though there are still some wasteful 'once through' systems in existence. The output from these systems is an effluent, even if there is no contamination (see legal definition of trade effluent above). Occasionally, a system may be dumped for thorough cleaning; this is becoming a more common event in view of the preventative measures now required to be taken against *Legionellosis*. Cooling waters should be low in terms of COD unless there is significant contamination with process materials; the volume will depend on the design of the particular system. Both parameters must be quantified with consideration also given to occasional events such as a system clean. The effect of dispersants and biocides on effluent treatment processes must be assessed though it is not normally significant.

15.2.5.2 *Movement of process materials.*   Deliveries of process materials to site may result in the need to wash road tankers. This practice should be avoided if at all possible. Empty drums may need to be washed before return or disposal. Substantial volumes and/or mass loads can arise and must be quantified. Procedures should exist for dealing with unplanned washdown and spillage that could arise during delivery or despatch, to avoid treatment plant failures, consent contraventions and prosecutions. Spillage must be quantified before washing down and this should be avoided entirely if the spillage can be isolated and pumped into containers for safe disposal. Movements of materials around a site give rise to similar potential problems.

15.2.5.3 *Contaminated storm water.*   Runoff from rain which falls on a site may become contaminated by spilt process materials, e.g. in drum storage and delivery areas etc. It is not uncommon for runoff from such areas to have COD concentrations of several hundred milligrams per litre or more and this cannot legally be discharged to a storm water drainage system. There are numerous prosecutions on record for infringements of pollution control legislation caused by such events. Consequently, it is necessary either to prevent the contamination from occurring, or if it is unavoidable, the drainage from these areas should be linked into the process effluent system. The volume and load contribution must again be quantified as part of the overall assessment of process effluent.

15.2.5.4 *Steam raising.*   This will not usually be a significant input either volumetrically or in terms of COD. The only input should be from boiler blowdown which is usually controlled from an energy efficiency point of view. However in certain cases, particularly on older plants, steam can be contaminated with process materials and if this is significant it will not be possible to return condensate to the boiler. This could give rise to hot water being dumped into the effluent system.

15.2.5.5 *Vehicle washing.* The washing of vehicles, even if it is only exterior washing, technically produces a trade effluent. Although weaker than process effluents many authorities will object to its presence in storm water systems and it may therefore be necessary to include it in the process effluent drainage system. Drainage from vehicle servicing, fuelling and parking areas may also become contaminated as a result of oil/fuel spillage or leakage and this will need to be assessed if these facilities are present on a site.

15.2.6 Drainage

Two things should be apparent from the above discussion: (i) that a variety of wastewaters may be produced from even a fairly simple processing operation and (ii) that as far as possible, rainwater runoff should be kept clean. The way in which the site drainage system is designed will have a major effect on the economics of effluent management, treatment and disposal. The worst possible situation is where all wastewater (process, foul and storm) enters a single combined drain. Unfortunately for many older sites this is the situation that will be inherited and it is invariably a costly exercise to put right.

Too many newer sites have compromised nominally separate drainage systems by making connections to the nearest available drain in short sighted attempts to save money. This practice is totally contrary to the principle of getting it right first time and it will certainly cost much more to re-engineer the drainage system, perhaps under threat of prosecution, than to do the job properly in the first place.

A modern processing plant built with longer term environmental and economic considerations in mind may need a number of separate drainage systems. Consider process effluent first. If a treatment plant is to be provided it does not follow that all the effluent needs to be treated before discharge to foul sewer. Weaker streams, such as cooling water and boiler blowdown, could be bypassed, combining with the treated effluent downstream of the treatment plant. The benefit is that those components of the treatment plant that are sized on the basis of hydraulic load will cost less to provide and operate. Also, some of these weaker effluents (in terms of COD) may contain chemicals that could have adverse effects on some treatment processes.

If a treatment plant is not going to be provided it may still be beneficial to separate strong and weak effluents as it can be cheaper to tanker away some high strength, low volume effluents than to discharge them to sewer. Certainly any waste stream with a COD level above about 50 000 mg/l should be considered as a potential candidate for segregation.

Foul drainage from toilets, washrooms and canteens must be kept completely separate from other drainage. If this is not achieved then apart from infringing likely discharge consent requirements, sampling of the other discharges becomes unpleasant and is more likely a major hygiene issue. If all drainage from the site will ultimately enter a foul sewer then process and foul effluent should not be combined above the final flow and quality monitoring point for the process effluent.

Storm drainage is potentially the most difficult to deal with. Totally uncontaminated water should be discharged to a surface water sewer or watercourse. Totally uncontaminated means just that; if the rainwater runs across a yard on which lorries or even cars are parked it will not be uncontaminated as it will probably pick up some oil at least. If oil is the only problem then a suitable oil interceptor may be all that is required; this must be designed to cope with the maximum likely flow rate and most importantly, it must be regularly emptied, otherwise the oil will simply build up until it is flushed through. The advice of the appropriate regulatory authority must be sought before installing equipment or making any discharge.

On a processing site it is likely that some areas of the site will become contaminated with process materials. There are three approaches to the problem. The first and best is to avoid surface contamination, i.e. clean up all spillages as they occur. This should be the objective anyway, whatever the sophistication of the drainage system, but in these days of staff reductions it is perhaps asking too much to expect this to be achieved at all times.

The second method is to keep spillage and stormwater apart. This can be achieved by roofing over areas that are likely to become contaminated, with roof downpipes sealed into the drain at ground level rather than discharging to open drains. This may be the most economical method where small areas are involved but is unlikely to be attractive for larger areas particularly if vehicular access is required.

The third method is to provide two drainage systems for stormwater. The first will receive all stormwater that is known to be uncontaminated and will discharge to river or surface water sewer. Drainage from roofs should be clean but bear in mind that factory extraction systems and vents can result in deposition of potentially polluting matter on roofs which can be washed off during storm conditions. It is common to mount cooling towers on roofs; the bleed from these must go to a process effluent drain. The second drainage system will collect drainage from areas known to be continuously or occasionally contaminated with process materials. If a treatment plant is to be provided the decision must be taken as to whether the polluting load of the runoff will justify treatment bearing in mind the hydraulic load that will be placed upon the treatment facility during storm conditions.

If the runoff is not treated the water company which receives the site discharge into its sewers may require it to be treated as a trade effluent and combined with the remaining process effluent (after treatment if provided) prior to the flow measurement and sampling point.

It is suggested that all stormwater systems be provided with a penstock or some other means of isolation so that a major spillage can be held pending a decision on its disposal. An accessible manhole and small sump immediately above the penstock will provide a suitable point from which the spillage and any flushing water can be pumped or tankered. It is equally as important that staff are aware of such facilities and how and when they are used. If this does not happen the management could end up dealing with either a major pollution or a flood!

15.2.7 Legal and cost aspects

Only an overview can be addressed here. Those who find themselves in the unfortunate position of being subjected to prosecution are advised to find themselves both a competent technical person and a solicitor who have relevant experience of the various statutes and their enforcement, particularly if the case is to be contested. The prosecution case is not always watertight and complex technical matters are often simplified for the benefit of magistrates. The effect of this is that negative aspects of the case can be overemphasized and a competent defence can ensure that this does not get out of hand.

Environmental legislation can be applied at three points, the environment itself, the discharge to the environment and the process creating the discharge. The thrust of recent and foreseeable legislation, particularly from the EC is towards the latter two. Historically, discharges to water were controlled on the ability of the receiving watercourse to absorb the discharge without adverse effects. The Wastewater Directive seeks to impose more blanket standards which have less regard to the ability of the receiving watercourse to accept the discharge. Integrated Pollution Control Legislation moves the point of control towards the process creating the discharge.

In terms of compliance, the basic principle is that unless the discharge is wholly uncontaminated rainwater, it cannot be discharged without statutory approval. In the United Kingdom there are two major statutes to consider: the Water Industry Act and the Water Resources Act, both passed in 1991. As well as containing primary provisions, these are also enabling Acts and subsequent Regulations have been and will continue to be made from time to time. These cover such matters as prescribed processes and substances and storage of chemicals and oil in particular circumstances. Insofar as effluent discharges are concerned, the Water Resources Act is concerned with the protection of the aquatic environment, i.e. rivers, lakes, coastal waters and groundwater with the National Rivers Authority as the enforcing body. The Water Industry Act is concerned with the protection of sewers and the sewage works into which they discharge.

The primary offences under the Water Resources Act are to cause or knowingly permit poisonous, noxious or polluting matter to enter relevant waters, or to discharge without a Consent or in contravention of the terms of a Consent. The maximum fine in a UK magistrates court is currently £20 000 per offence and an unlimited fine and/or imprisonment can be imposed by a Crown court.

There is a great legal debate over the terms 'cause' and 'knowingly permit'. However, it is generally not necessary for a positive action to occur for the offence to be proved. Failure to take a particular course of action can be deemed to be causing a discharge. Generally, companies are liable for the actions of their employees and depending on the circumstances, any contractors who may be on the site.

There is a personal as well as corporate liability under the Act and there appears to be an increasing tendency to target directors and other senior managers who, whilst not causing a discharge themselves, may knowingly permit it to occur. Managers who ignore these matters or give them low management and budgeting priority may be doing so at their own personal risk. Insurance companies are also beginning to exclude pollution events from general liability policies leaving many companies seriously exposed in the event of a pollution event.

The UK National Rivers Authority (NRA) are also empowered under the Act to require preventative measures to be taken where pollution is foreseeable or likely to occur. They can arrange for works to be undertaken themselves and can recover reasonable costs from the responsible person. It is likely that this provision will be used only in the most serious situations, and where there is no doubt about the consequences of not acting.

If it is desired to discharge directly to a river or other relevant waters (directly or via soakaways or irrigation) it is necessary to obtain a Consent well in advance of the need to start discharging. A detailed profile of the discharge will be required, hence the need for adequate preliminary assessments of both manufacturing and effluent treatment processes. Unsupported estimates are not usually acceptable and could be the first stage on the road to prosecution. If you do not know what you are losing from your manufacturing process the chances are that the effluent treatment process will be incorrectly designed or operated and the discharge will fail its Consent limits.

The Consent will normally contain volumetric and quality limits which, unless specified otherwise, must be complied with absolutely at all times. Statistically this is impossible to achieve but the authorities are generally unsympathetic to this argument. The Consent may also specify equipment that must be provided and maintained for monitoring purposes. Sampling and flow monitoring apparatus are usually the minimum requirements; these are briefly considered later. Continuous monitoring of key quality parameters may be specified for a major discharge though this is more likely to be a requirement where toxic or dangerous substances may be present.

Another major consideration is that details of the Consent, samples taken and actions arising will be held on a public register. Such registers are increasingly scrutinised by environmental interest groups and others with an interest in the aquatic environment such as angling clubs. It is possible for such organisations or indeed individuals to bring their own actions against a discharger, irrespective of any action or inaction by the NRA. Civil cases can also arise which may be more costly than criminal ones and the standard of proof required is less. It is not unknown for criminal and civil cases to proceed together and an example of this has occurred within the UK soft drinks industry quite recently.

The NRA (UK) will levy an annual charge to cover the costs of monitoring the discharge. A scoring system is used whereby points are awarded according to discharge volume, nature of components and type of receiving waters. The total is multiplied by a unit cost to derive the charge. Currently, typical charges range from several hundred to several thousand pounds.

The other main method of effluent disposal is discharge to a public foul sewer. In the United Kingdom the sewers and sewage works are owned by the privatised water companies, though most still have an agency agreement with the local authority covering the management of the sewerage system. Industrial discharges into sewers are controlled under the Water Industry Act of 1991 by the relevant water company.

Water companies need to treat as much industrial effluent as possible, particularly organically strong effluents from the food and drinks sector, because it attracts relatively high trade effluent charges. However they must not overload their sewage works because they would then fail their own river discharge Consents and risk prosecution by the NRA (in the UK). Hence there is a wide variation in the approach to industrial effluent control according to the degree of risk that a particular company is willing to take.

The only liquid that may be discharged without consent to a public foul sewer is domestic sewage. Stormwater may be discharged if there is no stormwater sewer available. This is a simplified summary and there are a few grey areas. There is no right to discharge any trade effluent into a public sewer, no matter what connections to the sewer may exist. An application for Consent must be made to the water company. The Consent will contain volumetric and quality conditions in a similar way to a river discharge Consent though parameters such as suspended solids and organic loads and concentrations may be much more relaxed, depending on local circumstances. Flow monitoring and sampling equipment will almost certainly be required for a discharge of any significance.

For many discharges to sewer from the juice/soft drinks industry, the most significant aspect of the Consent will be the trade effluent charges. These are usually calculated using the so-called 'Mogden formula' though there are various refinements of it and there may be attempts in the future to introduce alternative methods.

The Mogden formula calculates a cost per cubic metre of trade effluent discharged. For the food and drinks industry, the primary components of the cost are almost always the effluent strength factors which are assessed relative to the average strength of sewage treated by the water company. Hence if the strength of an effluent is ten times that of the average sewage strength the relevant unit cost factor is multiplied by ten. This can be seen from examination of the basic formula which is as follows.

$$C = R + V + \frac{Ot}{Os} \cdot B + \frac{St}{Ss} S$$

where C is the cost per cubic metre of effluent discharged, R is the reception (sewerage) charge in pence per m^3, V covers those elements of the sewage treatment process that are costed volumetrically in pence per m^3, B is the cost of biological treatment of sewage in pence per m^3, S is the cost of sludge treatment and disposal in pence per m^3, Ot and Os are the strengths of effluent and sewage respectively usually measured as mg/l settled COD and St and Ss are the suspended solids concentrations in mg/l of effluent and sewage, respectively.

The effluent strength and suspended solid levels are usually the mean values from a series of samples taken through a charging period, hence sampling technique is important and it is usually beneficial to devote adequate resources to this.

The water companies are empowered by statute to operate a charges scheme and there are obvious commercial advantages to them in maximising this income stream. It is often the issue of cost as much as Consent compliance that gives the incentive to examine the possibilities for on site treatment prior to a sewer discharge.

A sewer discharge should be treated with some respect. It is possible for concentrates, sugars and other process materials used within the juice and beverage industries to cause severe damage to, or even complete failure of sewage treatment processes. In such cases the water company could seek to recover the costs of recommissioning a treatment process which may run into many thousands of pounds. The discharge of effectively untreated sewage could cause severe damage to a watercourse and result in claims for damages from downstream interests, e.g. angling clubs, abstractors.

Sometimes a sewage treatment works or sewerage system may not have enough capacity to receive an increased or new discharge. In these circumstances the water company may be able to uprate its facilities to deal with the discharge and will require adequate recompense for doing so. This is usually achieved through some form of formal agreement which may replace or supplement a Consent. A number of options are possible. An initial capital contribution may be made for the provision of treatment plant and its availability for treating a specified quantity of the industrial effluent for an agreed period of time. Alternatively, the discharging company may agree to guarantee the water company a given level of revenue for a specified period, this being sufficient to provide the water company with an adequate rate of return on its investment in treatment plant.

15.2.8   Environmental Protection Act

In the UK relevant EC Directives are generally implemented through one or both of the Water Acts. One of the main issues at present is the effect of the Urban Wastewater Directive on Consent conditions and discharge costs. It is almost certain that some Consent conditions will be tightened as sewage works and other direct discharges need to meet higher standards. In the case of trade effluent discharges to sewer, the increased costs of meeting higher sewage works discharge standards are likely to be reflected in yet more substantial rises in trade effluent discharge costs. For those discharging to a watercourse, the Wastewater Directive will have a direct impact on their Consent conditions if they fall within one of the defined industry sectors. These include the manufacture of fruit and vegetable products, manufacture and bottling of soft drinks, breweries and the production of alcohol and alcoholic beverages.

The Environmental Protection Act (EPA) impinges upon undustrial effluent control in two main areas: prescribed processes and substances and sludge disposal. It is unlikely that any beverage or drinks process will be designated as a prescribed process in the foreseeable future, though a future EC Directive may change this by applying integrated pollution control to all food processes employing more than fifty people.

The provisions of the EPA do apply to the removal from a site of any sludge arising from an effluent treatment process or indeed to any other solid or liquid waste removed from the site other than via a permitted discharge to drain. The main regulations are those associated with duty of care which place responsibility upon the waste producers to ensure that the removal and ultimate disposal of waste is in accordance with the provisions of the Act.

The Act itself and the Duty of Care Regulations and code of practice should be consulted for full details. Only licensed contractors may be used to remove wastes and these must be taken to licensed disposal sites with very few exceptions. Wastes must be adequately described and records must be kept concerning details of the wastes disposed of.

Many food and drinks wastes may be applied to land provided that their benefit to the land can be positively demonstrated, usually by means of a soil growth test carried out by the Agriculture Development and Advisory Service. This route does not currently require a license, however a licensed contractor must still be used and if it is not carried out professionally there is a risk of nuisance, soil and crop damage or water pollution through runoff or infiltration. New disposal regulations applying from May 1994 put the practice into a decidedly grey area insofar as licensing requirements are concerned and with the water companies also hoping to increase sludge applications to land its availability may reduce markedly in years to come with resultant increases in disposal costs.

15.2.9   Flow and quality measurement

Whatever discharge route is selected it is likely that there will be a requirement for some form of flow measurement and, for significant discharges, suitable automatic sampling equipment. Even where this is not required as a condition of the Consent it may still be financially beneficial to provide it

The information gained from sampling and flow measurement will be used to assess compliance with Consents and to determine trade effluent charges in the case of discharges to sewer. The information, if collected and analysed properly, can also be a valuable management tool for preventing or reducing waste and therefore costs.

For smaller discharges to sewer the water companies generally do not require direct positive measurement of flow and samples are generally obtained by visiting officers who may take grab samples or sometimes a 24 h composite sample using an automatic sampler. In these situations, the discharge volume for charging purposes is estimated from water supply data by subtracting standard or agreed allowances for domestic usage and water lost into product or through evaporation etc. Although this avoids the expense of flow monitoring apparatus, the chargeable volume of trade effluent may be higher than it need be. This is because all water use that is outside of the allowances given is assumed to be returned to sewer as trade effluent. If a tap or hosepipe is left running by accident, say overnight, the volume will be charged for as trade effluent, at the prevailing strength of the trade effluent. Also it is unlikely that all losses of water will be fully assessed giving rise to further overcharging.

The next level of sophistication is to install water meters on individual process lines where all or most of the water is ultimately discharged as trade effluent. Apart from satisfying many Consent requirements this is an extremely useful management tool as it enables accountability for water and effluent to be delegated to individual areas of the site. The meters can also be used to measure losses such as cooling tower and boiler make-up water and product water. They are relatively cheap unless the pipe diameter is greater than about 50 mm but it should be remembered that they are prone to under-record as they age.

For large discharges to sewer and virtually all discharges to watercourse the Consent is likely to specify direct measurement. Open channel methods (which include partially filled pipes) are more common and are generally preferable as it is usually easy to check the calibration and function of the meter. Closed pipe methods may be acceptable where it is impractical to install an open channel device but it should be noted that some of the accuracy claims for these meters are rarely met in practice and their use should be considered with some caution.

Open channel methods are usually based on creating a channel condition so that the flow is proportional to the depth of liquid behind some structure in the channel, referred to as a primary device. The theory of this is covered in most standard texts on fluid mechanics. The most common devices for measuring effluent flows are weirs and flumes. The selection of an appropriate device requires careful consideration if acceptable accuracy is to be achieved. V notch weirs are generally the best for measuring low flows because at low liquid heads a small change in flow results in a larger change in head than is the case for say a flume. Horizontal weirs or flumes will be more appropriate for larger flows which can be measured with less headloss than a V notch.

One advantage of a flume is that it will tend to be more self cleansing than a weir as at zero flow there is no reservoir of liquid in which solids can settle. However this is not likely to be a significant issue in the drinks industry and the choice should be made primarily on flow characteristics and permissible headloss within the drainage system.

If a flume is installed it is vital to ensure that it cannot become submerged due to unsatisfactory downstream conditions. Such conditions may arise if the downstream flow is pumped and the pump operating levels are set too high or the channel narrows or its invert slope decreases. This can be assessed by calculating the modular limit for the flume which is the ratio of total downstream energy head to total upstream energy head. The limit is usually taken to be 0.7. In practice it is usually satisfactory to ensure that the downstream depth of flow does not exceed two-thirds of the upstream head.

Weirs must also not be operated in a way that can result in submergence. This is not usually a problem as it is common to provide some form of weir tank, with the price of increased headloss.

Other forms of primary device are available for specific flow conditions. It is best to contact a reputable manufacturing company who produce a range of types or to use a consultant when selecting a primary device.

The other component of a flow measuring installation is the secondary device which measures the head behind the primary device and converts it into flow units which may be displayed and recorded. Three methods of head measurement are common. These are ultrasonics, bubbler and conductance or capacitative probe. The ultrasonic method is becoming more popular and modern transducers claim to measure head to accuracies of 1 or 2 mm. The technique is temperature sensitive although modern transducers have built-in temperature correction.

Transducers should not be placed in sunlight or close to another heat source. The most common problem is where the effluent foams as the foam scatters the ultrasonic signal and renders the transducer useless. Foaming may occur even when the level of detergents present is low and may be a problem after biological treatment where various proteinaceous or bio-surfactant foams may be created. The best approach is to ensure that there is no weir or freefall of effluent for some distance above the transducer. Some success has also been achieved by mounting the transducer in a piece of pipe that extends below the liquid surface and for some distance above it. Floating solids and steam may also cause problems with ultrasonic measurement.

The bubbler method relies on a supply of air from a small compressor which is bubbled into the flow at a fixed height, behind the primary device. As the head of liquid changes, the back pressure within the air supply tube changes in proportion and it is this which is measured and converted to flow. It may be suitable in situations where ultrasonic transducers are not reliable. Regular maintenance of the system is essential, the output from the compressor must be constant and the air tube that dips below the liquid surface is prone to blockage resulting in gross over-recording of flow.

A conductivity or capacitance probe effectively measures the depth of flow directly. The problem is that gross solids will wrap themselves around the probe which is necessarily submerged within the flow to be measured. This eventually leads to complete fouling.

The flow meter itself is now usually microprocessor based and the chosen system should permit data to be downloaded periodically into suitable computer software for flow analysis. The flow recorder should continuously display the instantaneous flow rate and integrated total flow. Menu options on some instruments allow the total flow reading to be altered by the user; this is not acceptable if the unit is to be used to record flow for charging or Consent compliance purposes. It is wise to record the total flow figure on a daily basis and to download the recorded data to disk at intervals. In this way a historical record of flow is built up which will be essential in the event of any billing or Consent dispute arising.

Flow recorder calibration should be checked regularly, at no longer than three-month intervals. Siting the primary device too far below ground will make this a costly exercise as confined space entry procedures may then be required. The flow recorder should be capable of driving a sampling machine in a flow proportional manner; this is usually achieved by transmission of a volt-free impulse every time a specified number of flow units are recorded.

Sampling is at least as important as flow measurement, arguably more so since Consent compliance is largely based upon it. Also COD may comprise about 70% of a typical juice process trade effluent charge so adequate sampling facilities are important. An automatic sampler can be bought in various forms. The cheapest is a single bottle machine where all samples taken are added to and stored in a single container. This should suffice unless there is a requirement to assess the effluent quality profile, in which case a multi-bottle unit is needed. This permits samples for specified time periods (e.g. hourly) to be placed in separate containers for individual analysis.

Flow proportional sampling is most accurate but time-based sampling is usually nearly as good in most instances. Where time-based sampling is used the sampling interval should not exceed 15 min unless the effluent has a fairly stable composition. Installation and maintenance of sampling equipment is also important.

For significant discharges from the food and drinks industry it is worth considering the benefits of a refrigerated sampler. The first components of a 24 h sample will be in the container for at least this period of time before removal, and degradable components are likely to do just that during warm weather, particularly if the effluent is also warm. Even with refrigeration some ageing may occur, this usually manifesting itself by way of a drastic reduction in pH. This may result in Consent failures that are not real and written agreement with the controlling authority should be reached as to how pH compliance will be assessed.

There are few publications on the subject of instrumentation for effluent monitoring that are suitable for general information and reading. The book by Endress and Hauser (1992) provides introductory coverage of flow measurement, sampling and some on-line parameters and may be of use to those needing to know more about these methods.

Some basic analytical facilities are desirable though the costs of external analysis of COD and suspended solids should not exceed about £20 per sample. The main factor involved in the use of an external laboratory is time, and internal facilities are generally required if samples are used for process control.

15.2.10   Management schemes for effluent

A formal management scheme may be developed using many of the principles and techniques outlined so far. A process audit and mass balance exercise should define the general quantities and quality of effluent that must necessarily be discharged. A management scheme will help ensure that these levels are maintained as targets not to be routinely exceeded.

The benefits of this are fairly obvious. Process losses can be minimised, water use will be optimised and effluent treatment plants if they are required can be designed and operated on a sound basis. The overall result is better compliance with discharge standards and lower costs.

Successful implementation of a scheme requires commitment at all levels but especially from management as it is likely that there will be some initial costs, though these should be readily recouped as savings materialise. Most schemes work by setting budget targets for volume and quality of discharge from each defined effluent producing area. The targets must be realistic and should normally be related to production output to be fair.

To have credibility, target compliance must be assessed by positive measurement of discharge parameters. The costs of sampling, volume measurement and analysis will depend on the number of effluent-producing areas defined. Usually it will not be necessary to define each drain as an area, more likely a department or section will be so defined. For example the following activities may be defined as effluent-producing areas; each area may have multiple effluent discharge points.

— Raw materials preparation,
— Blending,
— Bottling,
— Materials storage tanks,
— Waste storage area.

It is important that each defined area should have an accountable manager.

Some third party involvement in drawing up a scheme may be useful to validate it from a technical point of view and to assist in training and awareness of staff.

At the end of each accounting period the overall site costs for water and effluent are apportioned to each defined area on the basis of its relative water consumption and effluent discharge as measured. This results in wastage having a direct impact on the departmental budget of the accountable manager giving a powerful incentive to monitor and improve.

Discussion of this topic would not be complete without reference to wider formal environmental standards. In the UK most people have a reasonable knowledge of the quality standard BS 5750; this is now complemented by the environmental quality standard BS 7750.

In outline, BS 7750 extends the principles of BS 5750 to the environmental area. The argument is that a quality product cannot be made at the expense of the environment. Description of and implementation of the standard are beyond the scope of this book; for those who wish or need to pursue it the publication by Rotheryl (1993) may be of help.

15.2.11   Effluent treatment

The reasons for treating effluent on site will be to meet Consent requirements and/or to reduce discharge costs. The subject is worthy of a book in itself and references have been given for those who are interested. Before examining the techniques available for treatment of effluents from the soft drinks/beverage industries it is worth considering one or two points that always arise when considering on-site treatment.

A realistic financial appraisal of any treatment proposal must be made, however it may be necessary to modify standard payback calculations. Consider the following example. A company wishes to install a new manufacturing facility that will discharge 1500 kg/d of BOD. The water company will only accept 500 kg/d into the existing sewage works but offers the additional required capacity for a capital contribution say of £850 000, with no reduction in trade effluent charges. The company can construct its own treatment plant for £500 000 but the cost of running the plant will almost equal the savings made from reduced trade effluent charges. A standard calculation on the capital cost of £500 000 would not give a reasonable payback since there is little positive cashflow. What has to be considered is the difference in capital outlay required to achieve compliance with Consent. Other factors such as ease of extending treatment capacity and future legislative changes need to be considered but are not easy to express in terms acceptable to an accountant.

The above example is typical of that which might arise in an area of the country where trade effluent charges are low. Where trade effluent charges are higher it should be possible to see some positive cashflow (savings in trade effluent costs—operating costs) which will give a return on capital employed. The payback period will obviously vary with local circumstances but may typically be between two and six years for treatment schemes removing the majority of COD.

A full assessment of likely operating costs must be made and this should be undertaken by someone who is independent of the company supplying the treatment plant. Costs to be considered include the following.

—Power,
—Chemicals,
—Sludge disposal,
—Maintenance,
—Operator costs,
—Management costs,
—Analytical costs.

Specialist help may be needed to operate treatment plants successfully and economically; a figure should be budgeted for consultant or other specialist help. Sludge disposal costs could account for 50% of operating costs of some plants, the treatment and disposal of sludge is perhaps the area that is most likely to be neglected with increasingly costly implications.

When assessing capital costs, necessary drainage modifications, pumping, sampling and flow monitoring requirements must not be overlooked. Where chemicals are used in the treatment process safe arrangements must be engineered for their storage and use. For most plants a laboratory area should be provided for process monitoring. The product quality control laboratory may not wish to handle effluent samples due to potential hygiene problems.

The effluent should be fully characterised with regard to the proposed treatment technology. Each effluent is different and this is one reason why table after table of characterisation and treatment performance data has not been included in this chapter. For those who wish to inform themselves in this matter there are many textbooks and papers available but it should not be assumed that a particular technique will work on your own effluent even if the basic characteristics (COD, solids etc.) appear to be similar to published data. Again, this is an area for the specialist.

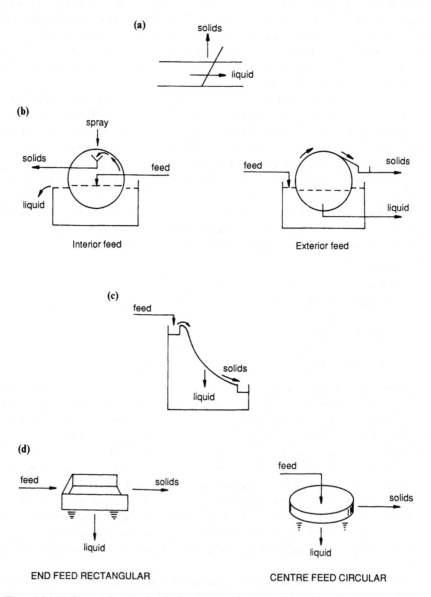

Figure 15.4 Various types of screen for effluent treatment. (a) In-channel bar screen, (b) rotary drum screeners, (c) tangential or static screen, (d) vibratory screens.

15.2.11.1 Treatment techniques. Treatment techniques can be classified into three areas, physical, chemical and biological though most treatment schemes will make use of at least two.

Physical treatment methods that can be applied to soft drink and beverage effluents will consist essentially of screening, settlement and flotation.

Screens come in a variety of formats (see Figure 15.4), commonly of a rotating drum type for industrial applications. The aperture should be as fine as possible consistent with adequate free flow. Apertures of 2 to 5 mm should be sufficient to remove gross solids and pieces of fruit etc. Any fatty or greasy materials can adhere to the screen, capturing solids and resulting in blockage. This should not be a major problem in this industry but is worth bearing in mind. It must be remembered that a headloss will occur whenever a screen is installed and it is wise to provide either a duplicate installation or an emergency (alarmed) bypass. The facility should be designed with regard to the method and frequency of screenings removal; vehicular access may be required.

Settlement and flotation in themselves are not likely to achieve a great deal with this type of effluent. Much of the organic load will be in solution and only that portion of it that can be precipitated can be removed by these techniques. Preliminary work can establish this proportion but it is likely to be low and would not justify the use of these processes on a stand-alone basis. They may be of use as a preliminary stage prior to a biological plant. One application for settlement might be the removal of grit, or possibly fruit seeds which could damage downstream mechanical equipment. Both of these should settle readily so that a relatively small surface area is required. Long settlement times will lead to degradation of the effluent with generation of organic acids. Whilst it is arguable that these may be more treatable than the primary carbohydrate a lowering of pH will occur which may have corrosive effects and some odour will be generated.

Flotation (see Figure 15.5) may be appropriate for pre-treatment of the effluent prior to a biological stage if there is free or emulsified oil or fat present. Chemicals will be required to crack any emulsions prior to flotation.

Chemical treatment may take various forms. The most common is perhaps pH correction. If there is no other treatment this should be applied downstream of any balancing tanks. If the effluent is held for any length of time after pH correction with no other treatment, the benefit of pH correction may be lost as acid-producing bacteria further degrade carbohydrates. Successful pH correction requires careful process design even though it may at first sight appear simple. Mixing, retention time and location of measuring probe are important parameters which must be optimised for each application.

Chemicals may also be applied to coagulate and flocculate the effluent prior to a solids removal process. The coagulant, which is commonly based on a metal salt, neutralises surface charge on small solids within the effluent enabling them to coagulate. The flocculant, which may have an opposite

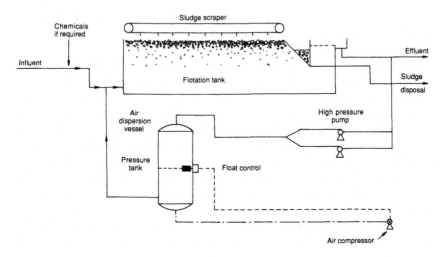

Figure 15.5 Dissolved air flotation for pre-treatment of effluent.

charge to the coagulant encourages bridging of coagulated solids resulting in floc growth. Flocculants are typically organic polymeric compounds of varying molecular weight and charge depending on the duty required. Such treatment is usually applied prior to flotation.

The other main use of chemicals will be to condition the effluent prior to biological treatment. It is common to add nitrogen and phosphorus-based chemicals as nutrients for biological growth as many effluents are deficient in one or both of these. Addition should be continuous and is usually achieved by injecting the chemical into the influent to the biological process through a dosing pump.

To achieve substantial removal of dissolved organic matter a biological treatment plant must be utilised. Although advances are being made in the application of membrane technology there is some way to go before it offers a cost effective alternative process. Biological methods rely on the use of the effluent as substrate for anaerobic or aerobic microbes. Some of the effluent components are therefore converted into additional biomass or sludge which must ultimately be removed from the process. The remainder is oxidised to provide energy for microbial growth. Aerobic microbes have an efficient metabolic pathway which can result in virtually complete utilisation of biodegradable effluent components under acceptable loading conditions. Hence an effluent of high quality can be produced.

Anaerobic microbes are rather less efficient; they have lower growth rates and the effluent is not completely utilised for substrate resulting in partial oxidation. The consequences of this are that aerobic processes produce higher quality effluents at the expense of more sludge production and in some configurations considerable energy input. If a river quality effluent, or almost complete removal of COD, is required then it is an aerobic process that must be used though it may of course be preceded by other technologies which can reduce the treatment load and hence its cost. Anaerobic processes will produce a poorer quality effluent but sludge production will be minimal. Also one of the products of anaerobic metabolism is methane which can be burned to provide energy for both the anaerobic treatment process itself, which should be maintained at 35°C, and possibly other processes.

A properly designed aerobic process should achieve at least 95% removal of COD though more than one process stage may be needed. Aerobic processes can be operated at high loadings when a poorer quality effluent is produced and sludge production increases through a lower rate of endogenous metabolism. An anaerobic process may be expected to remove between 60% and 85% of the COD load applied. Claims for greater efficiencies should be investigated carefully.

Aerobic processes may be based on fixed film or suspended processes. The traditional filter or bacteria bed is the most common fixed film process in which the microbes attach themselves to an inert support over which the effluent is trickled. Their operating costs are relatively low but the degree of process control over them is relatively limited and is really restricted to the application and recirculation rate. Due to the short contact time any shock loads tend to be transmitted directly to the final effluent.

The activated sludge method (Figure 15.6) is based on a flocculated suspension of microbes which is contacted with the effluent in one or more aeration tanks. The flocculated suspension is then separated in a downstream settlement tank and returned to the aeration process. Thus the retention time of microbes in the process exceeds the hydraulic retention time. High loads are adsorbed by the biomass up to a point and provided that there is a sufficient resting period there is less deterioration in effluent quality. This is however a simple summary and in practice a number of operational conditions can arise which affects the ability of the microbial suspension to settle. Poor settlement results in carryover of microbial solids in the final effluent. Specialist help is often necessary to diagnose and provide long term cures for these so called 'sludge bulking' episodes though a recently written and very

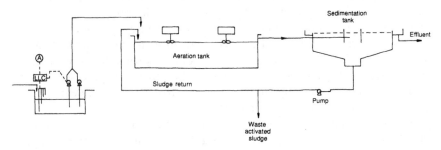

Figure 15.6 The activated sludge process.

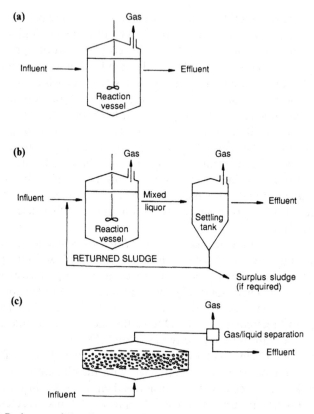

Figure 15.7 Basic types of plant for anaerobic treatment. (a) Conventional, (b) anaerobic contact, (c) anaerobic filter.

readable manual on the subject (Jenkins and Richard, 1993) provides excellent initial guidance for those with some knowledge of the process.

Anaerobic processes (Figure 15.7) are also based on fixed film or suspended processes. Loadings are generally higher with up to several kg BOD/m^3 per day being applied compared with typically one or less for an activated sludge process. In some applications it has been found adavantageous to separate the acid-producing and gas-producing stages of the process. There is less experience of anaerobic technology in the UK than elsewhere and there is some scepticism towards its use. Recent advances in the understanding of some of the finer aspects of the process microbiology should however remove some of the mystique surrounding it. Quite often anaerobic processes require more intensive monitoring and process control than equivalent aerobic processes but this is not a hard and fast rule.

All biological treatment processes tend to be designed on the basis of empirical criteria. This is because heterogenous populations of microbes are involved making the application of kinetic models difficult though recent refinements make this approach possible. However kinetic models in their basic form usually assume steady state conditions, something which rarely arises in effluent treatment.

Flow and strength balancing have been omitted from this discussion as it is not really a treatment as such. However, its use should be considered as it may enable treatment processes to be designed on the basis of mean hydraulic and quality parameters rather than maximum ones. Due to the sensitivity of anaerobic processes substantial balancing is almost always required. This can add substantially to the captial cost of treatment; it must also be designed carefully as stored effluent can deteriorate rapidly giving rise to odours.

For discharges to sewers, balancing may, in some cases, be all that is required though it is possible that pH correction may be needed prior to discharge. Generally, for economical use of chemicals, it will be found advantageous to correct pH after the balance tank rather than within it.

The above technologies may be combined in various ways according to the effluent characteristics and treatment standards required. A typical sequence for treatment prior to a sewer discharge might be screening, followed by flow/strength balancing and then biological treatment.

Chemical treatment and flotation might be incorporated if emulsified oils or fats were present in sufficient quantities to interfere with the biological stage. Fatty materials are best removed as early as possible in the treatment process; indeed if they originate from a single stream of effluent it may be worthwhile treating this stream separately from the main flow. A treatment scheme for a discharge to river would tend to follow the same pattern except that normally at least two biological treatment stages would be employed, the second of which must be aerobic. The effluent might then be polished in a tertiary stage such as sand filtration or lagoon according to Consent requirements.

Some new processes have been marketed in recent years, the two most prominent being reed beds and biological aerated filters. There is nothing fundamentally new about these; the microbiology is basically the same as for existing processes. As with most new processes, quite unjustified claims are initially made for their performance before the reality of operating experience forces a sober reassessment. These two and a few others rightfully have their place in the list of available processes but must not be seen as a final solution to all problems with existing processes. The number of applications of such new processes to the food and drink sector is limited and in many cases consists of pilot scale plants. Caution is therefore advised in their use and their application may be best undertaken through a development project rather than an off-the-shelf solution.

The correct application of treatment technology is always one of the major challenges, since every effluent is different. The effects of each chosen process on downstream processes must be carefully evaluated. For example whilst balancing is generally desirable, it is not unknown for changes to occur within the effluent making it more difficult to treat unless the requisite precautions are taken. Chemical treatment processes can remove substances that are essential nutrients for subsequent biological treatment stages, causing micro-biological imbalances and process failure unless this is taken into account at the design stage.

15.2.12 Sludge treatment

The vast majority of effluent treatment processes will result in the production of sludge. In simple terms the sludge consists of the impurities removed from the effluent plus the residues of any chemicals that might be used in precipi-tation reactions. The impurities may not be in their original form; if a biological process has been used the sludge will consist of the excess biomass that has grown upon the components of the effluent.

The term sludge can be misleading because it is quite common to obtain sludges from biological processes that are around 1% solids, i.e. 99% water. Tankering of sludge for off-site disposal will bear a cost which will vary depending on method of disposal and geographical area. Tankering away sludge of this consistency even for land injection, which is likely to be the cheapest disposal route, will probably give rise to 50% or more of the plant operating costs.

It is generally well worthwhile thickening sludge by some means. Increas-ing the solids content from 1% to just 2% will reduce the volume by 50%; this can sometimes be achieved by simple gravity settlement. Most sludges, particularly those derived from activated sludge processes will require mechanical dewatering to reach significantly above about 2% solids.

In many cases there are two options for sludge. It can be thickened to between about 5% and 8% solids if land injection is available as a disposal option. Otherwise it can be more fully dewatered to 20% solids or more if disposal is to landfill. Note however that many activated sludges when treated alone do not readily dewater above about 15% solids.

Flotation and centrifugation are common techniques for thickening sludge prior to land injection. Chemical conditioning of the sludge is usually required. Normally the water removed from the sludge needs to be recycled to the treatment plant; it may contain a significant load of COD and fine solids which must be taken into account in the design of the plant. Changes in the character of activated sludges, for example due to bulking problems, can affect the performance of thickening or dewatering processes; this may be offset to some extent by varying the chemical dosing regime.

Techniques for more intensive dewatering include vacuum filters, belt filter presses and plate and frame filters. Membrane technology is beginning to make inroads into the market though it is still considered to be unreliable and costly by many.

Other technologies being applied to sludge include composting and drying, both of which provide a recyclable material which may be suitable for horticultural as well as agricultural use. Sludges derived from food and drink manufacturing may be particularly suited to this, however substantial de-watering is required as a pre-treatment and the capital costs of drying are high at present. Incineration is being used more widely for sewage sludge but its capital cost and environmental issues are likely to make it generally unattractive to industry.

It is important to develop a strategy for sludge disposal at the earliest possible moment, particularly with regard to changing regulations and resultant pressures on disposal alternatives. Although land injection may be relatively cheap at present, this may not be the case in a few years time as more and more land is needed for the increasing volumes of sludge being generated as a result of tightening legislation. Landfill sites may become more reluctant to accept wet sludges and will certainly be charging much more to handle them. The implication is that it may become more economical to dewater sludge as fully as possible or to convert it via one of the processes outlined above into a usable and perhaps saleable product.

It is hoped that this chapter has at least given an outline of the issues involved in effluent management. It should be appreciated that a multi-disciplinary approach is required involving management, scientific and engin-eering skills, and that each factory's effluent will have its own distinctive characteristics. With these points in mind and a logical approach to the issues summarised above, answers to most problems should be forthcoming. If effluent management is integrated into the overall site management it should not become an uncontrollable problem.

References

Barnes, O., Forster, C. P. and Hruday, S. E. (eds.) (1984) *Surveys in Industrial Wastewater Treatment, Food and Allied Industries*, Pitman.

Dean Adams, V. (ed.) (1990) *Water and Wastewater Examination Manual*, Lewis.

Endress and Hauser (eds.) (1992) *Wastewater Measurement and Automation*, ISBN 3-9520220-1-2.

Horan, N. (ed.) (1990) *Biological Wastewater Treatment Systems, Theory and Operation*, Wiley.

Jenkins and Richard (1993) *Manual on the Causes and Control of Activated Sludge Bulking and Foaming*, 2nd edn., Daigger, Lewis.

Metcalfe and Eddy (eds.) (1991) *Wastewater Engineering, Treatment, Disposal and Reuse*, 3rd edn., McGraw-Hill.

Rotheryl, B. (1993) BS 7750, *Implementing the Environment Management Standard and the EC Eco Management Scheme*, Gower.

Tebbutt, T. H. Y. (ed.) (1992) *Principles of Water Quality Control*, 4th edn., Pergamon.

Index